LONDON MATHEMATICAL SOCIETY LECTURE NOTE SERIES

Managing Editor: Professor M. Reid, Mathematics Institute,
University of Warwick, Coventry CV4 7AL, United Kingdom

The titles below are available from booksellers, or from Cambridge University Press at
www.cambridge.org/mathematics

London Mathematical Society Lecture Note Series: 399

Circuit Double Cover
of Graphs

CUN-QUAN ZHANG
West Virginia University

CAMBRIDGE
UNIVERSITY PRESS

CAMBRIDGE
UNIVERSITY PRESS

University Printing House, Cambridge CB2 8BS, United Kingdom

One Liberty Plaza, 20th Floor, New York, NY 10006, USA

477 Williamstown Road, Port Melbourne, VIC 3207, Australia

314-321, 3rd Floor, Plot 3, Splendor Forum, Jasola District Centre, New Delhi - 110025, India

103 Penang Road, #05-06/07, Visioncrest Commercial, Singapore 238467

Cambridge University Press is part of the University of Cambridge.

It furthers the University's mission by disseminating knowledge in the pursuit of
education, learning and research at the highest international levels of excellence.

www.cambridge.org
Information on this title: www.cambridge.org/9780521282352

First published 2012

A catalogue record for this publication is available from the British Library

ISBN 978-0-521-28235-2 Paperback

Additional resources for this publication at http://math.wvu.edu/~cqzhang/

To

To my mentors, colleagues, and students –
who make research exciting

Contents

Foreword

When I use the term *multigraph decomposition*, I mean a partition of the edge set. In particular, a *cycle decomposition* of a multigraph is a partition of the edge set into cycles, where I am using *cycle* to indicate a connected subgraph in which each vertex has valency 2.

There is a short list of cycle decomposition problems that I view as important problems. At the top of my list is the so-called *cycle double cover conjecture* which is the underlying motivation for this book. My reasons for ranking it at the top are discussed next.

If a conjecture has been largely ignored, then longevity essentially is irrelevant, but when a conjecture has been subjected to considerable research, then longevity plays a significant role in its importance. The cycle double cover conjecture has been with us for more than thirty years and has received considerable attention including three special workshops devoted to just this single conjecture. Thus, just in terms of longevity the cycle double cover conjecture acquires importance.

There is a deep, but not well understood, connection with the structure of graphs for if a graph X contains no Petersen minor, then a vast generalization of the cycle double conjecture is true. Trying to understand what is going on in this realm adds considerably to the allure of the cycle double cover conjecture.

Another strong attraction of the conjecture is the connections with other subareas of graph theory. These include topological graph theory, graph coloring, and flows in graphs.

This book provides a thorough exploration of the general conjecture, the approaches that have been developed, connections with the subareas mentioned above, and is presented in a highly readable form. Anyone interested in the conjecture will welcome this addition to the mathematical literature.

I wish to close the foreword with a few words about the author, Cun-Quan Zhang, known to his friends as CQ. His story is a remarkable testament to perseverance.

He was born and raised in the People's Republic of China. When the cultural revolution began in May 1966, CQ was in the seventh grade. His school, like most, closed at that time and many schoolmates became Red Guards. Due to some kind of family background problem, he was considered unqualified to join them.

He was sent to a mountain village in 1969 in Guizhou Province in western China. One year later he started to pick up all the lost education of middle school and high school. He did this with no teacher and no help. His only possibility was to teach himself.

There was nothing to do in the rice fields during Winter so that was the best time to read books. During Spring, Summer, and Fall, when there was work to be completed in the fields, he was able to read only by candlelight.

Surprisingly, textbooks were not a problem as they were available in used-book stores in Shanghai (science was considered useless at that time). The cost of a set of calculus books was no more than a pack of cigarettes.

After he finished learning high school material, he started to teach himself basic university courses in physics, mathematics, chemistry, and electrical engineering. Mathematics was the subject he loved the most.

Universities opened in 1972, but only for children from revolutionary families. He did not qualify. Also, entrance exams were not required as government selection and family background were the keys for admission.

His real mathematics career started after the cultural revolution. Chairman Mao died in 1976 and two years later university re-opened for the general public. The only requirement for entrance was a federal entrance exam.

He took an entrance exam for a graduate program in 1978, and was admitted into a master's program in mathematics in 1978 (Qufu Normal College in the hometown of Confucius). His master's thesis adviser was Professor Yongjin Zhu in the Chinese Academy of Science. He was hired by the Chinese Academy of Science as a junior faculty in 1981. Six months later, in the Fall of 1982, he came to Simon Fraser University to begin a Ph.D. under my supervision.

He was the first student from mainland China to undertake graduate studies in mathematics at Simon Fraser University. He set a very high standard for the students that followed him and was a joy to supervise. Upon completion of his Ph.D., he took a position at West Virginia University where he remains to this day.

I have watched his career with great interest. He has constantly maintained excellent standards in his research, teaching and graduate supervision. I am happy to have known him and to be able call him a friend for almost thirty years. Well done, CQ!

Brian Alspach
Callaghan, Australia

Foreword

It has been more than thirty years since I first encountered the Circuit Double Cover Conjecture. A colleague had approached me in the hallway, with what he then referred to as "a nice little problem." Indeed, even back then, the problem had already been floating here and there for quite some time. Communication, however, was not remotely what it is today and new ideas spread around erratically and at a very slow pace.

It did not seem difficult. I thought, at first, that he meant it to be an exercise for our Graph Theory course, and was somewhat embarrassed, as I was not able to solve it right away. "Every edge doubled," I was thinking out loud, "that makes an Eulerian graph. ... bridgeless, so there should be a simple way to construct a circuit partition, that avoids both copies of an edge on the same circuit... Well, I will think about it."

Three decades and hundreds of related publications later, and I still think about it, and so do many others. Indeed, a fascinating "nice little problem."

No serious mathematical question, solved or unsolved, is as simple to state and as easy to understand. No background is required; nothing essential about graphs; not even basic arithmetic. Take the intuitive concept of a line joining two points, the idea of following such lines back to the starting point to form a circuit, the ability to count "one, two" and voilà, you have the Circuit Double Cover Conjecture.

Yet, at least until the conjecture is settled, we cannot be sure that any other problem is actually harder to solve.

That said, the CDC conjecture appears to lack the fame and glamour associated with some other, celebrated and publicly praised mathematical problems. When searching the literature and the electronic space for lists of open mathematical questions, the CDC conjecture scarcely comes up. It never does among "The most famous problems," (or "most important," or "most elegant"). It gets somewhat better when the search is restricted to Graph Theory, but even then, if found, the CDC conjecture is located far down on such lists. Whatever reasons are behind that situation, C.-Q. Zhang's book in hand can be a step toward changing it.

In terms of mathematical genealogy, the CDC conjecture descends from a dignified noble dynasty, the family of "Circuit Cover" problems. Its oldest ancestor is Leonard Euler's work, dated 1736, on the Seven Bridges of Königsberg, which many consider as the very start of graph theory. Family resemblance may be superficial, yet definitely apparent – Euler's work can well be titled "The Theory of Circuit SINGLE Covers."

Another prominent grandparent is the Four-Color Theorem, formerly, the long standing Four-Color Conjecture. Common family features here are more deeply hidden and harder to spot. A four-coloring of a planar map is equivalent to a circuit double cover, which can be double covered by three even subgraphs. That simple (but not straightforward) observation led W. T. Tutte to develop his theory of nowhere zero flows, which gave birth to some additional noteworthy members of the family, among them the 5-nowhere zero flow and the 3-nowhere zero flow conjectures, now standing open for over half a century. The assertion of the CDC conjecture can also be equivalently formulated in terms of nowhere zero flows of a certain kind.

Once the family circle is allowed to grow wider, it includes the entire theory of graph-coloring. To see that, one should adopt a matroid theoretical approach, by which nowhere zero flows and graph proper vertex colorings become one. When stated within matroid theory, the CDC conjecture (for graphs) is equivalent to its generalization to regular matroids. However, with the wider community of graph theory researchers, students and enthusiasts in mind, the author deliberately chose to avoid matroid theoretical concepts and terminology in his book.

The branch of that family, rooted in the seminal work of Tutte (matroids excluded), was the subject of C.-Q. Zhang's first book: *Integer Flows and Cycle Covers of Graphs*. In his current book, the author zooms in to focus on the rich body of results, methods and questions, directly and indirectly related to the CDC conjecture.

C.-Q. Zhang's fascination with nowhere zero flow theory, circuit covers and the CDC conjecture, alongside his expertise in the areas led to his becoming a center of activity in the field, as well as the holder of a knowledge data base and a living communication hub for those who share his fascination. Most of the related new questions, new results and new ideas are addressed first to him, for verification, and comparison against the vast, ever growing and updating body of knowledge that he possesses.

In his book, Zhang draws a comprehensive panoramic image of the continuing quest for a solution to the CDC conjecture. Hundreds of related research results have spread over time across a wide variety of journals, conference proceedings, lecture notes, internal reports and private correspondence. The most essential of these results (more than a few due to Zhang himself), not only are cited in the book, but are virtually rewritten (some meanwhile improved) and the fabric of connections among them revealed. Zhang has established a uniform framework in which most of the work done so far, as well as potential directions for future work are described and understood in a clear and systematic manner.

Among the various methods and lines of action, which may lead to a solution, an attentive reader can identify the author's own plan and vision. It appears to be a rather long and involved chain of arguments, that he mostly developed himself, with an affirmative answer to the CDC conjecture at the end. Although more than a sketchy schema, it is very clearly less than a proof (and certainly not claimed to be one). The mostly paved looking path is fragmented by obstacles which, while few in number, are significant. Can all these obstacles be bridged over to complete a valid proof?

One way or another, most experts regrettably agree that the days of reasonably long and elegant solutions to long standing open problems are over. A solution to the CDC conjecture is expected to be nothing but a long sequence of rather complex arguments, rich with case analysis and technical details and hard to follow and verify. This does not at all exclude the vital need for brilliant novel methods and sparkling new ideas. An intensive coordinated team effort may be the right tool to apply here. Some activity is indeed conducted with that prospect in mind, for example, the two weeks long workshop, solely devoted to the CDC conjecture, which was held in Vancouver, British Columbia, in August 2007.

When will the CDC conjecture finally be settled? Within one month? One year? A decade? Another century? Or was it already solved last week?

My Ph.D. advisor, the late professor Haim Hanani, published his own Ph.D. dissertation back in 1938. The thesis dealt with some aspects of the Four-Color Conjecture. Thirty eight years later, during my studies under his guidance, I once entered Hanani's office, to find him gazing at a complex looking diagram that he had previously sketched on his office blackboard. "What is it that you are solving," I humbly asked. "Not really solving," Hanani smiled at me while responding: "It is the Four-Color Conjecture again. I left it to rest for forty years. Now I can afford the pleasure to devote what years and strength still left in me to struggle with a problem that I will most likely not live to see solved." "You, however," he proceeded, waving a finger at me in warning, "you, do not dare come anywhere near that problem. It is poison, potentially lethal for young professional careers."

Ironically, when we had that conversation, the Four-Color Conjecture had already been solved, with the solution not yet announced. Hanani was there to see the proof published, a few months later, with his own old Ph.D. result on the reference list.

Poison for young careers? Not necessarily. I would not recommend the CDC conjecture as an obsession for young researchers, but it can well be a very stimulating and rewarding challenge to deal with, alongside other, more modest problems. The quest is not solely about the destination. It is mostly about the journey. Yes, they say there is a priceless precious trophy, hidden at the end of the road. We may reach it, or not. Meanwhile, let us calmly walk the trail and enjoy the magnificent scenery.

C.-Q. Zhang's book will make a very helpful companion and guide along that journey.

Michael Tarsi
Tel-Aviv, Israel

Preface

The Circuit (Cycle) Double Cover Conjecture (CDC conjecture) is easy to state: *For every 2-connected graph, there is a family $\mathcal{F}$ of circuits such that every edge of the graph is covered by precisely two members of $\mathcal{F}$.* As an example, if a 2-connected graph is properly embedded on a surface (without crossing edges) in such a way that all faces are bounded by circuits, then the collection of the boundary circuits will "double cover" the graph.

The CDC conjecture (and its numerous variants) is considered by most graph theorists to be one of the major open problems in the field. One reason for this is its close relationship with topological graph theory, integer flow theory, graph coloring and the structure of snarks.

This long standing open problem has been discussed independently in various publications, such as G. Szekeres (1973 [219]), A. Itai and M. Rodeh (1978 [119]), and P. D. Seymour (1979 [205]). According to Professor W. T. Tutte, *"the conjecture is one that was well established in mathematical conversation long before anyone thought of publishing it."* Some early investigations related to the conjecture can be traced back to publications by Tutte in the later 1940s.

Some material about circuit covers was presented in the book *Integer Flows and Cycle Covers of Graphs* (1997 [259]) by the author as an application of flow theory. There are several reasons why the author decided to write a follow-up book mainly on this subject. Of course, after more than a decade, some new progress and discoveries have been made. Furthermore, in the previous book, circuit cover is a secondary subject and its presentation is based mainly on the techniques and approaches of integer flow theory. However, not all techniques in this subject rely completely on flow theory, and most material can be approached and presented independent of flow theory. Without using flow theory, 3-edge-coloring of cubic graphs becomes one of the major techniques in this book. Since flow theory does provide some powerful tools and beautiful descriptions in this area, it will be covered in later chapters when readers (especially students) have gained sufficient familiarity with the other approaches that they are prepared for further depth. However, most of the book can be read with little or no knowledge of flow theory.

Most of the basic lemmas and theorems of integer flow theory in this book are presented (many without proof) in an appendix. Readers who are interested in further study of the theory of integer flow are referred to the book *Integer Flows and Cycle Covers of Graphs* [259] for more comprehensive coverage.

Since the main subject of this book is circuit double covers, and a smallest counterexample to the CDC conjecture is cubic, most graphs considered in this book are cubic, and the circuit coverage is restricted to 1 or 2 in most cases. Non-cubic graphs will be considered whenever flow theory is applied and in the presentation of a few other techniques.

While topological techniques offer promising approaches in this field, most results and techniques presented in this book are combinatorial. In order to keep the main theme of the book as focused as possible, the author decided to concentrate for the most part on results and techniques in structural graph theory. Topological results are presented only if either they can be obtained by combinatorial methods or they are needed as background for a combinatorial approach.

Most material presented in this book follows the pioneering survey papers by Jaeger (1985 [130]), and Jackson (1993 [121]), and covers the major topics discussed and presented at three historical workshops *(Barbados, February 25–March 4, 1990; IINFORM, Vienna, January 1991; PIMS at UBC, Vancouver, August 22–31, 2007)*.

After the publication of the first book by the author, some changes in notation and terminology were strongly suggested by colleagues. Circuits are commonly defined as connected 2-regular subgraphs, while cycles are defined as subgraphs with even degree at every vertex (even subgraphs). These definitions were originally adapted from matroid and flow theory because a vector of the cycle space of a graph is often called a cycle. In order to be consistent with other popular textbooks and avoid possible confusion, in this book a cycle is an even subgraph. Cycle (even subgraph) double cover, circuit double cover are technically the same (in most cases) because every even subgraph has a circuit decomposition.

A set of appendices collects some fundamental graph theoretical results which are useful in the study of circuit covers.

The Petersen graph has been at the core of much major research in structural graph theory. Families of graphs containing no subdivision of the Petersen graph have been proved to have various strong properties. However, for families of graphs with a Petersen subdivision, either some graphic properties do not hold, or determining whether they do is an extremely hard problem. Many circuit covering problems fall in one of these two categories. For the convenience of readers, an appendix (Section B.2), *A Mini Encyclopedia of the Petersen Graph*, is included in this book.

A section of exercises of varying difficulty is included at the end of each chapter. Some interesting results that are not presented in the main part of the chapters are included in the exercises. Appendix D, *Hints for Exercises*, appears at the end of the book.

For the convenience of readers, basic terminology and notation as well as some special terminology is listed in a glossary.

I would like to thank B. Alspach, M. Ellingham, G. Fan, H. Fleischner, J. Goldwasser, R. Gould, J. Hägglund, A. Hoffmann-Ostenhof, B. Jackson, T. Jensen, M. Kochol, H.-J. Lai, R. Luo, K. Markström, B. Sagan, P. Seymour, D. West, M. Tarsi, X. Zha, and other experts and colleagues in this area for their encouragement, valuable suggestions and information.

This book is developed based on a series of lectures that the author gave at various workshops and graduate courses at West Virginia University. I would like to express my appreciation to the WVU graduate students (in particular, W. Tang, K. Toth, Y. Wu, D. Ye, and others in my classes at West Virginia University). Their valuable comments helped me to polish the book.

Readers may send their suggestions via email to
cqzhang@math.wvu.edu
or by regular mail to me. Corrections and updated information will be available at the World Wide Web site:
http://math.wvu.edu/~ cqzhang/ .

Cun-Quan 'C.-Q.' Zhang
October 2011

1

Circuit double cover

Most terminology and notation in this book follow standard textbooks in graph theory (such as [18], [19], [41], [242], etc.). Some terminology is slightly different from those classical textbooks.

Definition 1.0.1 Let $G = (V, E)$ be a graph with vertex set V and edge set E.
  (1) A circuit is a connected 2-regular graph.
  (2) A graph (subgraph) is even if the degree of each vertex is even.
  (3) A bridge (or cut-edge) of a graph G is an edge whose removal increases the number of components of G (equivalently, a bridge is an edge that is not contained in any circuit of G).

Note that, in much of the literature related to circuit covers and integer flows, an even graph/subgraph is also called a cycle, which is adapted from matroid theory, and is different from what is used in many other graph theory textbooks. For the sake of less confusion, we will use *even graph/subgraph* instead of *cycle* in this book.

Definition 1.0.2 (1) A family $\mathcal{F}$ of circuits (or even subgraphs) of a graph G is called a circuit cover (or an even subgraph cover) if every edge of G is contained in some member(s) of $\mathcal{F}$.
  (2) A circuit cover (or an even subgraph cover) $\mathcal{F}$ of a graph G is a double cover of G if every edge is contained in precisely two members of $\mathcal{F}$.

Graphs considered in this book may contain loops or parallel edges. However, most graphs are bridgeless (since this is necessary for circuit covering problems).

1.1 Circuit double cover conjecture

The following is one of the most well-known open problems in graph theory, and is the major subject of this book.

The Circuit Double Cover (CDC) Conjecture. *Every bridgeless graph G has a family $\mathcal{F}$ of circuits such that every edge of G is contained in precisely two members of $\mathcal{F}$.*

The CDC conjecture was presented as an "open question" by Szekeres ([219] p. 374) for cubic graphs (as we will see soon in Section 1.2, it is equivalent for all bridgeless graphs). The conjecture was also independently stated by Seymour in [205] (Conjecture 3.3 on p. 347) for all bridgeless graphs. An equivalent version of the CDC conjecture was proposed by Itai and Rodeh (Open problem (ii) in [119] p. 298) that *every bridgeless graph has a family $\mathcal{F}$ of circuits such that every edge is contained in one or two members of $\mathcal{F}$*. (It is not hard to see that this open problem is equivalent to the circuit double cover conjecture. See Exercise 1.3.)

For the origin of the conjecture, some mathematicians gave the credit to Tutte. According to a personal letter from Tutte to Fleischner [237], he said, *"I too have been puzzled to find an original reference. I think the conjecture is one that was well established in mathematical conversation long before anyone thought of publishing it."* It was also pointed out in the survey paper by Jaeger [130] that *"it seems difficult to attribute the paternity of this conjecture;"* and also pointed out in some early literatures (such as [82]) that *"its origin is uncertain."* This may explain why the CDC conjecture is considered as "folklore" in [19] (Unsolved problem 10, p. 584).

An early work related to circuit double cover that we are able to find is a paper by Tutte published in 1949 [229], [237].

The circuit double cover conjecture is obviously true for 2-connected planar graphs since the boundary of every face is a circuit and the set of the boundaries of all faces forms a circuit double cover of an embedded graph. One might attempt to extend this observation further to all 2-connected graphs embedded on some surfaces. However, it is not true that *any embedding* of a 2-connected graph is free of a handle-bridge. That is, the boundary of some face may not be a circuit. So, *can we find an embedding of a 2-connected graph G on some surface Σ such that the boundary of every face is a circuit?* The existence of such an embedding is an even stronger open problem in topology (see Section 1.4 for more detail).

The circuit double cover conjecture is also true for the family of 3-edge-colorable cubic graphs since a 3-edge-colorable cubic graph is covered by three bi-colored 2-factors (see Section 1.3). This observation will be further extended to all bridgeless graphs (not necessarily cubic) that admit nowhere-zero 4-flows (see Chapter 7 and Appendix C).

1.2 Minimal counterexamples

In this section, we study some basic structures of a smallest counterexample to the circuit double cover conjecture.

Definition 1.2.1 Let G be a graph. The suppressed graph of G is the graph obtained from G by replacing each maximal subdivided edge with a single edge, and is denoted by $\overline{G}$.

Lemma 1.2.2 [205] *Let G be a bridgeless graph. If G has no circuit double cover, then there is a bridgeless cubic graph G' such that $|E(G')| \leq |E(G)|$ and G' has no circuit double cover either.*

Proof Let $v \in V(G)$ with $d(v) \geq 4$. By Theorem A.1.14 (the vertex splitting method), there is a pair of edges e_1, e_2 of G incident with v such that the graph $G_{[v;\{e_1,e_2\}]}$ obtained from G by splitting e_1 and e_2 away from v remains bridgeless. Let G' be the suppressed graph of $G_{[v;\{e_1,e_2\}]}$. It is evident that any circuit double cover of G' can be adjusted or modified to a circuit double cover of G. Thus, G' does not have a circuit double cover. Repeating this procedure, we obtain a bridgeless cubic graph G'' which is smaller than G and has no circuit double cover either. $\square$

Lemma 1.2.3 [205] *Let G be a bridgeless graph. If G has no circuit double cover, then G is contractible to an essentially 4-edge-connected graph G' such that G' has no circuit double cover either.*

Proof Let T be a non-trivial 2- or 3-edge-cut of G with components Q_1, Q_2. Let G_i be the graph obtained from G by contracting Q_i, for each $i = 1, 2$. If G_i has a circuit double cover $\mathcal{F}_i$ for each $i = 1, 2$, then $\mathcal{F}_1$ and $\mathcal{F}_2$ can be modified at edges of T so that the resulting family of circuits is a circuit double cover of G (see Figure 1.1). $\square$

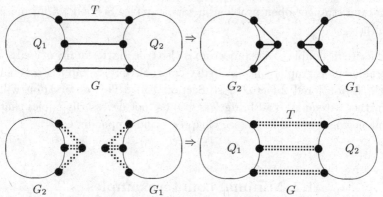

Figure 1.1 *CDC of G constructed from G_1 and G_2*

With some straightforward observations, the following theorem summarizes some structure of a minimal counterexample to the conjecture of circuit double cover.

Theorem 1.2.4 *If G is a minimum counterexample to the circuit double cover conjecture, then*

(1) G is simple, 3-connected and cubic (Lemmas 1.2.2 and 1.2.3);

(2) G has no non-trivial 2 or 3-edge-cut (Lemma 1.2.3);

(3) G is not 3-edge-colorable (see Section 1.3);

(4) G is not planar (see Sections 1.1 and 1.4).

There are more structural properties for a smallest counterexample to the conjecture. The following is a partial summary of some results we will discuss in this book.

(1) G contains no Hamilton path (Corollary 4.2.8).

(2) G contains a subdivision of the Petersen graph (Theorem 3.2.1).

(3) The girth of G is at least 12 (Theorem 9.1.1).

(4) For each edge $e \in E(G)$, the suppressed cubic graph $\overline{G - \{e\}}$ is not 3-edge-colorable (Theorem 2.2.4).

(5) $G \neq \overline{G' - \{e\}}$ for some 3-edge-colorable cubic graph G' and some $e \in E(G')$ (Theorem 4.2.3).

(6) G is of oddness at least 6 (Theorem 4.2.4).

1.3 3-edge-coloring and even subgraph cover

Recall some definitions. A subgraph H is even if the degree of every vertex is even (defined in Definition 1.0.1). A family $\mathcal{F}$ of even subgraphs is called an even-subgraph double cover of a graph G if every edge e of G is contained in precisely 2 members of $\mathcal{F}$ (defined in Definition 1.0.2).

Definition 1.3.1 An even-subgraph double cover $\mathcal{F}$ of a graph G is called a k-even-subgraph double cover if $|\mathcal{F}| \leq k$.

It is trivial that every circuit double cover is an even subgraph double cover. And, by Lemma A.2.2, it is also straightforward that every even subgraph double cover can be converted to a circuit double cover (by replacing each even subgraph with a set of circuits).

The following theorem was formulated by Jaeger [130]. (The equivalence of (1) and (2) was also applied in [205].)

Theorem 1.3.2 *Let G be a cubic graph. Then the following statements are equivalent:*
(1) G is 3-edge-colorable;
(2) G has a 3-even subgraph double cover.
(3) G has a 4-even subgraph double cover.

Proof (1) $\Rightarrow$ (2): Let $c : E(G) \to \mathbb{Z}_3$ be a 3-edge-coloring of G. Then

$$\{c^{-1}(\alpha) \cup c^{-1}(\beta) : \alpha, \beta \in \mathbb{Z}_3, \alpha \neq \beta\}$$

is a 3-even subgraph double cover of G.

(2) $\Rightarrow$ (1): Let $\{C_\mu : \mu \in \mathbb{Z}_2 \times \mathbb{Z}_2 - \{(0,0)\}\}$ be a 3-even subgraph double cover of G. Then let $c : E(G) \to \mathbb{Z}_2 \times \mathbb{Z}_2 - \{(0,0)\}$ be defined as follows: for each edge $e \in E(G)$, $c(e) = \mu' + \mu''$ if e is covered by even subgraphs $C_{\mu'}$ and $C_{\mu''}$. It is easy to see that c is a proper 3-edge-coloring of G.

(2) $\Rightarrow$ (3): Trivial, let $C_4 = \emptyset$.

(3) $\Rightarrow$ (2): Let $\{C_1, C_2, C_3, C_4\}$ be a 4-even subgraph double cover of G. Then $\{C_1 \triangle C_2, C_1 \triangle C_3, C_1 \triangle C_4\}$ is a 3-even subgraph double cover of G. $\qquad\square$

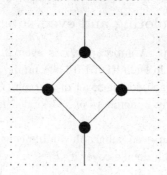

Figure 1.2 *A 2-cell embedding of K_4 on the torus that is not circular*

1.4 Circuit double covers and graph embeddings

All surfaces considered in this book are connected 2-manifolds (see [243] or [182] for an introduction to the topology of surfaces). The set of faces of an embedded graph G is denoted by $F(G)$.

Definition 1.4.1 Let G be a 2-connected graph and Σ be a surface. An embedding of G on Σ is called a 2-cell embedding if each face of the embedded graph is homeomorphic to the open unit disk.

Definition 1.4.2 Let G be a graph G that has a 2-cell embedding on a surface Σ.

(1) A coloring of the faces of the embedded graph G is proper if for each edge $e \in E(G)$, the faces on the two sides of e are colored differently.

(2) The embedded graph G is k-face-colorable on Σ if there is a proper face coloring of G requiring at most k colors.

Note that a 2-cell embedding of a graph G on a 2-manifold does not guarantee a proper face coloring of the graph G since some edge might be on the boundary of the same face (see Figure 1.2).

Thus, we have to consider an embedding with the following property.

Definition 1.4.3 A 2-cell embedding of a 2-connected graph G on a surface is circular if the boundary of each face is a circuit.

It is not hard to see that the definition of a circular 2-cell embedding of a graph on a surface has the following equivalent statements:

(1) each edge is on the boundaries of two distinct faces;

(2) the closure of each face is homeomorphic to the closed unit disk.

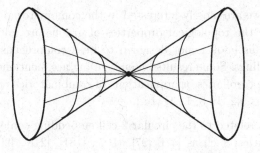

Figure 1.3 *A pinch point on a pseudo-surface*

For a 2-connected planar graph, the circuit double cover conjecture is obviously true since the family of all facial boundaries is a circuit double cover of the graph. This is also true for every 2-connected graph which has a circular 2-cell embedding on some surface. However, one cannot assume that every 2-connected graph has such an embedding, since that is an even stronger open problem.

Conjecture 1.4.4   *Every 2-connected graph has a circular 2-cell embedding on some 2-manifold.*

Conjecture 1.4.4 has also been considered as a "folklore" conjecture (see [185]) with possible origin due to Tutte in the mid-1960s (see [19], Conjecture 3.10 on p. 95, or [33]). It is also presented in the articles [97], [167] as an open question.

Note that the CDC conjecture and Conjecture 1.4.4 are equivalent for connected *cubic* graphs (Exercise 1.4). Let $\mathcal{F}$ be a circuit double cover of G. One can consider each circuit C of $\mathcal{F}$ as the boundary of a disk and join those disks at the edges of G; the resulting surface is certainly a 2-manifold.

For a non-cubic graph G, with the method we used in the above paragraph, the "surface" that we obtained from a circuit double cover may not be a 2-manifold. It is possible that some "pinch point" appears on the constructed "surface," which is sometime called a pseudo-surface (see Figure 1.3).

Some mathematicians believe that the circular 2-cell embedding is a feasible approach to the CDC conjecture, although Conjecture 1.4.4 is stronger. However, there is little progress yet in this direction.

In this book, we are mostly interested in the combinatorial structure of graphs, while the topological properties of graphs are not a major subject. Thus, this book is not designed to be a comprehensive study of graph embeddings. Some results on embedding are mentioned simply because they are corollaries or applications of combinatorics results (for example, Sections 9.2, 10.8, 13.1, etc.).

Partial results related to the circular 2-cell embedding conjecture can be found in articles, such as [47], [97], [167], [181], [218], [251], [252], [253], etc.

1.5 Open problems

Conjecture 1.5.1 (Strong CDC conjecture; Seymour, see [61] p. 237, also [62], [83]) *Let G be a bridgeless cubic graph and C be a circuit of G. The graph G has a circuit double cover $\mathcal{F}$ with $C \in \mathcal{F}$.*

A recent computer aided search [21] showed that Conjecture 1.5.1 holds for all cubic graphs of order at most 36.

Conjecture 1.5.2 (Sabidussi and Fleischner [60], and Conjecture 2.4 in [2] p. 462) *Let G be a cubic graph such that G has a dominating circuit C. Then G has a circuit double cover $\mathcal{F}$ such that C is a member of $\mathcal{F}$.*

Remark. (1) Conjecture 1.5.1 implies both the circuit double cover conjecture and Conjecture 1.5.2.

(2) Conjecture 1.5.2 is an equivalent version of the original Sabidussi Conjecture (Conjecture 10.5.2) for the compatible circuit decomposition problem.

Problem 1.5.3 (Seymour [210]) Can you prove that a smallest counterexample to the CDC conjecture does not have any cyclic 4-edge-cut?

1.6 Exercises

Exercise 1.1 Explain why the following "proof" to the circuit double cover conjecture is incorrect.

"Proof" Let G be embedded on a surface Σ. Then the collection of the boundaries of all faces (regions) forms a circuit double cover of the graph.

Exercise 1.2 Explain why the following "proof" to the circuit double cover conjecture is incorrect.

"Proof" Let $2G$ be the graph obtained from G by replacing every edge with a pair of parallel edges. Since the resulting graph $2G$ is an even graph, by Lemma A.2.2, the even graph $2G$ has a circuit decomposition $\mathcal{F}$. Is it obvious that $\mathcal{F}$ is a circuit double cover of the original graph G?

Exercise 1.3 Show that the circuit double cover conjecture is equivalent to the following statement.

"Every bridgeless graph has a family $\mathcal{F}$ of circuits such that every edge is contained in one or two members of $\mathcal{F}$." (Open problem (ii) in [119] p. 298.)

Exercise 1.4 Let G be a bridgeless cubic graph. The graph G has a circuit double cover if and only if G has a circular 2-cell embedding on some surface.

Exercise 1.5 Let T be a minimal edge-cut of a graph G with $|T| \leq 3$. Let Q_1, Q_2 be the two components of $G - T$ and for each $\{i, j\} = \{1, 2\}$, let $H_i = G/Q_j$ be the graph obtained from G by contracting all edges of Q_j. If both H_1, H_2 have k-even subgraph double covers (for some integer k), then G also has a k-even subgraph double cover.

Exercise 1.6 (Ding, Hoede and Vestergaard [42]) Let G be a 2-connected graph other than a circuit and with no loop. Assume that G and all of its bridgeless subgraphs have circuit double covers. Show that G has a circuit double cover $\mathcal{F}$ consisting of distinct circuits.

Definition 1.6.1 Let Γ be a group and $S \subset \Gamma$ such that $1 \notin S$ and $\alpha \in S$ if and only if $\alpha^{-1} \in S$. The Cayley graph $G(\Gamma, S)$ is the graph with the vertex set Γ such that two vertices x, y are adjacent in $G(\Gamma, S)$ if and only if $x = y\alpha$ for some $\alpha \in S$.

Exercise 1.7 (Hoffman, Locke and Meyerowitz [109]) Show that every Cayley graph with minimum degree at least 2 has a circuit double cover.

2
Faithful circuit cover

2.1 Faithful circuit cover

The concept of faithful circuit cover is not only a generalization of the circuit double cover problem, but also an inductive approach to the CDC conjecture in a very natural way.

Let $\mathbb{Z}^+$ be the set of all positive integers, and $\mathbb{Z}^\star$ be the set of all non-negative integers.

Definition 2.1.1 Let G be a graph and $w : E(G) \mapsto \mathbb{Z}^+$. A family $\mathcal{F}$ of circuits (or even subgraphs) of G is a faithful circuit cover (or faithful even subgraph cover, respectively) with respect to w if each edge e is contained in precisely $w(e)$ members of $\mathcal{F}$.

Figure 2.1 shows an example of faithful circuit covers of (K_4, w) where $w : E(K_4) \mapsto \{1, 2\}$. Here $w^{-1}(1)$ induces a Hamilton circuit and $w^{-1}(2)$ induces a perfect matching (a pair of diagonals).

It is obvious that *the circuit double cover is a special case of the faithful circuit cover problem that the weight w is constant 2 for every edge.*

Definition 2.1.2 Let G be a graph. A weight $w : E(G) \mapsto \mathbb{Z}^+$ is

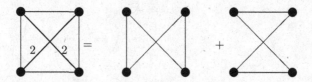

Figure 2.1 *Faithful circuit cover – an example*

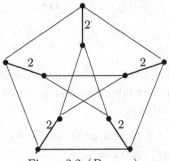

Figure 2.2 (P_{10}, w_{10})

eulerian if the total weight of every edge-cut is even. And (G, w) is called an eulerian weighted graph.

Definition 2.1.3 Let G be a graph. An eulerian weight $w : E(G) \mapsto \mathbb{Z}^+$ is admissible if, for every edge-cut T and every $e \in T$,

$$w(e) \leq \frac{w(T)}{2}.$$

And (G, w) is called an admissible eulerian weighted graph if w is eulerian and admissible.

If G has a faithful circuit cover $\mathcal{F}$ with respect to a weight $w : E(G) \mapsto \mathbb{Z}^+$, then the total weight of every edge-cut must be even since, for every circuit C of $\mathcal{F}$ and every edge-cut T, the circuit C must use an *even* number of *distinct* edges of the cut T. With this observation, the requirements of being *eulerian and admissible* are *necessary* for faithful circuit covers.

Problem 2.1.4 Let G be a bridgeless graph with $w : E(G) \mapsto \mathbb{Z}^+$. If w is admissible and eulerian, does G have a faithful circuit cover with respect to w?

Unfortunately, Problem 2.1.4 is *not* always true. The Petersen graph P_{10} with an eulerian weight w_{10} (see Figure 2.2) does not have a faithful circuit cover: where the set of weight 2 edges induces a perfect matching of P_{10} and the set of weight 1 edges induces two disjoint pentagons (Proposition B.2.26).

Definition 2.1.5 Let G be a bridgeless graph and w be an admissible eulerian weight of G. The eulerian weighted graph (G, w) is a contra pair if G does not have a faithful circuit cover with respect to the weight w.

For a given weight $w : E(G) \mapsto \mathbb{Z}^+$, denote

$$E_{w=i} = \{e \in E(G) : w(e) = i\}.$$

Some lemmas and definitions about eulerian weights are presented in this section to prepare for later discussions. Most of these results are straightforward observations.

Proposition 2.1.6 *Let G be a graph and $w : E(G) \mapsto \mathbb{Z}^+$. Then the following statements are equivalent.*
(1) The weight w is eulerian.
(2) The subgraph induced by all odd-weight edges is even.

Proof Exercise 2.3.   □

Definition 2.1.7 Let $\mathcal{F}$ be a set of circuits of G. The mapping $w_{\mathcal{F}} : E(G) \mapsto \mathbb{Z}^\star$ defined as follows is called the coverage weight of $\mathcal{F}$, or the weight induced by $\mathcal{F}$: for each edge $e \in E(G)$, $w_{\mathcal{F}}(e)$ is the number of members of $\mathcal{F}$ containing the edge e.

Proposition 2.1.8 *Let $\mathcal{F}$ be a set of circuits of G. The weight $w_{\mathcal{F}}$ induced by $\mathcal{F}$ is eulerian and admissible.*

Proposition 2.1.9 *Let G be a graph with a weight $w : E(G) \mapsto \{1, 2\}$. Then w is admissible if and only if G is bridgeless.*

Proof Exercise 2.2.   □

Definition 2.1.10 An eulerian weight $w : E(G) \mapsto \mathbb{Z}^+$ is an eulerian $(1, 2)$-weight of G if $1 \leq w(e) \leq 2$ for every $e \in E(G)$.

Since a smallest counterexample to the CDC conjecture is cubic, *most graphs G considered in this book are cubic, and most weights $w : E(G) \mapsto \mathbb{Z}^+$ are eulerian $(1, 2)$-weights.* By Proposition 2.1.9, we may omit the requirement of admissibility if the graph is bridgeless and the $(1, 2)$-weight is eulerian.

The following are some definitions that we will use later.

Definition 2.1.11 Let $w : E(G) \mapsto \mathbb{Z}^\star$. The support of w, denoted by $supp(w)$, is the set of edges e with $w(e) \neq 0$.

Definition 2.1.12 Let $w : E(G) \mapsto \mathbb{Z}^\star$ be an admissible eulerian weight of G. A family $\mathcal{F}$ of even subgraphs of G is called a faithful k-even subgraph cover if $\mathcal{F}$ is a faithful cover of (G, w) and $|\mathcal{F}| \leq k$.

2.2 3-edge-coloring and faithful cover

Like many mainstream research areas in graph theory, 3-edge-coloring plays a central role in the study of the faithful circuit cover problem. The following is one of the most frequently used lemmas in this field.

Lemma 2.2.1 (Seymour [205]) *Let G be a cubic graph and $w : E(G) \mapsto \{1, 2\}$ be an eulerian weight. Then the following statements are equivalent:*

(1) G is 3-edge-colorable;

(2) G has a faithful 3-even-subgraph cover with respect to w.

Proof (1) $\Rightarrow$ (2): Let $c : E(G) \mapsto \mathbb{Z}_3$ be a 3-edge-coloring of G, and let $C_{(\alpha, \beta)}$ be the 2-factor of G induced by all α-, β-colored edges, for each $\{\alpha, \beta\} \subset \mathbb{Z}_3$. Let $C_0 = G[E_{w=1}]$ which is an *even subgraph* (by Proposition 2.1.6). Then

$$\{C_0 \triangle C_{(0,1)}, C_0 \triangle C_{(0,2)}, C_0 \triangle C_{(1,2)}\} \tag{2.1}$$

is a faithful 3-even subgraph cover of G with respect to the eulerian $(1, 2)$-weight w.

(2) $\Rightarrow$ (1): Let $\{C_{\langle 1,0 \rangle}, C_{\langle 0,1 \rangle}, C_{\langle 1,1 \rangle}\}$ be a faithful 3-even subgraph cover of (G, w), and let $C_{\langle 0,0 \rangle} = G[E_{w=1}]$. Define an edge-coloring of G as follows, $c : E(G) \mapsto \mathbb{Z}_2 \times \mathbb{Z}_2 - \{\langle 0, 0 \rangle\}$ where, for each edge $e \in C_{\langle \alpha_1, \beta_1 \rangle} \cap C_{\langle \alpha_2, \beta_2 \rangle}$, let $c(e) = \langle \alpha_1 + \alpha_2, \beta_1 + \beta_2 \rangle$. It is easy to see that c is a 3-edge-coloring of the cubic graph G since $\{C_{\langle 0,0 \rangle}, C_{\langle 1,0 \rangle}, C_{\langle 0,1 \rangle}, C_{\langle 1,1 \rangle}\}$ is a 4-even subgraph double cover. $\square$

One may notice that it is possible that one member of Equation (2.1) could be an empty set if C_0 is a member of $\{C_{(0,1)}, C_{(0,2)}, C_{(1,2)}\}$. Thus, we have the following lemma for this special case.

Lemma 2.2.2 *Let G be a cubic graph and $w : E(G) \mapsto \{1, 2\}$ be an eulerian weight. Then the following statements are equivalent:*

(1) G has a faithful 2-even-subgraph cover with respect to w;

(2) G has a 3-edge-coloring $c : E(G) \mapsto \mathbb{Z}_3$ such that $E_{w=2}$ is monocolored (that is, $E_{w=2} = c^{-1}(i)$ for some $i \in \mathbb{Z}_3$).

⋆ Applications of Lemma 2.2.1

Since the 4-color theorem is equivalent to the 3-edge-coloring for all bridgeless cubic planar graphs (Theorem B.1.1), an immediate corollary of Lemma 2.2.1 is the following early result (Theorem 2.2.3) by Seymour.

An alternative proof of Theorem 2.2.3 (slightly stronger) is provided by Fleischner (see Theorem 10.3.1) without using 4-color theorem.

Theorem 2.2.3   (Seymour [205], and Fleischner [59], [62]) *If G is a planar, bridgeless graph associated with an eulerian weight $w : E(G) \mapsto \{1,2\}$, then G has a faithful circuit cover with respect to w.*

Theorem 1.3.2 shows that a 3-edge-colorable cubic graph has a circuit double cover. In the following theorem, the circuit double cover conjecture is verified for a family of "nearly" 3-edge-colorable graphs.

Theorem 2.2.4   (Jaeger [131]) *Let G be a bridgeless cubic graph. If G has an edge e_0 such that the suppressed cubic graph $\overline{G - e_0}$ is 3-edge-colorable, then G has a 5-even subgraph double cover.*

Proof It is sufficient to show the theorem for 2-edge-connected cubic graphs. Let $e_0 = x_0 y_0$. Since $\overline{G - e_0}$ is 3-edge-colorable, the suppressed cubic graph $\overline{G - e_0}$ remains bridgeless and connected. Hence, by Theorem A.1.3, there is a circuit C_0 of $G - e_0$ containing both x_0 and y_0. Define an eulerian weight $w_1 : E(\overline{G - e_0}) \mapsto \{1,2\}$ with

$$w_1(e) = \begin{cases} 1 & \text{if } e \in E(C_0) \\ 2 & \text{otherwise} \end{cases}$$

By Lemma 2.2.1, let $\mathcal{F}_1$ be a faithful 3-even subgraph cover of $(\overline{G - e_0}, w_1)$.

Let Q_1 and Q_2 be two segments of the circuit C_0 between x_0 and y_0. Then $\mathcal{F}_1 \cup \{Q_1 + e_0, Q_2 + e_0\}$ is a 5-even subgraph double cover of G. □

The main idea in the proof of Theorem 2.2.4 is that the graph G has two subgraphs, each of which is a subdivision of a 3-edge-colorable cubic graph, and their intersection is an even subgraph. The union of some circuit covers of these two subgraphs yields a double cover of G. This is one of the major approaches to the circuit double cover conjecture. This method will be further studied in, for example, Chapters 3, 4, 5, 6, etc.

2.3 Construction of contra pairs

In Proposition B.2.26, we have shown that (P_{10}, w_{10}) is a cubic contra pair. In this section, we are to construct some infinite families of cubic contra pairs.

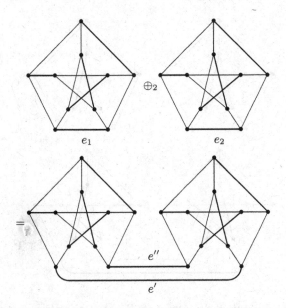

Figure 2.3 $\oplus_2$-sum of two copies of (P_{10}, w_{10}) at edges e_1 and e_2

$\star$ $\oplus_2$- and $\oplus_3$-sums

Definition 2.3.1 ($\oplus_2$-sum, see Figure 2.3) Let G_i be a cubic graph with an eulerian weight $w_i : E(G_i) \mapsto \{1, 2\}$ and $e_i = x_i y_i \in E_{w_i = \alpha}(G_i)$, for each $i = 1, 2$ and some $\alpha \in \{1, 2\}$. The $\oplus_2$-sum of (G_1, w_1) and (G_2, w_2) at edges e_1 and e_2, denoted by $(G_1, w_1) \oplus_2 (G_2, w_2)$, is the admissible eulerian weighted cubic graph constructed as follows: replace $\{e_1, e_2\}$ with two new edges $\{e', e''\}$ such that $e' = x_1 x_2$ and $e'' = y_1 y_2$ with $w(e') = w(e'') = \alpha$.

Lemma 2.3.2 *Let G_i be a cubic graph with an eulerian weight $w_i : E(G_i) \mapsto \{1, 2\}$ and $e_i = x_i y_i \in E_{w_i = \alpha}(G_i)$, for each $i = 1, 2$ and some $\alpha \in \{1, 2\}$. Then the $\oplus_2$-sum of (G_1, w_1) and (G_2, w_2) at e_1, e_2 is a contra pair if and only if one of (G_1, w_1) and (G_2, w_2) is a contra pair.*

Definition 2.3.3 ($\oplus_3$-sum, see Figure 2.4) Let G_i be a cubic graph with an eulerian weight $w_i : E(G_i) \mapsto \{1, 2\}$, for each $i = 1, 2$. And let $v_i \in V(G_i)$, for each $i = 1, 2$, with $N_{G_i}(v_i) = \{x_{i,1}, x_{i,2}, x_{i,3}\}$ and $w_1(v_1 x_{1,j}) = w_2(v_2 x_{2,j})$ for each $j \in \{1, 2, 3\}$. The $\oplus_3$-sum of (G_1, w_1) and (G_2, w_2) at v_1 and v_2, denoted by $(G_1, w_1) \oplus_3 (G_2, w_2)$, is the admissible eulerian weighted cubic graph constructed as follows: replace

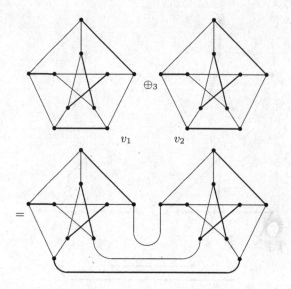

Figure 2.4 $\oplus_3$-*sum of two copies of* (P_{10}, w_{10}) *at vertices* v_1 *and* v_2

six edges $\{x_{i,j}v_i : i \in \{1,2\}, j \in \{1,2,3\}\}$ with three new edges $\{x_{1,j}x_{2,j} : j \in \{1,2,3\}\}$ with weight

$$w(x_{1,j}x_{2,j}) = w_1(x_{1,j}v_1) = w_2(x_{2,j}v_2).$$

Lemma 2.3.4 *Let* G_i *be a cubic graph with an eulerian weight* w_i : $E(G_i) \mapsto \{1,2\}$, *for* $i = 1, 2$. *Then the* $\oplus_3$-*sum of* (G_1, w_1) *and* (G_2, w_2) *at a pair of vertices of* G_1 *and* G_2 *is a contra pair if and only if one of* (G_1, w_1) *and* (G_2, w_2) *is a contra pair.*

⋆ Isaacs–Fleischner–Jackson product

The following method of constructing a cyclically 4-edge-connected contra pair was discovered by Jackson [121]. An equivalent version for the *compatible circuit decomposition problem* (see Section 10.6) was discovered independently by Fleischner [121].

Definition 2.3.5 Let (G_F, w_F) and (G_M, w_M) be a pair of cubic contra pairs. The Isaacs–Fleischner–Jackson product (or IFJ-product) of (G_F, w_F) and (G_M, w_M), denoted by $(G_F, w_F) \otimes_{IFJ} (G_M, w_M)$, is constructed as follows (see Figure 2.5).

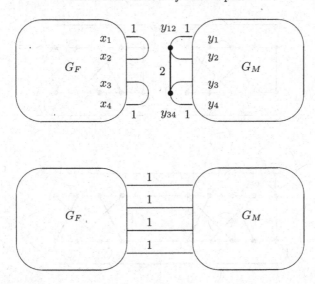

Figure 2.5 *IFJ-product*

Let $x_1x_2, x_3x_4 \in E_{w_F=1}$, $y_{12}y_{34} \in E_{w_M=2}$, and $y_{12}y_1$, $y_{12}y_2$, $y_{34}y_3$, $y_{34}y_4 \in E_{w_M=1}$. The eulerian weighted graph $(G_F, w_F) \otimes_{IFJ} (G_M, w_M)$ is obtained from (G_F, w_F) and (G_M, w_M) by deleting the edges x_1x_2, x_3x_4 and the vertices y_{12}, y_{34} and adding four new edges $e_i = x_iy_i$ for each $i = 1, 2, 3, 4$ with weight 1.

Figure 2.6 is an example of the IFJ-product of two copies of (P_{10}, w_{10}).

Lemma 2.3.6 (Jackson [121], also see [75]) *Let (G_F, w_F) and (G_M, w_M) be a pair of cubic contra pairs. Then the IFJ-product $(G_F, w_F) \otimes_{IFJ} (G_M, w_M)$ is also a contra pair.*

Proof Prove by contradiction. Suppose that the IFJ-product $(G_F, w_F) \otimes_{IFJ} (G_M, w_M)$ has a faithful circuit cover $\mathcal{F}_I$. Let $\mathcal{F}'$ be a subfamily of $\mathcal{F}_I$ such that each member of $\mathcal{F}'$ contains some edge of $\{e_1, \ldots, e_4\}$. Deleting all e_i from each member of $\mathcal{F}'$, we obtain a set of four paths $\{P'_F, P''_F, P'_M, P''_M\}$ where $P'_F, P''_F \subseteq G_F$ joining $\{x_1, \ldots, x_4\}$, and, P'_M, $P''_M \subseteq G_M$ joining $\{y_1, \ldots, y_4\}$. (See Figure 2.5.)

If y_1, y_2 are endvertices of P'_M, then (G_F, w_F) is not a contra pair. This contradicts that (G_F, w_F) is a contra pair.

If y_1, y_3 are endvertices of P'_M, then (G_M, w_M) is not a contra pair. This contradicts that (G_M, w_M) is a contra pair. $\square$

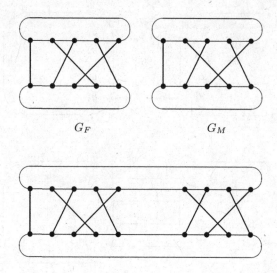

G_F G_M

Figure 2.6 *IFJ-product of two copies of* (P_{10}, w_{10})

Remark. In [63], [75], [121], [259], etc., it was conjectured that every contra pair, except for (P_{10}, w_{10}), has a cyclic edge-cut of size at most four. Some cyclic 5-edge-connected, cubic contra pair of order 36 was discovered in a recent computer aided search [21]. However, up to now, except for the three methods presented in this section, there is no general method for constructing cyclically highly connected contra pairs.

2.4 Open problems

Conjecture 2.4.1 (Seymour [205]) *Let* $w : E(G) \mapsto \mathbb{Z}^+$ *be an admissible eulerian weight of a bridgeless graph* G *such that* $w(e) \equiv 0 \bmod 2$ *for each edge* $e \in E(G)$. *Then* G *has a faithful circuit cover of* w.

Conjecture 2.4.2 (Goddyn [86]) *Let* $w : E(G) \mapsto \mathbb{Z}^+$ *be an admissible eulerian weight of a bridgeless graph* G. *If* $w(e) \geq 2$ *for every edge* e *of* G, *then* (G, w) *has a faithful circuit cover.*

Conjecture 1.5.1 (Strong circuit double cover conjecture; Seymour, see [61] p. 237, [62], also see [83]) *Let* $w : E(G) \mapsto \{1, 2\}$ *be an eulerian weight of a bridgeless cubic graph* G. *If the subgraph of* G *induced by weight one edges is a circuit, then* (G, w) *has a faithful circuit cover.*

Conjecture 2.4.3 [259] *For a graph G and a circuit cover $\mathcal{F}$ of G, let $w_{\mathcal{F}} : E(G) \mapsto \mathbb{Z}^{\star}$ where $w_{\mathcal{F}}(e)$ is the number of members of $\mathcal{F}$ containing the edge e, for every $e \in E(G)$.*

Suppose that G is a graph having a 4-circuit cover $\mathcal{F}_1 = \{C_1, \ldots, C_4\}$. Then there is a circuit cover $\mathcal{F}_2$ of G such that

$$w_{\mathcal{F}_2}(e) \leq w_{\mathcal{F}_1}(e)$$

and

$$w_{\mathcal{F}_2}(e) \equiv w_{\mathcal{F}_1}(e) \pmod 2, \quad and \quad 0 < w_{\mathcal{F}_2}(e) < 4$$

for each edge $e \in E(G)$.

Remark. Conjecture 2.4.3 implies the circuit double cover conjecture and will be further discussed in Section 12.2.

Problem 2.4.4 (Jackson [121]) Is the following problem NP-complete? *Let G be a bridgeless graph and w be an eulerian $(1, 2)$-weight of G. Does G have a faithful circuit cover with respect to w?*

2.5 Exercises

⋆ Admissible eulerian weights

Exercise 2.1 Let G be a graph with a weight $w : E(G) \mapsto \mathbb{Z}^{+}$. If $w(e)$ is odd for every edge e of G, then w is eulerian if and only if G is even.

Exercise 2.2 Let G be a graph with a weight $w : E(G) \mapsto \{1, 2\}$. Then w is admissible if and only if G is bridgeless.

Exercise 2.3 Let G be a graph and $w : E(G) \mapsto \mathbb{Z}^{+}$. The following statements are equivalent:

(1) w is eulerian;
(2) $\cup_{j=0} E_{w=2j}$ induces a parity subgraph of G;
(3) $\cup_{j=0} E_{w=2j+1}$ induces an even subgraph of G.

(Note that a spanning subgraph P is a *parity subgraph* of G if $d_P(v) \equiv d_G(v) \pmod 2$ for every vertex $v \in V(G)$.)

⋆ Faithful cover

Exercise 2.4 Let $w : E(G) \mapsto \mathbb{Z}^{+}$ be an admissible eulerian weight of a bridgeless graph G. Suppose that $T = \{e_1, \ldots, e_t\}$ is a non-trivial edge-cut of the graph G such that $w(e_1) = \sum_{j=2}^{t} w(e_j)$ (a critical cut). Let

Q_1 and Q_2 be the two components of $G - T$ and for each $\{i, j\} = \{1, 2\}$, let H_i be the graph obtained from G by contracting all edges of Q_j. If both (H_1, w) and (H_2, w) have faithful circuit covers, then (G, w) also has a faithful circuit cover.

Exercise 2.5   Let $w : E(G) \mapsto \mathbb{Z}^+$ be an admissible eulerian weight of a bridgeless graph G. Suppose that T is a non-trivial edge-cut of the graph G with $|T| \leq 3$. Let Q_1 and Q_2 be the two components of $G - T$ and for each $\{i, j\} = \{1, 2\}$, let H_i be the graph obtained from G by contracting all edges of Q_j. If both (H_1, w) and (H_2, w) have faithful circuit covers, then (G, w) also has a faithful circuit cover.

Exercise 2.6   Let $\mathcal{F} = \{C_1, C_2, \ldots, C_k\}$ be a k-even subgraph cover of a graph G. Then

$$[\mathcal{F} - \{C_1\}] \cup \{\triangle_{C \in \mathcal{F}'} C\},$$

for any subset $\mathcal{F}'$ of $\mathcal{F}$ with $C_1 \in \mathcal{F}'$, is a k-even subgraph cover of G. (Note that "$\triangle$" is the *symmetric difference* of two subgraphs, see Definition A.2.3.)

Exercise 2.7   Let G be a bridgeless cubic graph and $w : E(G) \mapsto \{1, 2\}$ be an eulerian weight. If $|V(G)| \leq 16$ and $E_{w=1}$ induces a circuit of G, then (G, w) has a faithful circuit cover.

⋆ Contra pairs

Exercise 2.8   (Hägglund [102]) An eulerian $(1, 2)$-weighted cubic graph (G, w) is called a **strong contra pair** if, for every $e \in E_{w=2}$, both (G, w) and $(G - e, w)$ are contra pairs. Find a strong contra pair.

Exercise 2.9   (Fleischner, Genest and Jackson [75] (revised)) If (G, w) is a cyclically 4-edge-connected, $(1, 2)$-weighted, contra pair with a cyclic 4-edge-cut T and $w(T) = 4$, then $(G, w) = (G_1, w_1) \otimes_{IFJ} (G_2, w_2)$ such that one of $\{(G_1, w_1), (G_2, w_2)\}$ is a contra pair.

Remark.   One may notice that we are only able to show that one of $\{(G_1, w_1), (G_2, w_2)\}$ is a contra pair, not both yet.

3
Circuit chain and Petersen minor

3.1 Weight decomposition and removable circuit

Definition 3.1.1 Let (G, w) be an eulerian weighted graph. A set of eulerian weighted subgraphs $\{(G_i, w_i) : i = 1, \ldots, t\}$ is called an eulerian weight decomposition of (G, w) if each G_i is a subgraph of G and $\sum_{i=1}^{t} w_i(e) = w(e)$ for every $e \in E(G)$ (where $w_i(e) = 0$ if $e \notin E(G_i)$). (See Figure 3.1 where $t = 2$.)

Definition 3.1.2 Let (G, w) be an admissible eulerian weighted graph. An eulerian weight decomposition $\{(G_i, w_i) : i = 1, \ldots, t\}$ of (G, w) is admissible if each (G_i, w_i) is admissible. (See Figure 3.1.)

Definition 3.1.3 Let (G, w) be an admissible eulerian weighted graph. If (G, w) has an admissible eulerian weight decomposition $\{(G_1, w_1), (G_2, w_2)\}$ such that G_2 is a circuit of G with a constant weight $w_2 \equiv 1$, then the circuit G_2 is called a removable circuit of (G, w). (See Figure 3.1.)

Definition 3.1.4 Let G be a bridgeless graph and $w : E(G) \mapsto \{1, 2\}$ be an eulerian weight of G. A contra pair (G, w) is minimal if,

(1) for every $e \in E_{w=2}$, the eulerian weighted graph $(G - e, w)$ has a faithful circuit cover, and

(2) (G, w) has no removable circuit.

Circuit chains, which first appeared in [205], play an important role in the study of minimal contra pairs and minimum counterexamples to the circuit double cover conjecture.

Definition 3.1.5 Let x_0 and y_0 be two vertices of a graph G. A family $\mathcal{P}$ of circuits of G is called a circuit chain joining x_0, y_0 if $\mathcal{P} = \{C_1, \ldots, C_p\}$ such that

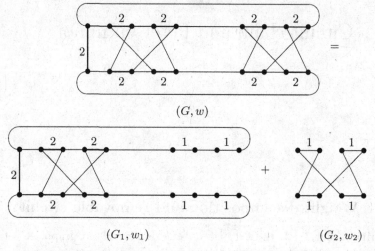

$$(G, w)$$

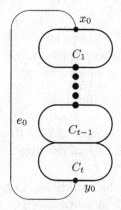

$$(G_1, w_1) \qquad\qquad\qquad (G_2, w_2)$$

Figure 3.1 *Weight decomposition* $(G, w) = (G_1, w_1) + (G_2, w_2)$ *where* G_2 *is a removable circuit*

Figure 3.2 *The circuit chain joining the endvertices of an edge* e_0

(1) $x_0 \in V(C_1)$ and $y_0 \in V(C_p)$,
(2) $V(C_i) \cap V(C_j) \neq \emptyset$ if and only if $i = j \pm 1$. (See Figure 3.2.)

3.2 Cubic minimal contra pair

The main theorem in this section is the following result.

Theorem 3.2.1 (Alspach and Zhang [3], also [4]) *Let G be a bridgeless cubic graph. If G contains no subdivision of the Petersen graph, then G has a faithful circuit cover with respect to every eulerian $(1,2)$-weight.*

Theorem 3.2.1 is proved in three steps. Part one is the following preparation, Part two is Lemma 3.2.3, and Part three is Theorem A.1.12 which is presented in Appendix A as one of the basic theorems in graph theory.

Preparation for the proof of Theorem 3.2.1. Let (G, w) be a counterexample to the theorem with the total weight $w(G) = \sum_{e \in E(G)} w(e)$ as small as possible. That is, (G, w) is a $(1,2)$-weighted cubic contra pair, but G contains no Petersen subdivision.

I. We claim that G *is 3-edge-connected and essentially 4-edge-connected.* Suppose that T is a 2-edge-cut of G with components Q_1 and Q_2 (see Figure 1.1). Then each suppressed weighted graph $(\overline{G/Q_i}, w)$ $(i = 1, 2)$ has a faithful circuit cover $\mathcal{F}_i$. By properly adjusting and joining some members of $\mathcal{F}_1 \cup \mathcal{F}_2$ at edges of T, we obtain a faithful circuit cover of the entire weighted graph (G, w). This is a contradiction.

It is similar if G has a non-trivial 3-edge-cut.

II. We claim that, *for every edge $e_0 \in E_{w=2}$, $(G - e_0, w)$ has a faithful circuit cover.* Note that, by I, $G - e_0$ remains bridgeless. Since the suppressed graph $(\overline{G - e_0}, w)$ is smaller, it has a faithful circuit cover.

III. We claim that (G, w) *has no removable circuit.* Suppose that (G, w) has a weight decomposition $(G, w) = (G_1, w_1) + (G_2, w_2)$ where G_2 is a circuit and w_2 is a constant 1 weight. The admissible eulerian weighted graph $(\overline{G_1}, w_1)$, which is smaller, has a faithful circuit cover $\mathcal{F}_1$. Now, $\mathcal{F}_1 \cup \{G_2\}$ is a faithful circuit cover of (G, w).

By I, II and III, the smallest counterexample to the theorem is a *minimal cubic contra pair* (see Definition 3.1.4). So, it is sufficient to prove the following theorem.

Theorem 3.2.2 (Alspach and Zhang [3], also [4]) *Every minimal $(1,2)$-weighted cubic contra pair must contain a subdivision of the Petersen graph.*

By applying Theorem A.1.12, Theorem 3.2.2 is an immediate corollary of the following lemma (part two of the proof of Theorem 3.2.1).

Lemma 3.2.3 (Alspach and Zhang [3], also [4]) *Every minimal $(1,2)$-weighted cubic contra pair is a permutation graph with $E_{w=2}$ as a perfect matching joining two chordless odd-circuits.*

Proof **I.** By Lemma 2.2.1, the cubic graph G is not 3-edge-colorable since (G, w) is a contra pair.

II. Let $e_0 = x_0 y_0$ be an *arbitrary* edge of $E_{w=2}$ and let $\mathcal{F}$ be a faithful circuit cover of $(G - e_0, w)$.

Let $\{C_1, \ldots, C_t\}$ ($\subseteq \mathcal{F}$) be a circuit chain joining the endvertices x_0 and y_0 of e_0. If $\mathcal{F} - \{C_1, \ldots, C_t\} \neq \emptyset$, then every member of $\mathcal{F} - \{C_1, \ldots, C_t\}$) is a removable circuit of (G, w). This contradicts that (G, w), as a minimal contra pair, has no removable circuit.

So, let $\mathcal{F} = \{C_1, \ldots, C_t\}$ be a circuit chain joining x_0 and y_0 where $C_\alpha \cap C_\beta \neq \emptyset$ if and only if $\alpha = \beta \pm 1$. We also notice that $E_{w=2}$ consists of edges of $\{e_0\} \bigcup_{i=1}^{t-1} [E(C_i) \cap E(C_{i+1})]$ and is, therefore, a *perfect matching* of the cubic graph G.

III. Let $c : E(G - e_0) \mapsto \{Red, Blue, Purple\}$ be an edge-coloring of $G - e_0$:

$$c(e) = \begin{cases} Red & \text{if } e \in [\bigcup_\mu E(C_{2\mu-1})] - [\bigcup_\mu E(C_{2\mu})] \\ Blue & \text{if } e \in [\bigcup_\mu E(C_{2\mu})] - [\bigcup_\mu E(C_{2\mu-1})] \\ Purple & \text{if } e \in [\bigcup_\mu E(C_{2\mu-1})] \cap [\bigcup_\mu E(C_{2\mu})]. \end{cases}$$

We notice that c is a proper 3-edge-coloring of the suppressed cubic graph $\overline{G - e_0}$, but not a proper edge-coloring of $G - e_0$ since x_0 (and y_0 as well) is incident with two same colored edges (both Red or both Blue).

IV. We claim that x_0 *and* y_0 *are contained in different components of* $(Red) \cup (Blue)$. Suppose not, let X be the component of $(Red) \cup (Blue)$ containing both x_0 and y_0. Alternating the colors (Red and Blue) along a segment of X between x_0 and y_0, and coloring the missing edge e_0 with *Purple*, we obtain a proper 3-edge-coloring of the entire graph G. This contradicts I that G is not 3-edge-colorable.

V. So, $(Red) \cup (Blue) = E_{w=1}$ is a 2-factor and x_0 and y_0 are contained in *two distinct* components X_1, X_2 of $(Red) \cup (Blue)$.

Note that the Red-Blue bi-colored circuits X_1 and X_2 are of *odd lengths* (by III) and all other components (if they exist) of $E_{w=1}$ are of *even lengths*. And

$$x_0 \in X_1, \quad y_0 \in X_2.$$

VI. We claim that $X_1 \cup X_2 = E_{w=1}$. That is, $E_{w=1}$ is the union of two vertex-disjoint odd length circuits X_1 and X_2. Suppose that X_3 is a third component of $E_{w=1}$.

Since the edge e_0 ($\in E_{w=2}$) was selected *arbitrarily*, by V, every edge e of $E_{w=2}$ must join the odd components X_1 and X_2 of $E_{w=1}$. Hence, there is no edge of $E_{w=2}$ incident with the component X_3. This contradicts II that $E_{w=2}$ is a perfect matching.

VII. In summary, $E_{w=2}$ is the perfect matching joining X_1 and X_2; and both X_1 and X_2 are chordless, odd-length circuits. □

3.3 Minimal contra pair

In Section 3.2, we dealt with faithful circuit cover for cubic graphs. The results for Petersen minor free cubic graphs (Theorem 3.2.1, Theorem 3.2.2) are further generalized in this section for general graphs (without the restriction of 3-regularity).

Theorem 3.3.1 (Alspach, Goddyn and Zhang [4]) *Let G be a bridgeless graph. If G contains no subdivision of the Petersen graph, then G has a faithful circuit cover with respect to every eulerian $(1,2)$-weight.*

With a similar argument as the proof of Theorem 3.2.1 (the preparation part), a smallest counterexample to the theorem is a minimal contra pair. Thus, Theorem 3.3.1 was already proved in Theorem 3.2.2 for cubic graphs. And, therefore, it is sufficient to verify the non-existence of high degree vertices in minimal contra pairs.

Lemma 3.3.2 (Alspach, Goddyn and Zhang [4]) *Every minimal $(1,2)$-weighted contra pair must be cubic.*

Outline of the proof of Lemma 3.3.2. Our goal is to verify the non-existence of high degree vertex. The following is a step-by-step outline of the proof.

Step 1.
    (1-1) The degree of every vertex is either 3 or 4.
    (1-2) Each degree 4 vertex is a cut-vertex of some component of $G[E_{w=1}]$, and every block of $G[E_{w=1}]$ is a circuit.
    (1-3) The even subgraph $G[E_{w=1}]$ has a unique circuit decomposition $\mathcal{X}$.

Step 2. The remaining part will be similar to the proof of Theorem 3.2.2.
    (2-1) The family $\mathcal{X}$ of circuits has precisely two members X_1, X_2 of odd lengths and all others (if they exist) are of even length.

(2-2) Every weight 2 edge joins X_1 and X_2.

Proof of Lemma 3.3.2. Let (G, w) be a smallest counterexample to the theorem. Similar to Theorem 3.2.1, G is 3-connected, essentially 4-edge-connected minimal contra pair (that is, (G, w) has no removable circuit and, for every $e \in E_{w=2}$, $(G - e, w)$ has a faithful cover).

I. Similar to the proof of Lemma 3.2.3, every faithful circuit cover of $(G - e_0, w)$ is a circuit chain joining x_0 and y_0.

Let $\mathcal{F} = \{C_1, \ldots, C_r\}$ be a faithful circuit cover of $(G - e_0, w)$ where $C_\alpha \cap C_\beta \neq \emptyset$ if and only if $\alpha = \beta \pm 1$. Hence, for every vertex v of G,

$$3 \le d(v) \le 4.$$

II. Let $S_1 = \cup_\mu E(C_{2\mu-1})$ and $S_2 = \cup_\mu E(C_{2\mu})$.

Here, $\{S_1, S_2\}$ is a faithful 2-even subgraph cover of $(G - e_0, w)$. And, each S_i is 2-regular since $\mathcal{F} = \{C_1, \ldots, C_r\}$ is a circuit chain.

III. By I, every vertex $v \in V(G) - \{x_0, y_0\}$ is in the intersection of precisely two members of the circuit chain $\mathcal{F}$. Hence,

(i) $d(v) = 3$ *if and only if*

$$|E(v) \cap E_{w=2}| = 1, \quad |E(v) \cap [S_1 - S_2]| = 1, \quad |E(v) \cap [S_2 - S_1]| = 1;$$

(ii) $d(v) = 4$ *if and only if*

$$E(v) \subseteq E_{w=1}, \quad |E(v) \cap S_1| = 2, \quad |E(v) \cap S_2| = 2;$$

(iii) and no vertex is of degree 5 or higher.

IV. Claim that, *for every faithful 2-even subgraph cover* $\{S_1', S_2'\}$ *of* $(G - e_0, w)$, *each* S_i' *must be 2-regular.*

Assume that $\{S_1', S_2'\}$ is a faithful 2-even subgraph cover of $(G - e_0, w)$ such that the maximum degree of $G[S_1']$ is 4. Let v be a vertex with $d(v) = 4$ and $E(v) \subseteq E(S_1')$.

Let $\mathcal{F}_i'$ be a circuit decomposition of S_i' for each $i = 1, 2$. The union $\mathcal{F}_1' \cup \mathcal{F}_2'$ forms a faithful circuit cover of $(G - e_0, w)$. By I, $\mathcal{F}_1' \cup \mathcal{F}_2'$ is a circuit chain $C_1', \ldots, C_{r'}'$ joining x_0 and y_0. Let $v \in C_j' \cap C_{j+1}'$. Note that C_j' and C_{j+1}' are edge-disjoint and both $\subseteq S_1'$. So, every vertex of the induced subgraph $G[C_j' \cup C_{j+1}']$ is of degree 2 or 4.

If v is the only vertex of $C_j' \cap C_{j+1}'$, then $\{x_0, v\}$ is a 2-cut of G, and $\{e_0\} \cup [E(v) \cap C_h']$ is a non-trivial 3-edge-cut of G for some $h \in \{j, j+1\}$. This contradicts that G is essentially 4-edge-connected.

Note that the induced subgraph $G[C_j' \cup C_{j+1}']$ is 2-connected. Let $u' \in C_j' \cap C_{j-1}'$ (or $u' = x_0$ if $j = 1$), and let $u'' \in C_{j+1}' \cap C_{j+2}'$ (or $u'' = y_0$

if $j + 1 = r$). Let C' be a circuit contained in the 2-connected subgraph $C'_j \cup C'_{j+1}$ and containing the vertices u' and u''. Then $[C'_j \cup C'_{j+1}] - C'$ is a removable even subgraph of (G, w). This contradicts that (G, w), as a minimal contra pair, has no removable circuit.

V. It is obvious that, *for each circuit C of $G[E_{w=1}]$, $\{S_1 \bigtriangleup C, S_2 \bigtriangleup C\}$ is also a faithful 2-even subgraph cover of $(G - e_0, w)$.*

VI. For each degree 4 vertex v, by III, let $\{e', e''\} = E(v) \cap S_1$ ($\subseteq E_{w=1}$). We claim that *no circuit of $G[E_{w=1}]$ contains both edges e' and e''.*

Suppose that C is a circuit of $G[E_{w=1}]$ containing both edges e' and e''. By V, $\{S_1 \bigtriangleup C, S_2 \bigtriangleup C\}$ is also a faithful 2-even subgraph cover of $(G - e_0, w)$. Note that the maximum degree of $S_2 \bigtriangleup C$ is 4. This contradicts IV.

Therefore, we have the following immediate conclusions about $E_{w=1}$.

Let v be a degree 4 vertex of G.

(VI-1) For each pair $\{e', e''\} = E(v) \cap S_i$ ($i = 1, 2$), the edges e' and e'' must be in different blocks of $G[E_{w=1}]$ (by VI).

(VI-2) The degree 4 vertex v must be a cut-vertex of some component of $G[E_{w=1}]$ (by VI-1).

(VI-3) The circuit decomposition of the even subgraph $G[E_{w=1}]$ must be unique (by VI-2).

Let $X_1, \ldots, X_t$ be the *unique circuit decomposition* of the induced even subgraph $G[E_{w=1}]$ (each X_i is a block of $G[E_{w=1}]$).

VII. Claim that *x_0 and y_0 are contained in two distinct components Q_1, Q_2 of $G[E_{w=1}]$.*

Suppose that x_0, y_0 are contained in the same component Q_1 of $G[E_{w=1}]$. Let P be a path of Q_1 joining x_0 and y_0. Then $\{S_1 \bigtriangleup [P \cup e_0], S_2 \bigtriangleup [P \cup e_0]\}$ is a faithful even subgraph cover of the contra pair (G, w). This is a contradiction.

So, let Q_1 and Q_2 be two *distinct* components of $G[E_{w=1}]$ and let $x_0 \in X_1$ and $y_0 \in X_2$ where X_j is a block of Q_j ($j = 1, 2$).

VIII. Claim that *the circuits X_1 and X_2 are of odd lengths, while all other X_i ($i > 2$) are of even lengths.*

Color the edges of $S_1 - S_2$ with red, edges of $S_2 - S_1$ with blue, edges of $S_1 \cap S_2 = E_{w=2}$ with purple. By III and VI-(1), each circuit X_i is of even length if $i \neq 1, 2$ since it is alternately colored with red and blue, while X_1 and X_2 are of odd lengths since each of x_0, y_0 is incident with two same colored edges.

IX. Claim that *every $e \in E_{w=2}$ must join X_1 and X_2.*

Since the edge $e_0 = x_0 y_0$ was selected arbitrarily from the matching $E_{w=2}$, all conclusions we have had above can be applied to every edge $e \in E_{w=2}$. That is, for *every $e = xy \in E_{w=2}$, $x \in X_1$, $y \in X_2$.*

X. Claim that *$G[E_{w=1}]$ has precisely two components $Q_1 = X_1$ and $Q_2 = X_2$.*

If Q_3 is a component of $G[E_{w=1}]$ other than Q_1 and Q_2, then, by IX, the component Q_3 is disconnected with other part of the graph G. This contradicts that G is connected.

If Q_1 has more than one block, then Q_1 must have a block X_3 other than X_1 that contains precisely one cut-vertex z of Q_1 (note that X_3 corresponds to a leaf in the block tree of Q_1). By IX, X_3 is not incident with any edge of $E_{w=2}$. Hence, z is also a cut-vertex of G, and, therefore, the even subgraph X_3 is a removable circuit of (G, w). This contradicts that (G, w), as a minimal contra pair, has no removable circuit.

XI. By X and III, G has no degree 4 vertex. This completes the proof of the lemma that G is cubic. $\qquad\qquad\qquad\qquad\qquad\qquad\qquad\square$

3.4 Structure of circuit chain

Most of Lemma 3.4.1 was inspired by or embedded in the proofs of Theorems 3.2.2 and 3.3.1. Beyond its applications for minimal contra pairs, it will be applied in further studies of contra pairs, circuit chain, and various subjects. So, we present it here as an independent lemma.

Note that the conclusions in the lemma are under somehow different conditions. Also notice that the graph G in the lemma is not necessarily *cubic.*

Lemma 3.4.1 *Let G be a bridgeless graph and $w : E(G) \mapsto \{1, 2\}$ be an eulerian weight of G and let $e_0 = x_0 y_0 \in E_{w=2}$. Assume that*

(a) (G, w) has no faithful circuit cover, and

(b) $(G - e_0, w)$ has a faithful circuit cover.

Then we have the following statements.

(1) For any faithful circuit cover $\mathcal{F}$ of $(G - e_0, w)$, x_0, y_0 are not contained in the same member of $\mathcal{F}$.

(2) If there is a faithful circuit cover $\mathcal{P}$ of $(G - e_0, w)$ which is also a circuit chain joining x_0 and y_0, then x_0 and y_0 are contained in different components of $E_{w=1}$ ($= \triangle_{C \in \mathcal{P}} C$).

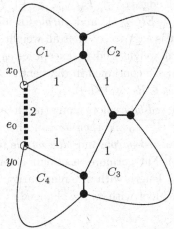

Figure 3.3 x_0 and y_0 are contained in the same component of $E_{w=1} = \triangle_{C \in \mathcal{P}} C$

(3) If there is a faithful circuit cover $\mathcal{P}$ of $(G - e_0, w)$ which is also a circuit chain joining x_0 and y_0, then there is another faithful circuit cover $\mathcal{P}^*$ of $(G - e_0, w)$ such that $\mathcal{P}^*$ contains a circuit chain $\mathcal{Q}^*$ joining x_0 and y_0 and

$$|\mathcal{Q}^*| \not\equiv |\mathcal{P}| \pmod 2.$$

(4) Let $\mathcal{F}$ be a faithful circuit cover of $(G - e_0, w)$ with $|\mathcal{F}|$ as large as possible, then

$$|\mathcal{F}| \geq 3.$$

(5) Every circuit of G containing e_0 is of length ≥ 5.

Proof **(1)** For otherwise, say $x_0, y_0 \in C$ ($C \in \mathcal{F}$). Then e_0 is a chord of C and $C + e_0$ has a circuit cover $\{C^*, C^{**}\}$ which covers e_0 twice. Replacing $\{C^*, C^{**}\}$ with C in $\mathcal{F}$, we obtain a faithful cover for (G, w).

(2) Let $\mathcal{P} = \{C_1, \ldots, C_t\}$ be a faithful circuit cover of $(G - e_0, w)$ which is also a circuit chain joining x_0 and y_0. Let $S_1 = \bigcup_\mu C_{2\mu-1}$, and $S_2 = \bigcup_\mu C_{2\mu}$. Here $E_{w=1} = S_1 \triangle S_2$. Assume that x_0 and y_0 are in the same component of $E_{w=1}$. Let P ($\subseteq E_{w=1}$) be a path joining x_0 and y_0. Then $\{S_1 \triangle [P \cup \{e_0\}], S_2 \triangle [P \cup \{e_0\}]\}$ is a faithful 2-even subgraph cover of (G, w). This contradicts that (G, w) is a contra pair. (See Figure 3.3.)

(3) Let $\mathcal{P} = \{C_1, \ldots, C_t\}$ be a faithful circuit cover of $(G - e_0, w)$ which is also a circuit chain joining x_0 and y_0. Color the edges of $G - e_0$ as follows: *Red* for all edges of $\bigcup_\mu E(C_{2\mu-1}) - \bigcup_\mu E(C_{2\mu})$, *blue* for all edges of $\bigcup_\mu E(C_{2\mu}) - \bigcup_\mu E(C_{2\mu-1})$ and *purple* for all edges of $[\bigcup_\mu E(C_{2\mu})] \cap [\bigcup_\mu E(C_{2\mu-1})]$. It is easy to see that all weight 1 edges are colored with *red and blue* and all weight 2 edges, except for e_0, are colored with *purple*. By (2), x_0 and y_0 are contained in different *red-blue* bi-colored circuits, say R_1 and R_2 ($x_0 \in R_1$, $y_0 \in R_2$).

Alternating the *red-blue* colors along the circuit R_2, we obtain another 3-edge-coloring of $\overline{G - e_0}$ and therefore another faithful cover $\mathcal{P}^*$ of $(G - e_0, w)$ consisting of *red-purple* and *blue-purple* bi-colored circuits (under the new coloring). Furthermore, a circuit chain $\mathcal{Q}^*$ ($\subseteq \mathcal{P}^*$) joining x_0 and y_0 must be of length with different parity from that of $\mathcal{P}$ since the colors of edges incident with y_0 are changed.

(4) Let (G, w) be a smallest counterexample to (4). Thus, the faithful cover $\mathcal{F}$ of $(G - e_0, w)$ must be a circuit chain $C_1, \ldots, C_r$ joining x_0 and y_0. Here $r \leq 2$ as G being a counterexample. By (1), $r \neq 1$.

By (3), $(G - e_0, w)$ has another faithful circuit cover $\mathcal{P}^*$ such that there is a circuit chain $\mathcal{Q}^*$ ($\subseteq \mathcal{P}^*$) joining x_0 and y_0 and $|\mathcal{Q}^*|$ is odd. Since the original chain $\mathcal{F}$ is of length $r = 2$, the new chain $\mathcal{Q}^*$ is of length ≥ 3 ($|\mathcal{Q}^*| > 1$, by (1)). This contradicts the choice that $|\mathcal{F}|$ is maximum.

(5) Let $\mathcal{F}$ be a faithful circuit cover of $(G - e_0, w)$ with $|\mathcal{F}|$ as large as possible. Let D be a circuit of G containing e_0 and of length at most four. Let $\mathcal{P} = \{C_1, \ldots, C_t\}$ be a circuit chain joining the endvertices of e_0 such that $C_i \in \mathcal{F}$ and $E(C_i) \cap E(D) \neq \emptyset$ for each $i = 1, \ldots, t$. Since D is of length at most four, we have that $t \leq 3$. By (4), $t > 2$ since $|\mathcal{F}|$ is maximum. Thus, $t = 3$. It is evident now that each edge of $D - e_0$ is covered only once by $\{C_1, C_2, C_3\}$ for otherwise, a shorter circuit chain would be obtained and it contradicts (4). So, endvertices of e_0 are contained in the same component of $C_1 \triangle C_2 \triangle C_3$. This contradicts (2) by considering the weighted graph $(\overline{G[\bigcup_{i=1}^3 C_i \cup \{e_0\}]}, w_{\{C_1, C_2, C_3\}})$ where $E_{w_{\{C_1, C_2, C_3\}}=1} = C_1 \triangle C_2 \triangle C_3$. $\qquad\square$

Some lemmas and exercises (such as Exercises 3.2, 3.3, 3.4, 3.5, 3.7, 3.8) also present useful structural information for circuit chains and contra pairs.

3.5 Open problems

Conjecture 3.5.1 (Goddyn [87]) (P_{10}, w_{10}) *is the only* minimal $(1,2)$-*weighted contra pair.*

Note that *minimal contra pair* is defined in Definition 3.1.4.

Definition 3.5.2 Let G be a bridgeless cubic graph and $w : E(G) \mapsto \{1,2\}$ be an eulerian weight of G. The eulerian weighted graph (G, w) is a critical contra pair if (G, w) has no faithful circuit cover and has no removable circuit.

By Lemma 3.2.3, a *minimal contra pair* must be a permutation graph. Hence, Conjecture 3.5.1 is equivalent to the following open problem.

Conjecture 3.5.3 (Goddyn [87]) *Let G be a permutation graph such that M is a perfect matching of G and $G - M$ is the union of two chordless circuits C_1 and C_2. If (G, w) is a critical contra pair with $E_{w=2} = M$ and $E_{w=1} = E(C_1) \cup E(C_2)$, then $(G, w) = (P_{10}, w_{10})$.*

Conjecture 3.5.4 (Goddyn [85], or see [87]) *Let (G, w) be an admissible eulerian $(1,2)$-weighted graph without removable circuit. If G is 3-connected and cyclically 4-edge-connected, then $(G, w) = (P_{10}, w_{10})$. (That is, (P_{10}, w_{10}) is the only 3-connected, cyclically 4-edge-connected, critical contra pair.)*

A recent computer aided search [21] showed that Conjectures 3.5.1, 3.5.3 and 3.5.4 hold for all cubic graphs of order at most 36.

3.6 Exercises

⋆ Structure of circuit chain

Exercise 3.1 Let $\mathcal{F}$ be a faithful circuit cover of an eulerian $(1,2)$-weighted cubic graph (G, w). If $\mathcal{F}$ is a circuit chain, then $E_{w=1}$ induces an even 2-factor of G.

Exercise 3.2 In Lemma 3.4.1-(4), the requirement of the maximality of $|\mathcal{F}|$ is necessary.

Exercise 3.3 Let G be a bridgeless graph, w be an eulerian $(1,2)$-weight of G, and $e = xy \in E_{w=2}$. Assume that
(1) $(G - e, w)$ has a faithful circuit cover,
(2) *every* faithful circuit cover of $(G-e, w)$ is a circuit chain connecting

x and y.

Let $\mathcal{F} = \{C_1, \ldots, C_k\}$ be a faithful circuit cover of $(G - e, w)$, which is a circuit chain with $x \in V(C_1)$ and $y, \in V(C_k)$. For each pair i, j with $1 \le i \le j \le k$ and $j - i < k - 1$, we have the following two conclusions.

(a) The circuit C_{j+1} (as well as C_{i-1}) intersects with *only one* component of $C_i \triangle \cdots \triangle C_j$ (where we let $C_0 = \{x\}$ and $C_{k+1} = \{y\}$.)

(b) If G is a permutation graph with $E_{w=1}$ inducing a pair of chordless circuits, then $C_i \triangle \cdots \triangle C_j$ is a *Hamilton circuit* of the subgraph $C_i \cup \cdots \cup C_j$.

Remark. The structure of a minimal contra pair is a permutation graph which is proved in Lemma 3.2.3.

$\star$ Girth for faithful cover

Applying Lemma 3.4.1, we have the following exercises dealing with short circuits in contra pairs.

Let (G, w) be an eulerian $(1, 2)$-weighted graph with $e_0 \in E_{w=2}$. In the following two exercises, we are to show that *if e_0 is contained in a "short" circuit C_0 (length ≤ 6) and C_0 contains "sufficiently many" edges of $E_{w=2}$, then $(G - e_0, w)$ having a faithful cover implies that (G, w) also has a faithful cover.*

Exercise 3.4 Let G be a cubic graph and $w : E(G) \mapsto \{1, 2\}$ be an eulerian weight. Let $e_0 = v_0 v_1 \in E_{w=2}$ be contained in a circuit $C_0 = v_0 v_1 \ldots v_{r-1} v_0$ and a path $P_0 \subseteq C_0 \cap E_{w=2}$ such that $P_0 = v_{r-1} v_0 v_1$ if $r = 5$, and $P_0 = v_{r-1} v_0 v_1 v_2$ if $r = 6$. If $r \le 6$ and $(G - e_0, w)$ has a faithful circuit cover, then (G, w) also has a faithful cover. (See Figure 3.4.)

Exercise 3.5 Let G be a cubic graph and $w : E(G) \mapsto \{1, 2\}$ be an eulerian weight. Let $e_0 = v_0 v_1 \in E_{w=2}$ be contained in a circuit $C_0 = v_0 v_1 \ldots v_5 v_0$ and a path $P_0 = v_5 v_0 v_1 \subseteq C_0 \cap E_{w=2}$. If there is another path $P_1 = v_2 v_3 v_4 \subseteq C_0 \cap E_{w=2}$ and $(G - e_0, w)$ has a faithful circuit cover, then (G, w) also has a faithful cover. (See Figure 3.5.)

Remark. Can we have some result similar to Exercises 3.4 and 3.5 for a short circuit C_0 with other distributions of $E_{w=2}$ edges around C_0? Possibly yes, but an exception exists. For example, in the contra pair (P_{10}, w_{10}) (see Figure 2.2), there is a 5-circuit containing a pair of

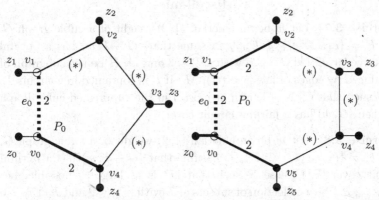

Figure 3.4 *A 5- or 6-circuit C_0 containing the edge $e_0 = v_0v_1$ (label (∗) is a weight of either 1 or 2)*

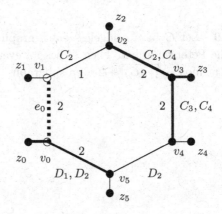

Figure 3.5 *A 6-circuit C_0 containing the edge $e_0 = v_0v_1$*

disjoint weight 2 edges. However, it remains open when $|E(C_0)| = 5$ and $|E_{w=2} \cap E(C_0)| = 1$. (See Exercise 3.6.)

Exercise 3.6 Let G be a cubic graph and $w : E(G) \mapsto \{1, 2\}$ be an eulerian weight. Let $e_0 = v_0v_1 \in E_{w=2}$ be contained in a circuit $C_0 = v_0v_1 \ldots v_4v_0$. Assume that $(G - e_0, w)$ has a faithful circuit cover $\mathcal{F}$, while (G, w) does not. Then either

$$|E_{w=2} \cap \{v_2v_3, v_3v_4\}| \geq 1$$

or $\mathcal{F}$ is not a circuit chain.

⋆ Miscellanies

Exercise 3.7   Let w be an eulerian $(1,2)$-weight of a cubic graph G, and $F = \{e_1, \ldots, e_t\} \subseteq E_{w=2}$. Assume that $(G - F, w)$ has a faithful circuit cover $\mathcal{F}$. Let $\mathcal{F}' \subseteq \mathcal{F}$ with $|\mathcal{F}'| \leq 3$ and let H be the subgraph of G induced by edges of $[\cup_{C \in \mathcal{F}'} E(C)] \cup F$. If F is contained in a circuit C_0 of H such that C_0 is of length at most 4 in the suppressed cubic graph $\overline{H}$, then (G, w) has a faithful circuit cover.

Exercise 3.8   Let w be an eulerian $(1,2)$-weight of a cubic graph G, and $F = \{e_1, \ldots, e_t\} \subseteq E_{w=2}$. Assume that $(G - F, w)$ has a faithful circuit cover $\mathcal{F}$. Choose $\mathcal{F}$ such that $|\mathcal{F}|$ is as large as possible. Let $\{\mathcal{F}_1, \ldots, \mathcal{F}_t\}$ be a collection of subsets of $\mathcal{F}$ with $|\mathcal{F}_j| \leq 2$ and $\mathcal{F}_i \cap \mathcal{F}_j = \emptyset$ (for $i \neq j$). Let H_j be the subgraph of G induced by $[\cup_{C \in \mathcal{F}_j} E(C)] \cup \{e_j\}$ for each $j \in \{1, \ldots, t\}$. If every H_j is bridgeless, then (G, w) has a faithful circuit cover.

Exercise 3.9   Let G be a bridgeless cubic graph containing no subdivision of the Petersen graph. Then G has an even subgraph cover $\mathcal{F}$ consisting of at most $\sqrt{|V(G)|} + 2$ even subgraphs.

4
Small oddness

4.1 k-even subgraph double covers

Recall Definition 3.1.1 for weight decomposition. Let (G, w) be an eulerian weighted graph. A set of eulerian weighted subgraphs $\{(G_i, w_i) : i = 1, \ldots, t\}$ is called an eulerian weight decomposition of (G, w) if each G_i is a subgraph of G and $\sum_{i=1}^{t} w_i(e) = w(e)$ for every $e \in E(G)$ (where $w_i(e) = 0$ if $e \notin E(G_i)$). (See Figure 3.1 where $t = 2$.)

Weight decomposition is an approach to the circuit double cover conjecture: if one can find a weight decomposition $\sum_{i=1}^{t}(G_i, w_i)$ of $(G, 2)$ such that each (G_i, w_i) has a faithful circuit cover (where the eulerian weight of $(G, 2)$ is the constant 2).

In this chapter, we will study how to find a *pair* of weighted graphs (G_1, w_1) and (G_2, w_2) for some families of cubic graphs G such that

(1) $(G_1, w_1) + (G_2, w_2)$ is a weight decomposition of $(G, 2)$,

(2) and each (G_j, w_j) has a faithful circuit cover.

In most cases, each suppressed cubic graph $\overline{G_j}$ is expected to be 3-edge-colorable and, therefore, a faithful cover is obtained by applying Lemma 2.2.1. (Some other results may also be used for faithful coverings of $(\overline{G_j}, w_j)$, such as, Theorem 3.2.1, Proposition B.2.27, etc. However, identifying suppressed subgraphs $\overline{G_j}$ with those structural requirements is technically much more challenging.)

The approach of weight decomposition will be re-visited in later chapters. In Chapter 5, $(G, 2) = (G_1, w_1) + (G_2, w_2) + (G_3, w_3)$ (three factors in the decomposition) and, similarly, each $\overline{G_j}$ is 3-edge-colorable. In Chapter 6, some circuit extension techniques will be applied so that the strong requirement of 3-edge-coloring will be relaxed with strong CDC results.

Recall Definition 1.3.1. Let G be a graph. A family $\mathcal{F}$ of even sub-graphs is called a k-**even subgraph double cover** of G if every edge e of G is contained in precisely 2 members of $\mathcal{F}$, and, $|\mathcal{F}| \leq k$.

Lemma 4.1.1 [257] *A cubic graph G has a 6-even subgraph double cover if and only if $(G, 2)$ has an eulerian weight decomposition $(G_1, w_1) + (G_2, w_2) = (G, 2)$ such that both suppressed cubic graphs $\overline{G_1}$ and $\overline{G_2}$ are 3-edge-colorable.*

Proof "⇒": Let $\{C_1, \ldots, C_6\}$ be a 6-even subgraph double cover of G. Let $G_1 = G[C_1 \cup C_2 \cup C_3]$ and $w_1 = w_{\{C_1, C_2, C_3\}}$, and, $G_2 = G[C_4 \cup C_5 \cup C_6]$ and $w_2 = w_{\{C_4, C_5, C_6\}}$ (note that w_1, w_2 are eulerian weights induced by the even subgraph covers, see Definition 2.1.7). By Lemma 2.2.1, both suppressed cubic graphs $\overline{G_1}$ and $\overline{G_2}$ are 3-edge-colorable.

"⇐": Since both suppressed cubic graphs $\overline{G_1}$ and $\overline{G_2}$ are 3-edge-colorable, by Lemma 2.2.1, for each $j = 1, 2$, the weighted graph (G_j, w_j) has a faithful 3-even subgraph cover $\mathcal{F}_j$. Hence, $\mathcal{F}_1 \cup \mathcal{F}_2$ is a 6-even sub-graph double cover of G.   □

If we further apply Lemma 2.2.2 to one of the terms G_1 and G_2, then we can obtain an analog of Lemma 4.1.1.

Lemma 4.1.2 [257] *A cubic graph G has a 5-even subgraph dou-ble cover if and only if $(G, 2)$ has an eulerian weight decomposition $(G_1, w_1) + (G_2, w_2) = (G, 2)$ such that both suppressed cubic graphs $\overline{G_1}$ and $\overline{G_2}$ are 3-edge-colorable, and $E_{w_1 = 2}$ is a mono-colored 1-factor under some 3-edge-coloring of $\overline{G_1}$.*

Proof See Exercise 4.1.   □

The following is an equivalent version of Lemma 4.1.2.

Lemma 4.1.3 (Hoffmann-Ostenhof [111]) *Let G be a 2-connected cubic graph. Then G has a 5-even subgraph double cover if and only if there is a matching M of G with the following two properties,*
   (1) $\overline{G - M}$ is 3-edge-colorable,
   (2) there is a pair of even subgraphs C_1, C_2 in G with $E(C_1) \cap E(C_2) = M$.

Proof See Exercise 4.4.   □

Remark. Generalizations of Lemmas 4.1.1 and 4.1.2 for non-cubic graphs are presented in Exercises 8.12 and 8.13.

Conjecture 4.1.4 (Celmins [31] and Preissmann [192]) *Every bridge-less graph has a 5-even subgraph double cover.*

Conjecture 4.1.4 is true for the Petersen graph (see Proposition B.2.24). But this conjecture has been verified for very few families of graphs. The most notable results concerning this conjecture are Theorems 4.2.3 and 4.2.4 for the family of cubic graphs with small oddness.

4.2 Small oddness

Definition 4.2.1 Let S be an even subgraph of a cubic graph G. A component C of S is odd (or even) if C contains an odd (or even, respectively) number of vertices of G.

Definition 4.2.2 Let G be a bridgeless cubic graph. For a spanning even subgraph S of G, the oddness of S, denoted by $odd(S)$, is the number of odd components of S. For the cubic graph G, the oddness of G, denoted by $odd(G)$, is the minimum of $odd(S)$ for all spanning even subgraph S of G.

Note that a spanning even subgraph mentioned in Definition 4.2.2 may not be a 2-factor, it may consist of a set of circuits and some isolated vertices.

The following are some straightforward observations.

Fact. A cubic graph G is 3-edge-colorable if and only if $odd(G) = 0$.

Fact. The oddness of every cubic graph must be even.

Note that determination of the oddness of a cubic graph is a hard problem since the determination of 3-edge-colorability of a cubic graph is an NP-complete problem [113].

For a 3-edge-colorable cubic graph G_1 and an edge $e \in E(G_1)$, it is obvious that the suppressed cubic graph $G_2 = \overline{G_1 - e}$ is of oddness at most 2. And, therefore, a bridgeless cubic graph containing a Hamilton path is also of oddness at most 2 (see Corollary 4.2.8).

Theorem 4.2.3 (Huck and Kochol [117] and [114], [144]) *Let G be a bridgeless cubic graph with oddness at most 2. Then G has a 5-even subgraph double cover.*

38 *Small oddness*

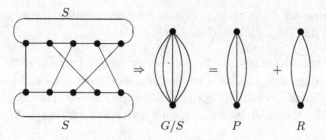

Figure 4.1 *Contracting a spanning even subgraph S of P_{10}*

Theorem 4.2.3 was further improved by Huck [116] (a computer assisted proof), and by Häggkvist and McGuinness [101], independently for oddness 4 graphs.

Theorem 4.2.4 (Huck [116], Häggkvist and McGuinness [101]) *Let G be a bridgeless cubic graph with oddness at most 4. Then G has a circuit double cover.*

Because the proof of Theorem 4.2.4 is very complicated, only the proof of Theorem 4.2.3 is presented here.

Definition 4.2.5 Let G be a graph. A spanning subgraph P is a *parity subgraph* of G if

$$d_P(v) \equiv d_G(v) \pmod 2$$

for every $v \in V(G)$.

Here, we first present an outline of the proof of Theorem 4.2.3. Let S be a spanning even subgraph of G with precisely two odd components. By Lemma 4.1.2, it is sufficient to show the following statements.

(1) There is a partition $P \cup R$ of the edge set $E(G) - E(S)$ such that P is a bridgeless parity subgraph of the contracted graph G/S (that is, $G_1 = \overline{S \cup P}$ is of at most oddness 2, $G_2 = \overline{S \cup R}$ is of oddness zero); and
(2) both $G_1 = \overline{S \cup P}$ and $G_2 = \overline{S \cup R}$ are 3-edge-colorable. (See Figures 4.1 and 4.2.)

Theorem 4.2.3 will be proved when Lemma 4.1.2 is applied to the weight decomposition $(G, 2) = (G_1, w_1) + (G_2, w_2)$ where $w_j(e) = 1$ if $e \in E(S)$ and, $w_j(e) = 2$ otherwise for each $j = 1, 2$.

The following lemmas from [257] were originally inspired by the proof

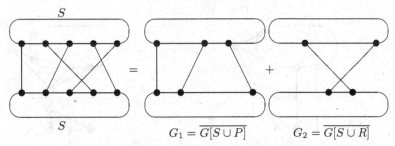

$$G_1 = \overline{G[S \cup P]} \qquad G_2 = \overline{G[S \cup R]}$$

Figure 4.2 *Both G_1 and G_2 are 3-edge-colorable*

in [117], and are presented here for the preparation of the final proof of Theorem 4.2.3. (In the following lemma, the bridgeless graph H can be viewed as the graph obtained from an oddness 2 cubic graph G by contracting a spanning even subgraph S.)

Lemma 4.2.6 *Let H be a bridgeless graph which contains only two odd degree vertices x_1 and x_2. If P is a smallest bridgeless parity subgraph of H, then P is the union of three edge-disjoint paths joining x_1 and x_2 and every edge of P is contained in some 3-edge-cut of the subgraph P (which separates the vertices x_1 and x_2).*

Proof Since the parity subgraph P has only two odd vertices x_1 and x_2, any edge-cut of even size cannot separate x_1 and x_2 (by Lemma A.2.8). By Menger's Theorem (Theorem A.1.3), let Q_1, Q_2 and Q_3 be three edge-disjoint paths joining x_1 and x_2 in P. Since the union of Q_1, Q_2 and Q_3 is also a bridgeless parity subgraph of P, by the minimality of P, we have that $P = Q_1 \cup Q_2 \cup Q_3$.

If there is an edge e of P such that e is not contained in any 3-edge-cut of P, then the local edge-connectivity between x_1 and x_2 in $P - \{e\}$ remains at least 3, and, therefore, $P - \{e\}$ contains another set of three edge-disjoint paths Q_1', Q_2' and Q_3' joining x_1, x_2. This is a contradiction since the union of Q_1', Q_2' and Q_3' induces a bridgeless parity subgraph of P which is smaller than P. □

Lemma 4.2.7 *Let G be a bridgeless cubic graph of oddness 2 and S be a spanning even subgraph of G with only two odd components. Let P be a smallest bridgeless parity subgraph of the contracted graph G/S. Then the suppressed graph $\overline{G[E(S) \cup E(P)]}$ is 3-edge-colorable.*

Proof Let G be a smallest counterexample to the lemma. Thus, $G/S = P$ and, $E(G) = E(S) \cup E(P)$. Let x_1 and x_2 be the two odd vertices of

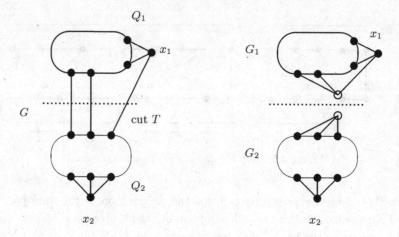

Figure 4.3 *Graph G and contracted graphs $G_1 = G/Q_2$, $G_2 = G/Q_1$*

P, which are created by contracting two odd components X_1 and X_2 of S in G.

I. We claim that *there is no non-trivial 2- or 3-edge-cut T of G such that $T \subseteq E(P)$ and, T separates x_1 and x_2 in G/S.*

Since $\{x_1, x_2\}$ is the set of all odd degree vertices of G/S, every edge-cut separating x_1 and x_2 must be of odd size (by Lemma A.2.8). Suppose that $T \ (\subseteq E(P))$ is a non-trivial 3-edge-cut of G separating x_1 and x_2 with components Q_1 and Q_2 where $x_i \in Q_i$ for $i = 1, 2$.

Then, for each $\{i, j\} = \{1, 2\}$, let G_i be the cubic graph obtained from G by contracting Q_j (see Figure 4.3), S_i be the spanning even subgraph of G_i obtained from S by shrinking all components of S contained in Q_j as a single vertex component, and $P_i = G_i/S_i$ be obtained from P by deleting all edges of Q_j. Obviously, for each $\{i, j\} = \{1, 2\}$, each pair of G_i and S_i satisfies the description of the lemma, and P_i remains as a smallest bridgeless parity subgraph of G_i/S_i.

Thus, the cubic graph G_i is 3-edge-colorable since it is smaller than the smallest counterexample G. Properly renaming the colors of G_1 (if necessary), a combination of 3-edge-colorings of G_1 and G_2 yields a 3-edge-coloring of G. This contradicts that G is a counterexample.

II. Note that G/S satisfies Lemma 4.2.6 (by considering G/S as the graph H). Since P is a smallest bridgeless parity subgraph of G/S, by Lemma 4.2.6, we have that

(1) P (P = G/S) is the union of three edge-disjoint paths joining x_1 and x_2 in G/S,

(2) every edge of P (P = G/S) is contained in a 3-edge-cut of P = G/S separating x_1 and x_2.

III. By applying I and II, we have following description of the structure of the graph G.

(a) By II-(1), $|E_{G/S}(x_i)| = 3$, for each $i \in \{1, 2\}$.

(b) By I, each odd component $X_i = \{x_i\}$ is a single vertex component of S. That is, x_i is a vertex of G.

(c) By II-(2) and I, *every* edge of P is contained in a trivial 3-edge-cut $E_{G/S}(x_i) = E_G(x_i)$, for some $i \in \{1, 2\}$. Therefore,

$$E(P) = E(G/S) = E_G(x_1) \cup E_G(x_2).$$

(d) Since G is cubic, by (c), every vertex of $G - \{x_1, x_2\}$ is incident with one edge of $E(P) = E(G/S) = E_G(x_1) \cup E_G(x_2)$, and two edges of S.

Therefore, by (d),

$$|V(G) - \{x_1, x_2\}| \le |E(x_1)| + |E(x_2)| = 6.$$

That is, the bridgeless cubic graph G has at most eight vertices. By Proposition B.1.11, G is 3-edge-colorable. □

Proof of Theorem 4.2.3. The theorem is to be proved by applying Lemma 4.1.2. Let S be a spanning even subgraph of G with only two odd components in G. Let P be a smallest bridgeless parity subgraph of the contracted graph G/S.

By Lemma 4.2.7, $\overline{G_1} = \overline{G[E(S) \cup E(P)]}$ is 3-edge-colorable. Hence, the eulerian weighted graph $(G, 2)$ has a weight decomposition $(G, 2) = (G_1, w_1) + (G_2, w_2)$ where $G_2 = G[E(G) - E(P)]$, $G_1 = G[E(S) \cup E(P)]$, and the eulerian weight w_j with $w_j(e) = 1$ if $e \in E(S)$ and $w_j(e) = 2$ otherwise for each $j = 1, 2$. All of these satisfy the description in Lemma 4.1.2. This completes the proof the theorem. □

Since every cubic graph containing a Hamilton circuit is 3-edge-colorable, the following is a corollary of Theorem 4.2.3.

Corollary 4.2.8 (Tarsi [221], or see [84], [83], [117], and Exercise 6.2 in this book) *Every bridgeless graph containing a Hamilton path has a circuit double cover.*

Proof By Theorem 4.2.3, we need only to show that G has a 2-factor S which contains at most two odd components in G.

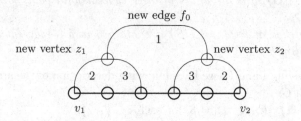

Figure 4.4 $c^{-1}(2) \cup c^{-1}(3)$ *is an even 2-factor in G', and is also a 2-factor with oddness ≤ 2 in G*

Let Q be a Hamilton path of G with endvertices v_1 and v_2 and end-edges $e_1 = v_1 u_1$, $e_2 = v_2 u_2$. (See Figure 4.4.) Construct a cubic graph G' from G as follows: for each $i = 1, 2$, insert a vertex z_i into an edge $f_i \in E_G(v_i) - \{e_i\}$, and add an edge $f_0 = z_1 z_2$. The Hamilton path Q is therefore extended to a Hamilton circuit Q' of G' by adding a path $v_1 z_1 z_2 v_2$.

Let $c' : E(G') \mapsto \{1, 2, 3\}$ be a 3-edge-coloring of G' with $c'^{-1}(1) \cup c'^{-1}(2) = E(Q')$. Without loss of generality, let $c'(f_0) = 1$. Let $S' = c'^{-1}(2) \cup c'^{-1}(3)$ which is a 2-factor of G' with all even components. Note that S' corresponds to a 2-factor of G with at most two odd components (each of which contains one of $\{v_1, v_2\}$). Thus, the corollary is proved by applying Theorem 4.2.3.   □

The definition of oddness for cubic graphs (Definition 4.2.2) can be extended for general graphs without the requirement of 3-regularity.

Definition 4.2.9 Let G be a bridgeless graph. For a spanning even subgraph S of G, the *oddness* of S, denoted by $odd(S)$, is the number of components of S that contains an odd number of odd degree vertices of G. And, the *oddness* of G, denoted by $odd(G)$, is the minimum of $odd(S)$ for all spanning even subgraph S of G.

Theorem 4.2.3 can be further extended for all bridgeless graphs with oddness at most 2 (see Exercise 4.6).

By Lemma 4.1.2 and some ideas that we used in the proof of Theorem 4.2.3, Conjecture 4.1.4 can be proved by verifying the following statement.

Conjecture 4.2.10 *Let G be a cyclic 4-edge-connected cubic graph.*

Then there is a spanning even subgraph S of G with the least number of odd components and there is a minimal bridgeless parity subgraph P in the contracted graph G/S such that the suppressed cubic graph $\overline{G[E(S) \cup E(P)]}$ is 3-edge-colorable.

The following conjecture is a weak version of the above conjecture, which implies the circuit double cover conjecture, but not Conjecture 4.1.4.

Conjecture 4.2.11 *Let G be a cyclic 4-edge-connected cubic graph. Then there is a spanning even subgraph S of G and there is a minimal bridgeless parity subgraph P in the contracted graph G/S such that the suppressed cubic graph $\overline{G[E(S) \cup E(P)]}$ has an even subgraph double cover containing S as an element.*

4.3 Open problems

Conjecture 4.1.4 (Preissmann [192] and Celmins [31]) *Every bridgeless graph has a 5-even subgraph double cover.*

Conjecture 4.2.10 *Let G be a cyclic 4-edge-connected cubic graph. Then there is a spanning even subgraph S of G with the least number of odd components and there is a minimal bridgeless parity subgraph P in the contracted graph G/S such that the suppressed cubic graph $\overline{G[E(S) \cup E(P)]}$ is 3-edge-colorable.*

Conjecture 4.2.11 *Let G be a cyclic 4-edge-connected cubic graph. Then there is a spanning even subgraph S of G and there is a minimal bridgeless parity subgraph P in the contracted graph G/S such that the suppressed cubic graph $\overline{G[E(S) \cup E(P)]}$ has an even subgraph double cover containing S as an element.*

Conjecture 4.3.1 (Hoffmann-Ostenhof [111]) *Let C be a circuit in a 2-connected cubic graph G. Then there is a 5-even subgraph double cover of G such that C is a subgraph of one of these five even subgraphs.*

Similar to Lemma 4.1.3, an equivalent statement of Conjecture 4.3.1 is proved in [111] (see Exercise 4.5).

4.4 Exercises

Exercise 4.1   (Lemma 4.1.2) A graph G has a 5-even subgraph double cover if and only if $(G, 2)$ has a weight decomposition $(G_1, w_1) + (G_2, w_2) = (G, 2)$ such that both suppressed cubic graph $\overline{G_1}$ and $\overline{G_2}$ are 3-edge-colorable, and $E_{w_1=2}$ is a mono-colored 1-factor under some 3-edge-coloring of $\overline{G_1}$.

Exercise 4.2   For a 3-edge-colorable cubic graph G_1 and an edge $e \in E(G_1)$. Show that the suppressed cubic graph $G_2 = \overline{G_1 - e}$ is of oddness at most 2.

Exercise 4.3   (Goddyn [83]) Let $3K_2$ be the graph with two vertices and three parallel edges. If a graph G contains a spanning subgraph H which is a subdivision of $3K_2$, then G has a 6-even subgraph double cover.

Remark.   See Chapter 5 about Kotzig frames which is an approach to the CDC conjecture generalized from Exercise 4.3.

Exercise 4.4   (Hoffmann-Ostenhof [111]) Let G be a 2-connected cubic graph. Then G has a 5-even subgraph double cover if and only if there is a matching M of G with the following two properties,
   (1) $\overline{G - M}$ is 3-edge-colorable,
   (2) there is a pair of even subgraphs C_1, C_2 in G with $E(C_1) \cap E(C_2) = M$.

Exercise 4.5   (Hoffmann-Ostenhof [111]) Let G be a 2-connected cubic graph and C be a circuit G. Then G has a 5-even subgraph double cover $\mathcal{F}$ with C as a subgraph of a member of $\mathcal{F}$ if and only if there is a matching M of G with the following three properties,
   (1) $\overline{G - M}$ is 3-edge-colorable,
   (2) there is a pair of even subgraphs C_1, C_2 in G with $E(C_1) \cap E(C_2) = M$,
   (3) $C \subseteq C_1$.

Exercise 4.6   (Huck and Kochol [117]) Let G be a bridgeless graph with oddness at most 2. Then G has a 5-even subgraph double cover.

5

Spanning minor, Kotzig frames

We continue the study of weight decomposition.

Recall Definition 3.1.1 for weight decomposition. Let (G, w) be an eulerian weighted graph. A set of eulerian weighted subgraphs $\{(G_i, w_i) : i = 1, \ldots, t\}$ is called an eulerian weight decomposition of (G, w) if each G_i is a subgraph of G and $\sum_{i=1}^{t} w_i(e) = w(e)$ for every $e \in E(G)$ (where $w_i(e) = 0$ if $e \notin E(G_i)$). (See Figure 3.1 where $t = 2$.)

In this chapter, we will further investigate how to find a set of three weighted graphs $\{(G_1, w_1), (G_2, w_2), (G_3, w_3)\}$ for some families of cubic graphs G such that

(1) $(G_1, w_1) + (G_2, w_2) + (G_3, w_3)$ *is a weight decomposition of* $(G, 2)$,
(2) and each (G_j, w_j) *has a faithful circuit cover.*

In most cases, some suppressed cubic graph $\overline{G_j}$ is expected to have a Hamilton circuit (and, therefore, is 3-edge-colorable). A CDC of such a graph can be obtained by applying arguments similar to those of Lemmas 4.1.1 and 4.1.2.

5.1 Spanning Kotzig subgraphs

Definition 5.1.1 A cubic graph H is called a Kotzig graph if H has a 3-edge-coloring $c : E(H) \mapsto \{1, 2, 3\}$ such that $c^{-1}(i) \cup c^{-1}(j)$ is a Hamilton circuit of H for every pair $i, j \in \{1, 2, 3\}$. (Equivalently, H is a Kotzig graph if it has a 3-circuit double cover.) The coloring c is called a Kotzig coloring of H.

Obviously, $3K_2$, K_4, Möbius ladders M_{2k+1} for every $k \geq 0$ (Fig-

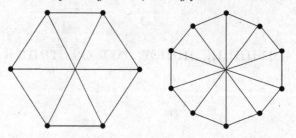

Figure 5.1 *Möbius ladders* M_3 *($M_3 = K_{3,3}$) and* M_5 *($M_5 = V_{10}$)*

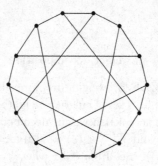

Figure 5.2 *Heawood graph*

ure 5.1), the Heawood graph (Figure 5.2), and the dodecahedron graph (Figure 5.3) are examples of Kotzig graphs.

The study of CDC for graphs containing some spanning subgraphs that are subdivisions of Kotzig graphs was initially started in [83]. Later, it was further generalized in [99].

Definition 5.1.2   A graph H is a spanning minor of another graph G if

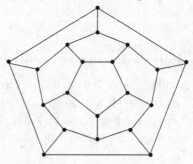

Figure 5.3 *Dodecahedron graph*

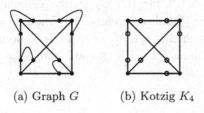

(a) Graph G (b) Kotzig K_4

Figure 5.4 K_4 *is a Kotzig spanning minor of the graph* G

G has a spanning subgraph that is a subdivision of H. (If H is a Kotzig graph, then, we say, G has a spanning Kotzig minor. See Figure 5.4)

Theorem 5.1.3 (Goddyn [83], also see [99]) *If a graph* G *has a Kotzig graph as a spanning minor, then* G *has a 6-even subgraph double cover.*

Before the proof of Theorem 5.1.3, we present an useful lemma, which is an analog of Lemmas 4.1.1 and 4.1.2.

A 2-factor F of a cubic graph G is even if every component is an even length circuit.

Lemma 5.1.4 (Häggkvist and Markström [99]) *Let* H *be a spanning subgraph of a cubic graph* G. *If* H *has a 3-even subgraph double cover* $\{C_1, C_2, C_3\}$ *such that*

(1) the matching $E(G) - E(H) = M$ *has a partition* $\{M_1, M_2, M_3\}$ *and,*

(2) for each $\mu \in \{1, 2, 3\}$, *the suppressed cubic graph* $\overline{G[E(C_\mu) \cup M_\mu]}$ *is 3-edge-colorable,*

then G *has a 9-even subgraph double cover. Furthermore, if, for each* $\mu \in \{1, 2, 3\}$, C_μ *corresponds to an even 2-factor in the suppressed cubic graph* $\overline{G[E(C_\mu) \cup M_\mu]}$, *then* G *has a 6-even subgraph double cover.*

Proof Since the suppressed cubic subgraph $\overline{G[E(C_\mu) \cup M_\mu]}$ is 3-edge-colorable ($\mu \in \{1, 2, 3\}$), by Lemma 2.2.1, it has a 3-even subgraph $(1, 2)$-cover $\mathcal{F}_\mu$ that covers $E(C_\mu)$ once and all other edges of M_μ twice. So, G has a 9-even subgraph double cover $\cup_{\mu \in \{1,2,3\}} \mathcal{F}_\mu$.

Furthermore, if C_μ corresponds to an even 2-factor in the suppressed cubic graph $\overline{G[E(C_\mu) \cup M_\mu]}$, then by Lemma 2.2.2, the suppressed cubic subgraph $\overline{G[E(C_\mu) \cup M_\mu]}$ has a 2-even subgraph $(1, 2)$-cover $\mathcal{F}_\mu$ that covers $E(C_\mu)$ once and all other edges of M_μ twice. So, G has a 6-even subgraph double cover. $\square$

Proof of Theorem 5.1.3. Let $\{C_1, C_2, C_3\}$ be a circuit double cover of $\overline{H}$ (consisting of three Hamilton circuits). Partition $M = E(G) - E(H)$ into $\{M_1, M_2, M_3\}$ such that $e \in M_j$ only if e is a chord of C_j for each $j \in \{1, 2, 3\}$. By Lemma 5.1.4 (the second part), the graph G has a 6-even subgraph double cover. $\qquad\square$

⋆ Generalizations of Kotzig graphs

The concept of spanning Kotzig minor is further generalized in [83] and [99].

Definition 5.1.5 Let H be a cubic graph with a 3-edge-coloring c : $E(H) \mapsto \mathbb{Z}_3$ such that

(∗)   edges in colors 0 and μ ($\mu \in \{1, 2\}$) induce a Hamilton circuit.

Let F be the even 2-factor induced by edges in colors 1 and 2. If, for *every* even subgraph $S \subseteq F$, switching colors 1 and 2 of the edges of S yields a new 3-edge-coloring having the same property (∗), then the 3-edge-coloring c is called a semi-Kotzig coloring. A cubic graph H with a semi-Kotzig coloring is called a semi-Kotzig graph.

If F has t components, then H is said to be a semi-Kotzig graph of rank t. (A semi-Kotzig graph of rank at most 2 is also called a switchable-CDC graph in [99]).

The following is a subfamily of semi-Kotzig graphs.

Definition 5.1.6 An iterated Kotzig graph H is a cubic graph constructed recursively as follows.

(1) Let $\mathcal{K}_0$ be a set of all Kotzig graphs with a Kotzig coloring c : $E(G) \mapsto \mathbb{Z}_3$.

(2) For $i > 0$, a cubic graph $H \in \mathcal{K}_{i+1}$ is constructed recursively from a graph $H'' \in \mathcal{K}_i$ and a graph $H' \in \mathcal{K}_0$: let $e = xy \in E(H'), f = uv \in E(H'')$, both colored with 0, replace e and f with a pair of 0-colored edges xu and yv (equivalently, H is a $\oplus_2$-sum of H' and H'' at a pair of 0-colored edges – see Section B.1.2 for the definition of $\oplus_2$-sum).

Members of the family $\bigcup_{i \geq 0} \mathcal{K}_i$ are called iterated Kotzig graphs.

Figure 5.5 is an example of iterated Kotzig graphs. It is evident that an iterated Kotzig graph has a semi-Kotzig coloring and hence is a semi-Kotzig graph. But a semi-Kotzig graph is not necessarily an iterated Kotzig graph. For example, the semi-Kotzig graph in Figure 5.6 [249] is

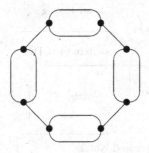

Figure 5.5 *An example of iterated Kotzig graphs – "string of pearls"*

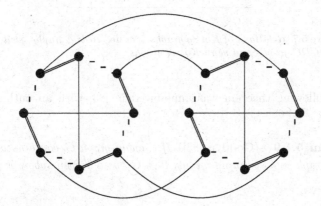

Figure 5.6 *A semi-Kotzig graph, but not iterated Kotzig*

not an iterated Kotzig graph. Hence we have the following relations (see Figure 5.7).

$$\{\text{Kotzig}\} \subset \{\text{iterated Kotzig}\} \cap \{\text{switchable CDC}\},$$
$$\{\text{iterated Kotzig}\} \cup \{\text{switchable CDC}\} \subset \{\text{semi-Kotzig}\}.$$

Analogies and stronger versions of Theorem 5.1.3 have been obtained for those generalizations.

Theorem 5.1.7 (Goddyn [83], also see [99]) *If a cubic graph G contains an iterated Kotzig graph as a spanning minor, then G has a 6-even-subgraph double cover.*

Theorem 5.1.8 (Häggkvist and Markström [99]) *If a cubic graph G contains a switchable CDC graph (a semi-Kotzig graph with rank at most 2) as a spanning minor, then G has a 6-even-subgraph double cover.*

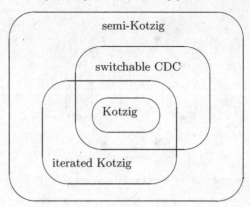

Figure 5.7 *Relations of Kotzig graphs, iterated Kotzig graphs, switchable CDC graphs and semi-Kotzig graphs*

The following theorem was announced in [83] with an outline of a proof.

Theorem 5.1.9   (Goddyn [83]) *If a cubic graph G contains a semi-Kotzig graph as a spanning minor, then G has a 6-even-subgraph double cover.*

Theorems 5.1.7, 5.1.8 and 5.1.9 will be proved together with a generalized result (Theorem 5.2.6).

⋆ Various spanning minors

In the following theorem, a spanning subgraph may not be a subdivision of a Kotzig graph. Those graphs are proved to have a circuit double cover.

Theorem 5.1.10   (Häggkvist and Markström [100]) *If a cubic graph G has a spanning subgraph H such that $\overline{H}$ is a 2-connected, cubic, simple graph with at most 10 vertices, then G has a circuit double cover.*

Theorem 5.1.10 generalizes some early results by Goddyn [83] that circuit double cover conjecture is true for a cubic graph G if G has a spanning subgraph H which is a subdivision of the Petersen graph or Möbius ladders $V_8 = M_4$.

5.2 Kotzig frames

If a cubic graph G has an even 2-factor, then the graph G has many nice properties: G *is 3-edge-colorable, G has a circuit double cover,* etc. Inspired by the structure of even 2-factors, Häggkvist and Markström [99] introduced the following concept, which extends the investigation of *connected* spanning minors to *disconnected* cases.

Definition 5.2.1 Let G be a cubic graph. A spanning subgraph H of G is called a frame of G if G/H is an even graph.

Definition 5.2.2 Let G be a cubic graph. A frame H of G is called a Kotzig frame (or *iterated Kotzig frame, switchable CDC frame, semi-Kotzig frame,* etc.) of G if, for each non-circuit component H_j of H, the suppressed graph $\overline{H_j}$ is a Kotzig graph (or iterated Kotzig graph, switchable CDC, semi-Kotzig graph, etc. respectively).

In Section 5.1, we have discussed cubic graphs with connected Kotzig frames (Theorem 5.1.3) and some of its generalizations (Theorems 5.1.7, 5.1.8, and 5.1.9). Those are results about frames with only one component. In this section, graphs with disconnected frames will be further studied.

The following is a generalization of Theorem 5.1.3.

Theorem 5.2.3 (Häggkvist and Markström [99]) *If a cubic graph G has a Kotzig frame that contains at most one non-circuit component, then G has a 6-even subgraph double cover.*

Similar to results in Section 5.1, Theorem 5.2.3 is further generalized as follows for iterated Kotzig frames, semi-Kotzig frames.

Theorem 5.2.4 (Häggkvist and Markström [99]) *If a cubic graph G contains an iterated Kotzig frame with at most one non-circuit component, then G has a 6-even subgraph double cover.*

Theorem 5.2.5 (Häggkvist and Markström [99]) *If a cubic graph G contains a switchable CDC frame with at most one non-circuit component, then G has a 6-even subgraph double cover.*

Theorem 5.2.6 (Ye and Zhang [249]) *If a cubic graph G contains a semi-Kotzig frame with at most one non-circuit component, then G has a 6-even subgraph double cover.*

The proof of Theorem 5.2.6 is presented at the end of this section.

The following conjecture about semi-Kotzig frames is a generalization of most results in this and the previous section, which was originally proposed in [99] for Kotzig frames, iterated Kotzig frames, and switchable CDC frames (see Conjecture 5.5.1).

Conjecture 5.2.7 *Let G be a cubic graph with semi-Kotzig frame. Then G has a circuit double cover.*

The following are some partial results for Conjecture 5.2.7 (and Conjecture 5.5.1).

Theorem 5.2.8 (Zhang and Zhang [262]) *Let G be a bridgeless cubic graph. If G contains a Kotzig frame H such that G/H is a tree if parallel edges are identified as a single edge (equivalently, the circumference of G/H is at most 2), then G has a 6-even subgraph double cover.*

Theorem 5.2.9 (Cutler and Häggkvist [38]) *Let G be a bridgeless cubic graph. Assume G contains a semi-Kotzig frame H such that H has precisely two components Q_1, Q_2 and $\overline{Q_1}$ is Kotzig while $\overline{Q_2}$ is semi-Kotzig. Then G has an even subgraph double cover.*

⋆ Proof of Theorem 5.2.6

Definition 5.2.10 Let H be a bridgeless graph. A mapping $c : E(H) \mapsto \mathbb{Z}_3$ is called a parity 3-edge-coloring of H if, for each vertex $v \in H$ and each $\mu \in \mathbb{Z}_3$,

$$|c^{-1}(\mu) \cap E(v)| \equiv |E(v)| \pmod 2.$$

It is obvious that if H itself is cubic, then a parity 3-edge-coloring is a proper 3-edge-coloring (traditional definition).

Preparation of the proof. Let $H_0, \ldots, H_t$ be components of the frame H such that H_0 is a subdivision of a semi-Kotzig graph and each H_i, $1 \leq i \leq t$, is a circuit of even length. Let $M = E(G) - E(H)$, and $H^* = H - H_0$.

Define an initial semi-Kotzig coloring $c_0 : E(\overline{H}_0) \to \mathbb{Z}_3$ of $\overline{H}_0$, where $F_0 = c_0^{-1}(1) \cup c_0^{-1}(2)$ is a 2-factor of $\overline{H}_0$ and $c_0^{-1}(0) \cup c_0^{-1}(\mu)$ is a Hamilton circuit of $\overline{H}_0$ for each $\mu \in \{1, 2\}$.

The semi-Kotzig coloring c_0 of $\overline{H}_0$ can be considered as an edge-coloring of H_0: each induced path is colored with the same color as its corresponding edge in $\overline{H}_0$ (note, this edge-coloring of H_0 is a parity 3-edge-coloring, which may not be a proper 3-edge-coloring).

The strategy of the proof is to show that G can be covered by three subgraphs $G(0,1), G(0,2)$ and $G(1,2)$ such that each $G(\alpha, \beta)$ has a 2-even-subgraph cover which covers edges of $M \cap E(G(\alpha, \beta))$ twice and the edges of $E(H) \cap E(G(\alpha, \beta))$ once. In order to prove this, we are going to show that the three subgraphs $G(\alpha, \beta)$ have the following properties.

(i) The suppressed cubic graph $\overline{G(\alpha, \beta)}$ is 3-edge-colorable (so that Lemma 2.2.2 can be applied to each of them).

(ii) $c_0^{-1}(\alpha) \cup c_0^{-1}(\beta) \subseteq G(\alpha, \beta)$ for each pair $\alpha, \beta \in \mathbb{Z}_3$.

(iii) The even subgraph H^* has a decomposition, H_1^* and H_2^*, each of which is even, (here, for technical reason, let $H_0^* = \emptyset$), such that $H_\alpha^* \cup H_\beta^* \subseteq G(\alpha, \beta)$, for each $\{\alpha, \beta\} \subset \mathbb{Z}_3$.

(iv) Each $e \in M = E(G) - E(H)$ is contained in precisely one member of $\{G(0,1), G(0,2), G(1,2)\}$.

(v) And most importantly, the subgraph $c_0^{-1}(\alpha) \cup c_0^{-1}(\beta) \cup H_\alpha^* \cup H_\beta^*$ corresponds to an even 2-factor of $\overline{G(\alpha, \beta)}$.

Can we decompose H^* and find a partition of $M = E(G) - E(H)$ to satisfy (v)? One may also notice that the *initial* semi-Kotzig coloring c_0 may not be appropriate. However, the color-switchability of the semi-Kotzig component H_0 may help us to achieve the goal. The properties described above in the strategy will be proved in the following claim.

We claim that G has the following property.

(∗) *There is a semi-Kotzig coloring c_0 of $\overline{H}_0$, a decomposition $\{H_1^*, H_2^*\}$ of H^* and a partition $\{N_{(0,1)}, N_{(0,2)}, N_{(1,2)}\}$ of M such that, let $C_{(\alpha, \beta)} = c_0^{-1}(\alpha) \cup c_0^{-1}(\beta)$,*

(1) for each $\mu \in \{1, 2\}$, the even subgraph $C_{(0,\mu)} \cup H_\mu^*$ corresponds to an even 2-factor of $\overline{G(0,\mu)} = \overline{G[C_{(0,\mu)} \cup H_\mu^* \cup N_{(0,\mu)}]}$, and

(2) the even subgraph $C_{(1,2)} \cup H^*$ corresponds to an even 2-factor of $\overline{G(1,2)} = \overline{G[C_{(1,2)} \cup H^* \cup N_{(1,2)}]}$.

Proof of (∗). Let G be a minimum counterexample to (∗). Let $c : E(H) \mapsto \mathbb{Z}_3$ be a parity 3-edge-coloring of H such that

(1) the restriction of c on $\overline{H}_0$ is a semi-Kotzig coloring, and

(2) $E(H^*) \subseteq c^{-1}(1) \cup c^{-1}(2)$ (a set of mono-colored circuits).

Let

$$F = c^{-1}(1) \cup c^{-1}(2) = E(H) - c^{-1}(0).$$

Label (or partition) the matching M as follows. For each edge $e = xy \in M$, $xy \in M_{(\alpha, \beta)}$ ($\alpha \leq \beta$ and $\alpha, \beta \in \mathbb{Z}_3$) if x is incident with two α-colored edges and y is incident with two β-colored edges. So, the matching M is partitioned into six subsets:

$$M_{(0,0)}, M_{(0,1)}, M_{(0,2)}, M_{(1,1)}, M_{(1,2)} \text{ and } M_{(2,2)}.$$

Note that this partition will be adjusted whenever the parity 3-edge-coloring c is modified.

The final semi-Kotzig coloring c_0 of $(*)$ will be obtained from this given parity coloring c by switching colors 1 and 2 along some components of $c^{-1}(1) \cup c^{-1}(2)$. (Following the Rule $(**)$ on p. 55.)

Claim 1. $M_{(0,\mu)} \cap G[V(H_0)] = \emptyset$, *for each* $\mu \in \mathbb{Z}_3$.

Suppose that $e = xy \in M_{(0,\mu)}$ where x subdivides a 0-colored edge of $\overline{H}_0$. Then, in the graph $\overline{G - e}$, the spanning subgraph H retains the same property as itself in G. Since $\overline{G - e}$ is smaller than G, $\overline{G - e}$ satisfies $(*)$: $\overline{H}_0$ has a semi-Kotzig coloring c_0 and $M - e$ has a partition $\{N_{(0,1)}, N_{(0,2)}, N_{(1,2)}\}$ and H^* has a decomposition $\{H_1^*, H_2^*\}$. In the semi-Kotzig coloring c_0, without loss of generality, assume that y subdivides a 0- or 1-colored edge of $\overline{H}_0$. For the graph G, add e into $N_{(0,1)}$ and $G(0,1)$. This revised partition $\{N_{(0,1)}, N_{(0,2)}, N_{(1,2)}\}$ of M and the resulting subgraphs $G(\alpha, \beta)$ satisfy $(*)$. This contradicts that G is a counterexample. $\diamond$

Since $c^{-1}(0) \subseteq H_0$ (each component of $H - H_0 = H^*$ is mono-colored with 1 or 2), for every edge $e \in M_{(0,\mu)}$ ($\mu \in \{1,2\}$), by Claim 1, the edge e has one endvertex belonging to $V(H - H_0) = V(H^*)$. That is,

$$M_{(0,0)} = \emptyset, \quad \text{and} \quad M_{(0,1)} \cup M_{(0,2)} \subseteq (H_0, H^*).$$

Let

$$G' = \overline{G - M_{(0,1)} - M_{(0,2)}}. \tag{5.1}$$

Here $F = c^{-1}(1) \cup c^{-1}(2)$ is a 2-factor of G' since all 0-colored edges induce a matching in G'.

Claim 2. *The contracted graph* G'/F *is acyclic.*

Note that, by (5.1), $E(G'/F) \subseteq M_{(1,1)} \cup M_{(1,2)} \cup M_{(2,2)}$.

Suppose to the contrary that G'/F contains a circuit Q (including loops). In the graph $\overline{G - E(Q)}$, the spanning subgraph H remains as a semi-Kotzig frame.

Then the smaller graph $\overline{G - E(Q)}$ satisfies $(*)$: $\overline{H}_0$ has a semi-Kotzig coloring c_0, and $M - E(Q)$ has a partition $\{N_{(0,1)}, N_{(0,2)}, N_{(1,2)}\}$, and H^* has a partition $\{H_1^*, H_2^*\}$. So add all edges of $E(Q)$ into $N_{(1,2)}$ and

$G(1,2)$. This revised partition $\{N_{(0,1)}, N_{(0,2)}, N_{(1,2)}\}$ of M and its resulting subgraphs $G(\alpha, \beta)$ also satisfy $(*)$ since $C_{(1,2)} \cup H^*$ corresponds to an even 2-factor of $\overline{G(1,2)} = \overline{G[C_{(1,2)} \cup H^* \cup N_{(1,2)}]}$. This is a contradiction. So Claim 2 follows. $\Diamond$

By Claim 2, each component T of G'/F is a tree. Along the tree T, we can modify the parity 3-edge-coloring c of H as follows:

$(**)$ *properly switch colors for some circuits in F so that every edge of T is incident with four same colored edges.*

Note that Rule $(**)$ is feasible since G'/F is acyclic (by Claim 2). Furthermore, under the modified parity 3-edge-coloring c, $M_{(1,2)} = \emptyset$. So

$$M = M_{(0,1)} \cup M_{(0,2)} \cup M_{(1,1)} \cup M_{(2,2)}.$$

The colors of all H_i ($i \geq 1$) give a decomposition $\{H_1^*, H_2^*\}$ of H^* where H_μ^* consists of all circuits of H^* mono-colored by μ for $\mu = 1$ and 2.
Let

$$G'' = G/H$$

where $E(G'') = M$. Then G'' is even since H is a frame. For a vertex w of G'' corresponding to a component H_i with $i \geq 1$, there is one $\mu \in \{1,2\}$ such that all edges incident with w belong to $M_{(0,\mu)} \cup M_{(\mu,\mu)}$. Here, define

$$N_{(0,\mu)} = M_{(0,\mu)} \cup M_{(\mu,\mu)}.$$

Hence, a vertex of G'' corresponding to H_i with $i \geq 1$ either has the degree in $G''[N_{(0,\mu)}]$ the same as its degree in G'' or has degree zero (by Rule $(**)$). So every vertex of $G''[N_{(0,\mu)}]$, other than the one corresponding to H_0, has an even degree. Since every graph has an even number of odd-degree vertices, it follows that $G''[N_{(0,\mu)}]$ is an even subgraph.
For each $\mu \in \{1,2\}$, let $G(0,\mu) = N_{(0,\mu)} \cup (c^{-1}(0) \cup c^{-1}(\mu))$. Since $G''[N_{(0,\mu)}]$ is an even subgraph of G'', the even subgraph $c^{-1}(0) \cup c^{-1}(\mu)$ corresponds to an even 2-factor of $G(0,\mu)$. And let $G(1,2) = F = c^{-1}(1) \cup c^{-1}(2)$. So G has the property $(*)$, a contradiction. This completes the proof of $(*)$. $\square$

Proof of Theorem 5.2.6. Let G be a graph with a semi-Kotzig frame. Then G satisfies $(*)$ and therefore is covered by three subgraphs $G(\alpha, \beta)$ ($\alpha, \beta \in \mathbb{Z}_3$ and $\alpha < \beta$) as stated in $(*)$.
Applying Lemma 2.2.2 to the three graphs $\overline{G(\alpha, \beta)}$, each $G(0,\mu)$ has

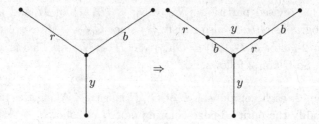

Figure 5.8 *Method 5.1: $(Y \rightarrow \triangle)$-operation*

a 2-even-subgraph cover $\mathcal{F}_{(0,\mu)}$ which covers the edges of $C_{(0,\mu)} \cup H_\mu^*$ once and the edges in $N_{(0,\mu)}$ twice, and $G(1,2)$ has a 2-even-subgraph cover $\mathcal{F}_{(1,2)}$ which covers the edges of $C_{(1,2)} \cup H^*$ once and the edges in $N_{(1,2)}$ twice. So $\bigcup \mathcal{F}_{(\alpha,\beta)}$ is a 6-even-subgraph double cover of G. This completes the proof.   □

5.3 Construction of Kotzig graphs

Kotzig graphs were originally studied in [152]. Well-known examples of Kotzig graphs are: K_4, Möbius ladders M_{2k+1} for every $k \geq 0$ (Figure 5.1), the Heawood graph (Figure 5.2), and the dodecahedron graph (Figure 5.3).

Some operations for constructing Kotzig graphs were introduced in [152], [153], [154], [155].

Method 5.1 [153] The $(Y \rightarrow \triangle)$-operations on a vertex of a Kotzig graph (see Figure 5.8).

Method 5.2 [153] The $\oplus_3$-sum of two Kotzig graphs. (See Section B.1.3, and Figure B.2.)

Remark. Method 5.1 can be considered as a special case of Method 5.2 as a $\oplus_3$-sum of a Kotzig graph and a copy of K_4.

Method 5.3 [153] Let G be a Kotzig graph, and e_1, e_2 be two edges of G colored differently under some Kotzig coloring of G, say, red and blue. Let H be a graph constructed by first subdividing each e_j twice, see Figure 5.9, then identify the pair of new vertices and edges between

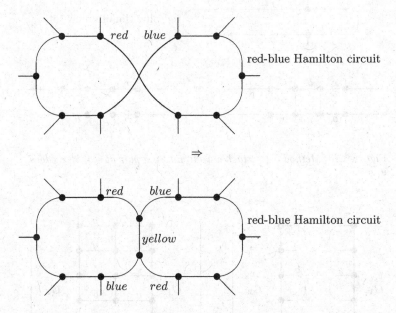

red-blue Hamilton circuit

red-blue Hamilton circuit

Figure 5.9 *Method 5.3: crossing merging along a red-blue Hamilton circuit*

them as the red-blue Hamilton circuit crossing itself at the new edges. Then the new cubic graph H is Kotzig.

Remark. Method 5.1 (the $(Y \to \triangle)$-operation) is a special case of Method 5.3.

Method 5.4 [99] Let G be a Kotzig graph and let e_1 and e_2 be two edges with different colors in some Kotzig coloring of G. The graphs G_1 and G_2 obtained by using the constructions illustrated in Figure 5.10 on e_1 and e_2 are Kotzig graphs.

Remark. The first case of Method 5.4 (adding a square) is a special case of Method 5.3.

Method 5.5 For each $\mu \in \{1,2\}$, let G_μ be a Kotzig graph and $e_{\mu,1}, e_{\mu,2}, e_{\mu,3}$ be three edges of G_μ colored differently under some Kotzig coloring $c : E(G_\mu) \mapsto \mathbb{Z}_3$. Say $c(e_{\mu,j}) = \mu + j \pmod{3}$. Let H be a graph constructed by first subdividing each $e_{\mu,j}$ twice, see Figure 5.11, then merge each pair of new edges (between two new vertices) inside $e_{1,j}$ and $e_{2,j}$ for each $j = 1, 2, 3$. Then the new cubic graph H is Kotzig.

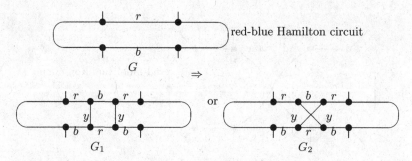

Figure 5.10 *Method 5.4: attach a 4-circuit on a pair of red, blue edges*
e_1, e_2

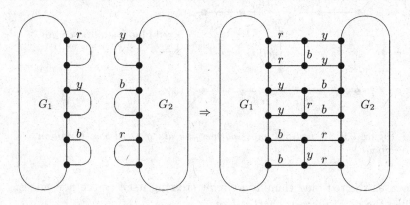

Figure 5.11 *Method 5.5: two Kotzig graphs are attached to each other
on three pairs of edges*

Some additional construction will be presented as exercises.

5.4 Three-Hamilton circuit double covers

Kotzig graphs have the property that they have a circuit double cover
consisting of three Hamilton circuits. There are several other closely related families of cubic graphs having some very similar properties to
Kotzig graphs and playing important roles in the study of circuit covering problems. For example, uniquely 3-edge-colorable cubic graphs ([55],
[56], [78], [93], [106], [225], [226], [235], [244] etc.), Hamilton weighted
graphs ([165], [256], [261] etc. and see Section 16.2).

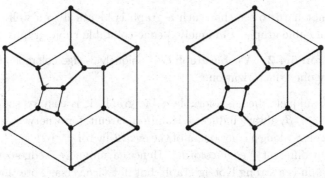

Bold edges form a Hamilton circuit
induced by $c^{-1}(1) \cup c^{-1}(2)$

Bold edges form an even 2-factor
induced by $c^{-1}(1) \cup c^{-1}(2)$

Figure 5.12 *A Kotzig graph, but not strong Kotzig*

⋆ Strong Kotzig graphs

Definition 5.4.1 A 3-edge-colorable cubic graph G is strong Kotzig if every 3-edge-coloring is Kotzig (equivalently, for *every* 3-edge-coloring $c : E(G) \mapsto \mathbb{Z}_3$ and for every pair $\{\alpha, \beta\} \subset \mathbb{Z}_3$, $c^{-1}(\alpha) \cup c^{-1}(\beta)$ induces a Hamilton circuit).

Graphs, such as, K_4, $K_{3,3}$, the Heawood graph (Figure 5.2), and the dodecahedron graph (Figure 5.3), are examples of strong Kotzig graphs (see Exercise 5.10).

A straightforward fact is that a cubic graph G is strong Kotzig if and only if every even 2-factor is connected (Exercise 5.8).

It is evident that every strong Kotzig graph is Kotzig, but not vice versa. The planar graph in Figure 5.12 is an example [247]. Other non-planar examples are Möbius ladders M_{2k+1} for every $k \geq 2$ (Figure 5.1) which are Kotzig, but not strong Kotzig (Exercise 5.9).

And some expansion operations defined in Section 5.3 preserve the property of being Kotzig, but not strong Kotzig (especially if the graph has more than one Kotzig coloring).

⋆ Uniquely 3-edge-colorable graphs

Note that a strong Kotzig graph may have more than one Kotzig coloring (Exercise 5.10). We may further consider a subfamily of strong Kotzig graphs that have precisely *one* 3-edge-coloring.

It is not hard to see that such a graph is identical to a well-studied family of cubic graphs – uniquely 3-edge-colorable cubic graphs.

Definition 5.4.2   A cubic graph G is uniquely 3-edge-colorable if it has precisely one 1-factorization.

For a uniquely 3-edge-colorable cubic graph, it is easy to show that $c^{-1}(\alpha) \cup c^{-1}(\beta)$ always induces a Hamilton circuit, for otherwise, switching the colors along a component of the even 2-factor $c^{-1}(\alpha) \cup c^{-1}(\beta)$, one obtains a different 3-edge-coloring. Hence, a uniquely 3-edge-colorable cubic graph is a strong Kotzig graph, but not vice versa. Note that $K_{3,3}$ ($K_{3,3} = M_3$) (Figure 5.1), the Heawood graph (Figure 5.2), and the dodecahedron graph (Figure 5.3) are examples of strong Kotzig graphs, but are not uniquely 3-edge-colorable (see Exercise 5.10).

⋆ Hamilton weighted graphs

It is not hard to see that a uniquely 3-edge-colorable cubic graph G contains precisely three Hamilton circuits, and, furthermore, for each Hamilton circuit C_0, the graph G has a circuit double cover containing the given circuit C_0 and consisting of three Hamilton circuits (equivalently, a strong CDC consisting of three Hamilton circuits). Is it possible that G has a CDC $\mathcal{F}$ containing the given one C_0 but some member of $\mathcal{F}$ may not be a Hamilton circuit? The answer is "yes": the generalized Petersen graph $P(9, 2)$ (Figure 16.25) is uniquely 3-edge-colorable [235] but some of its strong CDC consists of more than three members (see Exercise 16.4).

Let us consider a subfamily of uniquely 3-edge-colorable cubic graphs as follows:

a cubic graph G with a Hamilton circuit C_0 has the property of strong *$3H$-CDC if every CDC of G containing C_0 must consist of three Hamilton circuits.*

It is easy to see that a cubic graph G with the property of strong $3H$-CDC has only one strong CDC containing the given circuit C_0. Furthermore, a cubic graph G with the property of strong $3H$-CDC is a special case of cubic graphs admitting Hamilton weights (see Section 16.2). (It is proved in Exercise 16.2 that a Hamilton weighted graph (G, w) is uniquely 3-edge-colorable if and only if $E_{w=1}$ induces a Hamilton circuit.) In Chapter 16, we will see that cubic graphs with the property of strong $3H$-CDC play a central role in the structural study of circuit chain.

Remark. Denote the family of Kotzig graphs by $\mathcal{K}$, strong Kotzig graphs by $\mathcal{SK}$, uniquely 3-edge-colorable graphs by $\mathcal{U3EC}$, and graphs with strong $3H$-CDC property by $\mathcal{SD3HC}$.

By the discussion above,

$$\mathcal{K} \supset \mathcal{SK} \supset \mathcal{U3EC} \supset \mathcal{SD3HC}.$$

5.5 Open problems

⋆ From frames to CDC

The following conjecture combines several open problems proposed in various papers.

Conjecture 5.5.1 *Let G be a bridgeless cubic graph and let H be a spanning subgraph of G. Then G has a circuit double cover if H is*

(1) (Häggkvist and Markström [99], p. 186) *a Kotzig frame, or*

(2) (Häggkvist and Markström [99], p. 186) *an iterated Kotzig frame, or*

(3) (Häggkvist and Markström [99], p. 186) *a switchable CDC frame, or*

(4) (Ye and Zhang [249]) *a semi-Kotzig frame.*

Remark. Some partial results for Conjecture 5.5.1 can be found in Sections 5.1 and 5.2.

Conjecture 5.5.2 (Häggkvist and Markström [99], p. 205) *Let G be a cubic graph containing a spanning subgraph H such that $\overline{H}$ is an even Möbius ladder M_{2k}. Then G has an h-even subgraph double cover for some integer h.*

Remark. Conjecture 5.5.2 is true for M_2 since $M_2 = K_4$ which is a Kotzig graph (included in Theorem 5.1.3). And it is also proved in [83] for $M_4 = V_8$.

Conjecture 5.5.3 (Häggkvist and Markström [100], p. 777) *If a cubic graph G has a spanning subgraph H such that $\overline{H}$ is a 2-connected cubic graph with at most 10 vertices, then G has a 6-even subgraph double cover.*

Remark. Conjecture 5.5.3 strengthens Theorem 5.1.10, in which we only know the existence of a *circuit double cover*. However, the number of even subgraphs in the covering remains unknown.

Conjecture 5.5.4 (Häggkvist and Markström [99]) *Every cubic graph with a 3-edge-colorable spanning minor has a 6-even subgraph double cover.*

Remark. Exercise 5.13 proves the existence of such spanning minor, and, therefore, Conjecture 5.5.4 is equivalent to the following conjecture.

Conjecture 5.5.5 *Every bridgeless cubic graph has a 6-even subgraph double cover.*

Remark. Conjecture 5.5.5 is a weak version of Conjecture 4.1.4.

⋆ Existence of Kotzig frames

By Theorem 5.1.3, the following conjecture implies the circuit double cover conjecture.

Conjecture 5.5.6 *Every 3-connected, cyclically 4-edge-connected, cubic graph contains a spanning subgraph that is a subdivision of a Kotzig graph.*

This conjecture was originally proposed by Häggkvist and Markström in [99] (p. 205) for all 3-connected cubic graphs. The above version has an additional requirement: "cyclically 4-edge-connected", which was suggested by Hoffmann-Ostenhof (based on his counterexamples in [111]).

Note that the 3-connectivity in Conjecture 5.5.6 cannot be replaced with 2-connectivity. The graph in Figure 5.13 is a 2-connected counterexample to Conjecture 5.5.6 (discovered in [99]). Furthermore, Hoffmann-Ostenhof discovered more counterexamples for 3-connected cubic graphs [111], and he pointed out that an extra requirement of higher cyclic edge-connectivity is necessary for the existence of such frames.

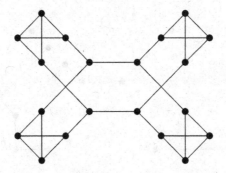

Figure 5.13 *A 2-connected graph without a Kotzig frame*

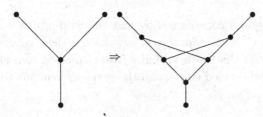

Figure 5.14 *Kotzig graph expansion by the method of Exercise 5.4*

5.6 Exercises

⋆ Constructions

Exercise 5.1   Show that graphs constructed using Methods 5.2, 5.3, 5.4 and 5.5 are Kotzig graphs.

Exercise 5.2   Show that Method 5.1 (the $(Y \to \triangle)$-operation) is a special case of Method 5.3.

Exercise 5.3   Show that the first case of Method 5.4 (adding a square) is a special case of Method 5.3.

Exercise 5.4   [99] Let G be a Kotzig graph and v be a vertex of G. Show that the new graph obtained from G by using the expansion method illustrated in Figure 5.14 is also a Kotzig graph.

Remark.   The method in Exercise 5.4 is a special case of the second part of Method 5.4.

Exercise 5.5   [99] Let G be a Kotzig graph, and let e_1, e_2 and e_3 be three edges in G with different colors in some Kotzig coloring of G.

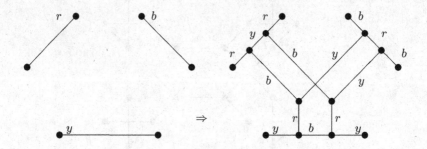

Figure 5.15 *Add two independent vertices*

Let H be a graph constructed by first subdividing the edges e_i twice, see Figure 5.15, then adding two new independent vertices v_1, v_2 and adding edges so that both v_1 and v_2 are connected to a new vertex on each of the subdivided edges, and the graph H remains cubic. Then H is also a Kotzig graph.

Exercise 5.6 Show that the method in Exercise 5.5 is a special case of Method 5.5: combine a Kotzig graph with $3K_2$.

⋆ Examples, counterexamples

Exercise 5.7 [99] Show that the Möbius ladder M_{2k+1} is Kotzig.

Exercise 5.8 Show that a cubic graph G is strong Kotzig if and only if every even 2-factor is connected.

Exercise 5.9 Show that Möbius ladders M_{2k+1} for every $k \geq 2$ are Kotzig, but not strong Kotzig.

Exercise 5.10 Show that $K_{3,3}$ ($K_{3,3} = M_3$) (Figure 5.1), the Heawood graph (Figure 5.2), and the dodecahedron graph (Figure 5.3) are all strong Kotzig graphs, but not uniquely 3-edge-colorable.

Exercise 5.11 Let $\langle \mathcal{K}_4 \rangle$ be the family of all cubic graphs obtained from $3K_2$ via a series of $(Y \to \triangle)$-operations. Show that each member of $\langle \mathcal{K}_4 \rangle$ has all the following properties: Kotzig, strong Kotzig, uniquely 3-edge-colorable, strong $3H$-CDC.

⋆ Spanning minors

Exercise 5.12 [99] Let H be an even subgraph of a cubic graph G. If G/H contains a spanning Kotzig graph, show that (1) G contains a spanning subgraph which is a subdivision of a Kotzig graph and (2) therefore, G has a circuit double cover.

Note, in Exercise 5.12, this spanning subgraph itself is a Kotzig graph, not a subdivision of a Kotzig graph.

Exercise 5.13 Every cyclically 4-edge-connected cubic graph contains a connected 3-edge-colorable cubic graph as a spanning minor.

6
Strong circuit double cover

6.1 Circuit extension and strong CDC

Note that in the contra pair (P_{10}, w_{10}) (see Figure 2.2), $E_{w_{10}=1}$ induces two disjoint circuits. How about an eulerian $(1,2)$-weighted graph (G, w) that $E_{w=1}$ induces a single circuit? The following is an open problem that addresses possible faithful covers for such weighted graphs.

Conjecture 1.5.1 (Strong circuit double cover conjecture, Seymour, see [61] p. 237, and [62], also see [83]) *Let w be an eulerian $(1,2)$-weight of a 2-edge-connected, cubic graph G. If the subgraph of G induced by weight 1 edges is a circuit, then (G, w) has a faithful circuit cover.*

Conjecture 1.5.1 has an equivalent statement.

Let G be a 2-edge-connected cubic graph and C be a circuit of G, then the graph G has a circuit double cover $\mathcal{F}$ with $C \in \mathcal{F}$.

Definition 6.1.1 Let C be a circuit of a 2-edge-connected cubic graph G. A strong circuit (even subgraph) double cover of G with respect to C is a circuit (even subgraph) double cover $\mathcal{F}$ of G with $C \in \mathcal{F}$. (As an abbreviation, $\mathcal{F}$ is called a strong CDC of G with respect to C.)

Conjecture 1.5.1 is obviously stronger than the circuit double cover conjecture. Conjecture 1.5.2 (Sabidussi Conjecture) is a special case of Conjecture 1.5.1 that the given circuit is dominating.

Problem 6.1.2 (Seymour [209], also see [66], [148]) For a 2-edge-connected cubic graph G and a given circuit C of G, does G contain a circuit C' with $V(C) \subseteq V(C')$ and $E(C) \neq E(C')$?

Definition 6.1.3 A circuit C of a graph G is extendable if G contains another circuit C' such that $V(C) \subseteq V(C')$ and $E(C) \neq E(C')$. And the circuit C' is an extension of C (or simply, a C-extension).

Problem 6.1.2 proposes a possible recursive approach to Conjecture 1.5.1.

Proposition 6.1.4 (Kahn, Robertson, Seymour [137], also see [33], [209], [211]) *If Problem 6.1.2 is true for every circuit in every 2-edge-connected cubic graph, then Conjecture 1.5.1 is true.*

Proof Induction on $|E(G)|$. Let C' be an extension of C and $G' = G - (E(C) - E(C'))$. In G', C remains as a circuit.

Since $E(C) - E(C')$ is a matching, the suppressed graph $\overline{G'}$ is cubic. In order to apply induction to the smaller graph $\overline{G'}$, we need to show that $\overline{G'}$ *is 2-edge-connected.* Suppose that $\overline{G'}$ has a bridge e_0 with components Q_1 and Q_2. Let T be the edge-cut of G separating $V(Q_1)$ and $V(Q_2)$. Note that $T - \{e_0\} \subseteq E(C) - E(C')$. Thus, by Lemma A.2.8, $T \cap E(C') = \emptyset$ since $|T \cap E(C')| \leq |\{e_0\}| = 1$. That is, one of Q_1 and Q_2, say Q_1, contains some edges of C but no edge of C'. This contradicts that $V(C) \subseteq V(C')$.

By induction, the suppressed cubic graph $\overline{G'}$ has a circuit double cover $\mathcal{F}'$ with $C' \in \mathcal{F}'$. Thus,

$$\mathcal{F}' - C' + \{C \triangle C', C\}$$

is a circuit double cover of G. $\qquad\qquad\Box$

In next a few sections, we will discuss Problem 6.1.2: partial results, special and useful methods, counterexamples, and possible modifications and approaches.

6.2 Thomason's lollipop method

Although the subject of this section may not seem very close to the main topic (Problem 6.1.2) of this chapter, it provides some partial results and some special methods for the circuit extension problem.

Theorem 6.2.1 (C. A. B. Smith, see [228], [225], or [236] p. 243) *Let G be a cubic graph. For every edge e of G, the number of Hamilton circuits of G containing e is even.*

The following proof of this theorem is provided by Thomason [225], and is popularly called the lollipop method.

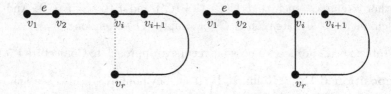

Figure 6.1 *Lollipop detour*

Definition 6.2.2 Let $P' = v_1 v_2 \ldots v_r$ be a path of a cubic graph. Let $v_i \in N(v_r) \cap \{v_2, \ldots, v_r\}$. The subgraph $P'' = v_1 v_2 (P') v_i v_r (\overleftarrow{P'}) v_{i+1} = P' \cup \{v_r v_i\} - \{v_i v_{i+1}\}$ is a path obtained from P' via a lollipop detour. (See Figure 6.1.)

Note, considering $v_1, v_2, \ldots, v_r$ as an ordering of the path P', the path $v_1 v_2 (P') v_i$ is a segment of P' along this direction, while $v_r (\overleftarrow{P'}) v_{i+1}$ is a segment of P' along the opposite direction.

Proof of Theorem 6.2.1. Let $e = v_1 v_2 \in E(G)$. Construct an auxiliary graph A as follows. Each vertex of A corresponds to a Hamilton path of G starting at the vertex v_1 and the edge $v_1 v_2$. Two vertices of A are adjacent to each other if and only if one of their corresponding Hamilton paths is obtained from the another corresponding Hamilton path via a *lollipop detour*.

It is evident that there are two cases for a Hamilton path $P^* = v_1 v_2 \ldots x$.

Case 1. If $v_1 \notin N(x)$, then the corresponding vertex in the auxiliary graph A is of degree 2.

Case 2. If $v_1 \in N(x)$, then the corresponding vertex in the auxiliary graph A is of degree 1. In this case, this Hamilton path is extendable to a Hamilton circuit of G containing the edge $v_1 v_2$.

Hence, each component of A is either a path or a circuit. Note that the number of odd vertices is even in A. Also note that there is a one-to-one correspondence between the set of degree 1 vertices of A and the set of Hamilton paths of G extendable to Hamilton circuits containing the given edge $v_1 v_2$.   □

The following modified lollipop lemma is useful in further studies (for cases of non-Hamilton circuits).

Lemma 6.2.3 *Let G be a cubic graph of order n and $C = v_1 v_2 \ldots v_r v_1$ be a circuit of G. Then*

(1) either there is a C-extension $C' = v_1 v_2 \ldots v_1$ containing the edge $v_1 v_2$ ($V(C) \subseteq V(C')$ and $E(C) \neq E(C')$),

(2) or there is a path $P = v_1 v_2 \ldots z$ starting at the vertex v_1 and the edge $v_1 v_2$ and $V(P) = V(C) \cup \{z\}$ for some vertex $z \notin V(C)$.

Proof The lollipop method applied in the proof of Theorem 6.2.1 is adapted in this proof.

Construct an auxiliary graph A for all paths P of G starting at the vertex v_1 and the edge $v_1 v_2$ with $V(P) = V(C)$. And P_1 and P_2 are adjacent to each other in A if and only if P_1 is obtained from P_2 via a lollipop detour.

The path $P' = v_1 v_2 \ldots v_r = C - \{v_r v_1\}$ is a vertex in the auxiliary graph A. Since the component of A containing the vertex P' must have another degree 1 vertex $P'' = v_1 v_2 \ldots x$, similar to the discussion in the proof of Theorem 6.2.1, there are two possibilities:

(1) $v_1 \in N(x)$ (thus, with this additional edge $v_1 x$, the path P'' is closed to be a circuit C'', which is a C-extension), or

(2) $N(x)$ contains a vertex z not in P', which extends the path P'' with an extra vertex z. □

Almost Hamilton circuit

For a cubic graph of order n, Problem 6.1.2 has been verified if a given circuit C is of length at least $n - 1$ in the following pioneering theorem.

Theorem 6.2.4 (Fleischner [64], also see [72]) *Let G be a cubic graph and C be a circuit of G. If $|V(G) - V(C)| \leq 1$, then C is extendable.*

Proof Let $w \in V(G) - V(C)$ and x be a neighbor of w. Let $N(x) = \{w, y_1, y_2\}$ where $x, y_1, y_2 \in V(C)$. Consider the suppressed cubic graph $G' = \overline{G - \{xw\}}$. C is a Hamilton circuit of G' containing the edge $e = y_1 y_2$. By Theorem 6.2.1, let C' be another Hamilton circuit of G' containing the edge $y_1 y_2$. Since x is a degree 2 vertex of $G - \{xw\}$ and is contained in the induced path $y_1 x y_2$, the circuit C' is a C-extension in G. □

Using the approach of Proposition 6.1.4, we have the following corollary related to circuit double cover.

Corollary 6.2.5 (Fleischner [64], also see [72]) *Let C be a circuit of a cubic graph G. If $|V(G) - V(C)| \leq 1$, then G has a strong CDC (containing C).*

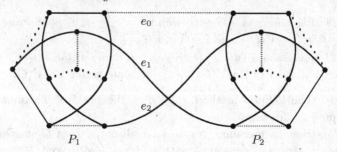

Figure 6.2 *The circuit C (of length $16 = n - 2$) is not extendable*

6.3 Stable circuits

Note that not every circuit is extendable, the graph in Figure 6.2 is an example discovered by Fleischner ([59], [61], [64], [66]) in which a circuit C does not have an extension (or, see Figure D.4(a) on p. 296 for a different drawing).

Definition 6.3.1 Let G be a 2-edge-connected, cubic graph and C be a circuit of G. If C is not extendable, then C is called a **stable circuit** of G.

Note that the graph G illustrated in Figure 6.2 is a $\oplus_3$-sum of two copies $G/P_1, G/P_2$ of the Petersen graphs (where P_1, P_2 are two components of $G - \{e_0, e_1, e_2\}$).

Proposition 6.3.2   (Fleischner) *The circuit C illustrated in Figure 6.2 is stable.*

Proof Assume that C' is a circuit G with $V(C') \supseteq V(C)$. We show that $C' = C$. Since the Petersen graph G/P_i does not have a Hamilton circuit, the graph G does not have any circuit of length ≥ 17. Thus,

$$V(C') = V(C).$$

Applying Proposition B.2.6 to the Petersen graph G/P_2, the circuit C' must contain the edge e_1. Symmetrically, the circuit C' must contain the edge e_2 in G/P_1. Applying Proposition B.2.6 again, we must have that $C'/P_i = C/P_i$ $(i = 1, 2)$ in G/P_i since $e_1, e_2 \in$ both $C'/P_i = C/P_i$ and $e_0 \notin C'/P_i = C/P_i$. □

In [66] and [148], infinite families of stable circuits are constructed by Fleischner and Kochol. Some of them are cyclically 4-edge-connected snarks [148].

Recently, a computer aided search [21] discovered stable circuits for some cyclically 4-edge-connected snarks of order n, for every even integer $n \in \{22, \ldots, 36\}$. However, the existence of stable circuits does not disprove the strong CDC conjecture for those snarks: the computer aided proof further verifies the strong CDC conjecture for all of those small snarks (cyclically 4-edge-connected snarks of order ≤ 36). Note that 3-edge-colorable graphs are not counterexamples to any faithful cover problem (Lemma 2.2.1), and graphs with non-trivial 2- or 3-edge-cut can be reduced to graphs of smaller orders.

Proposition 6.3.3 (Brinkmann, Goedgebeur, Hägglund and Markström [21]) *The strong circuit double cover conjecture holds for all bridgeless cubic graphs of order ≤ 36.*

Note that, for the stable circuit C illustrated in Figure 6.2, $|V(G) - V(C)| = 2$. Although it is not extendable, it is not a counterexample to the strong circuit double cover conjecture (see Exercise 6.8). However, for all cubic graphs, the strong CDC conjecture remains open if a circuit C is of length $n - 2$.

6.4 Extension-inheritable properties

In this chapter, by a pair (G, C), we shall always mean *a 2-edge-connected cubic graph G and a circuit C of G.*

One may notice from the induction proofs of Proposition 6.1.4 and Corollary 6.2.5, that the strong circuit double cover conjecture can be proved for a pair (G_1, C_1) (C_i is a circuit of G_i) if the following processing can be carried on until nothing is left:

find a C_1-extension C_2 in G_1 and let $G_2 = \overline{G_1 - (E(C_1) - E(C_2))}$,

find a C_2-extension C_3 in G_2 and let $G_3 = \overline{G_2 - (E(C_2) - E(C_3))}$,

$$\ldots, \ldots,$$

find a C_j-extension C_{j+1} in G_j and let $G_{j+1} = \overline{G_j - (E(C_j) - E(C_{j+1}))}$,

$$\ldots\ldots,$$

until $G_t = C_t$ for some t, then

$$\{C_1, C_1 \triangle C_2, C_2 \triangle C_3, \ldots, C_{t-1} \triangle C_t, C_t\}$$

is a circuit double cover of G_1 containing C_1.

For a pair (G, C) (C is a circuit of G), if one is able to find a C-extension C' that inherits certain property of (G, C), then one will be able to carry on the processing until reaching a complete circuit double cover of G. Based on this observation, we present the following definition (and propositions) which generalizes the idea from Corollary 6.2.5 and Proposition 6.1.4.

Definition 6.4.1 A given property $\mathcal{P}$ is extension-inheritable, if, for any pair (G, C) with property $\mathcal{P}$,

(1) the property $\mathcal{P}$ guarantees the existence of a C-extension C', and

(2) the reduced pair $(\overline{G - (E(C) - E(C'))}, C')$ also has the same property $\mathcal{P}$.

By Theorem 6.2.1 and Corollary 6.2.5, it is easy to see that *circuit of length at least* $n - 1$ is an example of an extension-inheritable property. With the same approach as for Proposition 6.1.4, we have the following lemma.

Lemma 6.4.2 *If a pair* (G, C) *has some extension-inheritable property* $\mathcal{P}$, *then the graph* G *has a circuit double cover containing* C.

In the remaining part of this section, following the approach in [72], some extension-inheritable properties are summarized.

Definition 6.4.3 A spanning tree T of a graph H is called a Y-tree if T consists of a path $x_1 \ldots x_{t-1}$ and an edge $x_{t-2}x_t$. A Y-tree is called a small-end Y-tree if $d_H(x_1) \leq 2$.

A Hamilton path $x_1 \ldots x_t$ of H is called a small-end Hamilton path if $d_H(x_1) \leq 2$.

The following is a list of some known extension-inheritable properties where G is a 2-edge-connected cubic graph and C is a circuit of G.

(1) $\mathcal{P}_0$: C *is a Hamilton circuit of* G *(Theorem 6.2.1).*

(2) $\mathcal{P}_1$: $|V(G) - V(C)| \leq 1$ *(Corollary 6.2.5).*

(3) $\mathcal{P}_2^*$ (Fleischner and Häggkvist [72]): $|V(G) - V(C)| \leq 2$ *and, in the case of* $|V(G) - V(C)| = 2$, *the distance between two vertices of* $V(G) - V(C)$ *is 3 (Proposition 6.4.4, see Figure 6.3).*

(4) $\mathcal{P}_{conn-4}$ (Fleischner and Häggkvist [72]): $|V(G) - V(C)| \leq 4$ *and* $G - V(C)$ *is connected.*

(5) $\mathcal{P}_{conn-5}$: $|V(G) - V(C)| \leq 5$ *and* $G - V(C)$ *is connected (Exercise 6.12).*

(6) $\mathcal{P}_{HP-s}$ (Fleischner and Häggkvist [98]): $H = G - V(C)$ *has a small-end Hamilton path (Proposition 6.4.5, see Figure 6.4).*

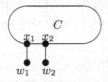

Figure 6.3 *An extendable circuit missing two vertices*

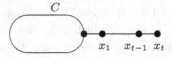

Figure 6.4 *An extendable circuit missing a small-end Hamilton path*

(7) $\mathcal{P}_{Y-tree-s}$: $H = G - V(C)$ *has either a small-end Hamilton path or a small-end Y-tree (Proposition 6.4.5, see Figures 6.4 and 6.5).*

Here are the proofs of some of these extension-inheritable properties.

Proposition 6.4.4 (Fleischner and Häggkvist [72]) *For a pair (G, C) with $|V(G) - V(C)| \leq 2$, and, in the case of $|V(G) - V(C)| = 2$, the distance between two vertices of $V(G) - V(C)$ is 3 (see Figure 6.3), then G has a circuit double cover containing C.*

Proof Let $x_j w_j \in E(G)$ $(j = 1, 2)$ where $x_1 x_2 \in E(C)$ and $\{w_1, w_2\} = V(G) - V(C)$. Then, in the suppressed cubic graph $\overline{G - \{x_1 w_1, x_2 w_2\}}$, by Proposition 6.2.1, there is a C-extension C' $(V(C') = V(C)$ and $E(C') \neq E(C))$. In the original graph G, the circuit C' corresponds to a circuit C' with $V(C') \supseteq V(C)$ and $E(C') \neq E(C)$. So, either C' inherits the property of C if $V(C') = V(C)$, or $V(G) - V(C')$ is smaller (missing at most one vertex). In the first case, the processing will be carried on. In the second case, Corollary 6.2.5 is applied. $\qquad\square$

The first part of the following proposition about small-end Hamilton path is proved by Fleischner and Häggkvist [98].

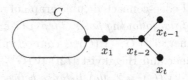

Figure 6.5 *An extendable circuit missing a small-end Y-tree*

Proposition 6.4.5   *For a pair (G, C), if $H = G - V(C)$ has either a small-end Hamilton path or a small-end Y-tree, then G has a circuit double cover containing C.*

Proof Induction on $|V(G)|$. Let $C = v_1 v_2 \ldots v_r v_1$ be the circuit and $H = G - V(C)$. Let $x_1 \ldots x_t$ be a small-end Hamilton path of H or $x_1 \ldots x_{t-1} + x_{t-2} x_t$ be a small-end Y-tree of H such that $x_1 v_1 \in E(G)$ (since $d_H(x_1) \leq 2$). By Lemma 6.2.3, either G has a circuit C' with $V(C) = V(C')$ and $E(C) \neq E(C')$, or G has a path $P = v_1 v_2 \ldots v_j x_h$ with $V(P) = V(C) \cup \{x_h\}$, which is further extended to a longer circuit $C' = v_1 v_2 \ldots v_j x_h \ldots x_1 v_1$. In either case, the reduced pair

$$(\overline{G - (E(C) - E(C'))}, C')$$

inherits the same property from (G, C): a small-end Hamilton path or a small-end Y-tree outside of C'.

By applying induction, let $\mathcal{F}'$ be a circuit double cover of the suppressed graph $\overline{G - (E(C) - E(C'))}$ with $C' \in \mathcal{F}'$. Hence, $\mathcal{F}' - C' + \{C' \triangle C, C\}$ is a strong CDC of G.   □

Exercises 6.10, 6.11 and 6.12 are some other extension-inheritable properties.

6.5 Extendable circuits

Although not every circuit is extendable (see the example in Section 6.3), characterizations of extendable circuits and stable circuits remain as one of the hopeful approaches to the strong CDC conjecture (at least for some families of graphs).

In Section 6.4, we presented some graphic properties that are inherited in the reduced pair $(\overline{G - (E(C) - E(C'))}, C')$ where C is a circuit of G and C' is a C-extension. Adding to those inheritable properties, in this section we will characterize some extendable circuits (although they may not have an extension-inheritable property).

Let G be a 2-edge-connected cubic graph of order n and C be a circuit of G. *If one of the following holds, then a C-extension exists.*

(1) (Fleischner [64]) $G - V(C)$ *is connected.*

(2) (Fleischner and Häggkvist [72]) $|V(G) - V(C)| \leq 2$, *and, in the case of* $|V(G) - V(C)| = 2$, *the distance between two vertices of* $V(G) - V(C)$ *is* 3 *(Proposition 6.4.4, see Figure 6.3).*

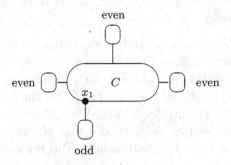

Figure 6.6 *G − V(C) has only one odd component*

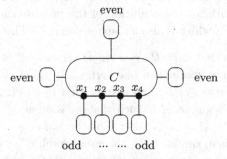

Figure 6.7 *All odd components of G−V(C) are adjacent to a segment of C*

(3) (Chan, Chudnovsky and Seymour [33]) $G - V(C)$ *has at most one odd component* (see Figure 6.6).

The following result is a combination and generalization of those early results.

Proposition 6.5.1 *Let G be a 2-edge-connected cubic graph of order n and C be a circuit of G and $H_1, \ldots, H_r$ be components of $G - V(C)$ where $H_1, \ldots, H_{t_o}$ are of odd orders, and others are of even orders. If C contains a path $x_1 x_2 \ldots x_{t_o}$ such that x_j is adjacent to some vertex of H_j for each $j \in \{1, \ldots, t_o\}$, then the pair (G, C) has a C-extension (see Figure 6.7).*

Proof Let G be a counterexample to the proposition with $|E(G)|$ as small as possible. Let $H_1, \ldots, H_r$ be components of $G - V(C)$ where $H_1, \ldots, H_{t_o}$ are of odd orders, and others are of even orders. Let $x_0 x_1$

$\ldots x_{t_o} x_{t_o+1}$ be a segment of C that each x_j is adjacent to some vertex of H_j (for every $j = 1, \ldots, t_o$).

I. We claim that G *does not contain any 2-edge-cut with separating $V(C)$ and some other vertex.*

Suppose that T is a 2-edge-cut with components Q_1 and Q_2, and $C \subseteq Q_1$. Let $Q_2 \subseteq H_j$ for some $j \in \{1, \ldots, r\}$. Since G is cubic and T is a 2-edge-cut, by Lemma A.2.8, $|V(Q_2)|$ must be even. Therefore, $\overline{G/Q_2}$ satisfies the proposition and is smaller than G. A C-extension in $\overline{G/Q_2}$ can be expanded to a C-extension in G. This is a contradiction.

II. We claim that *each H_j is acyclic* $(j \in \{1, \ldots, r\})$. Suppose not, let e be an edge of H_j that is not a cut-edge of H_j. By I, $G - e$ remains 2-edge-connected. Since $|V(\overline{H_j - e})| \equiv |V(H_j)| \pmod 2$, the cubic graph $G' = \overline{G - e}$ satisfies the condition of the proposition. Hence, G' has a C-extension C', which is also a C-extension in G. This is a contradiction.

III. We claim that *each H_j is a single vertex* $(j \in \{1, \ldots, r\})$. Suppose that H_j contains more than one vertex. By II, H_j is a tree. Let v_1, v_2 be two leaf vertices of H_j and assume that $v_1 \notin N(x_j)$ if $j \leq t_o$ (where $x_1 \ldots x_{t_o}$ is a segment of C such that x_j is adjacent to some vertex of H_j for each $j \in \{1, \ldots, t_o\}$). Let e be the leaf-edge of H_j incident with the leaf v_1. Then, similar to II, the cubic graph $G'' = \overline{G - e}$ satisfies the proposition and a C-extension exists in both G'' and G.

IV. By II and III, $G - V(C)$ is an independent set $\{w_1, \ldots, w_r\}$ where $H_j = \{w_j\}$. Consider the cubic graph $G''' = \overline{G - \{x_1 w_1, \ldots, x_r w_r\}}$. By Theorem 6.2.1, the edge $x_0 x_{r+1}$ of G''' is contained in another circuit which is a C-extension of G. $\qquad \square$

6.6 Semi-extension of circuits

If some circuit C is not extendable (such as the examples in Section 6.3), the graph G may still have a strong CDC containing C. In this section, we present a relaxed definition for circuit extendibility (introduced in [48]), by which the strong CDC conjecture (Conjecture 1.5.1) is true if every circuit of 2-connected cubic graphs has a semi-extension (Theorem 6.6.4 and Conjecture 6.6.3).

Before the introduction of the new concept of semi-extension, we first introduce the definition of Tutte bridge.

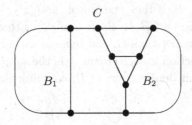

Figure 6.8 *Two Tutte bridges B_1, B_2 of a circuit C*

Definition 6.6.1 Let H be a subgraph of G. A **Tutte bridge** of H is either a chord e of H ($e = xy \notin E(H)$ with both $x, y \in V(H)$) or a subgraph of G consisting of one component Q of $G - V(H)$ and all edges joining Q and H (and, of course, all vertices of H adjacent to Q).

For a Tutte bridge B_i of H, the vertex subset $V(B_i) \cap V(H)$ is called the **attachment** of B_i, and is denoted by $A(B_i)$. (See Figure 6.8.)

Definition 6.6.2 Let C and D be a pair of distinct circuits of a 2-connected cubic graph G. Let $J_1, \ldots, J_p$ be the components of $C \triangle D$. The circuit D is a **semi-extension** of C if, for every Tutte bridge B_i of $C \cup D$,

(1) either the attachment $A(B_i) \subseteq V(D)$,
(2) or $A(B_i) \subseteq V(J_j)$ for some $j \in \{1, \ldots, p\}$.

Note that a semi-extension D of C may not contain all vertices of C.

It is easy to see that the concept of circuit semi-extension is a generalization of circuit extension: for a C-extension D, every Tutte bridge B_i has its attachment $A(B_i) \subseteq V(D)$ (since each J_j contains no vertex of $V(C) - V(D)$).

Conjecture 6.6.3 (Esteva and Jensen [48]) *For every 2-connected cubic graph G, every circuit C of G has a semi-extension.*

Theorem 6.6.4 (Esteva and Jensen [48]) *If Conjecture 6.6.3 is true for every 2-connected cubic graph, then the strong circuit double cover conjecture is true.*

Proof Induction on $|V(G)|$. Let D be a semi-extension of C in G. Let $\{J_1, \ldots, J_p\}$ be the set of all components of $C \triangle D$, and, for the sake of convenience, let $J_0 = D$. For each $i = 0, \ldots, p$, let $\{\mathcal{B}_0, \ldots, \mathcal{B}_p\}$ be a partition of the set of all Tutte bridges of $C \cup D$ such that, for each

$i \in \{0, \ldots, p\}$, $B_\mu \in \mathcal{B}_i$ if the attachment $A(B_\mu) \subseteq V(J_i)$. And let H_i be the union of J_i and all $B_\mu \in \mathcal{B}_i$. It is obvious that each suppressed cubic graph $\overline{H_i}$ is smaller than G, and remains 2-connected.

Applying the induction to each circuit J_i in the suppressed cubic graph $\overline{H_i}$, let $\mathcal{F}_i$ be a circuit double cover of H_i containing J_i. Then

$$\{C\} \cup \bigcup_{i=0}^{p} [\mathcal{F}_i - \{J_i\}]$$

is a circuit double cover of G containing C. □

Remark. For the cubic graph illustrated in Figure 6.2, although the circuit C does not have an extension, it is proved in [48] that it does have a semi-extension, and furthermore, the graph of order 18 does have a strong CDC containing C (Exercise 6.9). It is also proved in [49] that some families of stable circuits constructed in [148] also have semi-extensions.

Further generalizations

Similar to Definition 6.6.2 and Theorem 6.6.4, the concept of semi-extension can be further generalized as follows.

Definition 6.6.5 Let G be a 2-connected cubic graph, C be a circuit and D be a non-empty even subgraph of G with components $D_1, \ldots, D_q$. Let $J_1, \ldots, J_p$ be the components of $C \triangle D$. The even subgraph D is a weak-semi-extension of C if, for every Tutte bridge B_i of $C \cup D$,
   (1) either the attachment $A(B_i) \subseteq V(J_j)$ for some $j \in \{1, \ldots, p\}$,
   (2) or $A(B_i) \subseteq V(D_h)$ for some $h \in \{1, \ldots, q\}$.

Conjecture 6.6.6 *For every 2-connected cubic graph G, every circuit C of G has a weak-semi-extension.*

With a similar proof to that of Theorem 6.6.4, we have the following result.

Proposition 6.6.7 *If Conjecture 6.6.6 is true for every 2-connected cubic graph, then the strong circuit double cover conjecture is true.*

Remark. A refinement of semi-extension was further proposed by Jensen in [135].

Furthermore, Conjecture 6.6.3 and Theorem 6.6.4 can be extended for graphs without the regularity restriction.

Definition 6.6.2′ *Let C and D be a pair of connected even subgraphs of a 2-edge-connected graph G such that $C \neq D$ and $D \neq \emptyset$. Let $J_1, \ldots, J_p$ be the components of $C \triangle D$. The even subgraph D is a* semi-extension *of C if, for every Tutte bridge B_i of $C \cup D$,*

(1) either the attachment $A(B_i) \subseteq V(D)$,

(2) or $A(B_i) \subseteq V(J_j)$ for some $j \in \{1, \ldots, p\}$.

Conjecture 6.6.3' (Esteva and Jensen [48]) *For every 2-edge-connected graph G, every connected even subgraph of G has a semi-extension.*

The following is a generalization of Theorem 6.6.4.

Theorem 6.6.8 (Esteva and Jensen [48]) *If Conjecture 6.6.3' is true for every 2-edge-connected graph, then every bridgeless graph has an even subgraph double cover containing a given connected even subgraph.*

Proof See Exercise 6.14. □

6.7 Circumferences

The following theorem is an example of applications of some results about strong CDC problem.

Theorem 6.7.1 (Brinkmann, Goedgebeur, Hägglund and Markström [21]) *Let G be a 2-connected cubic graph of order n. If the circumference of G is at least $(n - 10)$, then G has a circuit double cover.*

Note that the CDC obtained in this theorem may not be a strong CDC.

Proof Let C be a longest circuit of G and $W = G - V(C)$. Here $|V(W)| \leq 10$. Let R be the set of all chords of C.

I. Assume that W contains an isolated vertex w (that is, $N(w) \subseteq V(C)$). Let $G_1 = G - (W - \{w\})$, and $G_2 = G - R - \{w\}$.

Here $\overline{G_2}$ is of order at most 36 since C is chordless in $\overline{G_2}$ and there are at most 9 vertices in $W - \{w\}$. By Proposition 6.3.3, G_2 has a circuit double cover $\mathcal{F}_2$ containing the circuit C.

For the graph $\overline{G_1}$, by Theorem 6.2.4, $\overline{G_1}$ has a circuit double cover $\mathcal{F}_1$ containing C. Hence $[\mathcal{F}_1 - \{C\}] \cup [\mathcal{F}_2 - \{C\}]$ is a CDC of G.

II. Now, we have that every component of W is of order ≥ 2. Let $G_3 = G - W$ and $G_4 = G - R$.

In the graph G_4, the number of edges between W and C is at most $3|V(W)| - 2[|V(W)| - \lfloor \frac{|V(W)|}{2} \rfloor] \le 20$ since no component of W is a singleton. Hence, the length of C is at most 20 in $\overline{G_4}$, and therefore, $|V(\overline{G_4})| \le |V(W)| + |V(C)| \le 10 + 20 = 30$. By Proposition 6.3.3, G_4 has a circuit double cover $\mathcal{F}_4$ containing the circuit C.

Since C is a Hamilton circuit of the suppressed cubic graph $\overline{G_3}$, by Lemma 2.2.1, $\overline{G_3}$ has a circuit double cover $\mathcal{F}_3$ containing C. Hence $[\mathcal{F}_3 - \{C\}] \cup [\mathcal{F}_4 - \{C\}]$ is a CDC of G.   □

Theorem 6.7.1 improves an earlier result for circumference of $(n-8)$ in [248] and another computer aided result for circumference of $(n-9)$ in [103].

6.8 Open problems

Conjecture 1.5.1 (Strong CDC conjecture, Seymour, see [61] p. 237, [62], also see [83]) *Let G be a 2-edge-connected cubic graph and C be a circuit of G. The graph G has a circuit double cover $\mathcal{F}$ with $C \in \mathcal{F}$.*

Conjecture 1.5.2 (Sabidussi and Fleischner [60], and Conjecture 2.4 in [2]) *Let G be a cubic graph such that G has a dominating circuit C. Then G has a circuit double cover $\mathcal{F}$ such that the given circuit C is a member of $\mathcal{F}$.*

See Sections 10.5 and 10.7 for further discussion about this conjecture.

Conjecture 6.8.1 (Fleischner [69]) *Let G be a cubic graph of order n with circumference at least $n - 1$. Then, for any given circuit C of G, the graph G has a CDC containing C.*

Remark. By Lemma 2.2.1 and Theorem 6.2.4, Conjecture 6.8.1 is true *(1) if G is 3-edge-colorable, and (2) if $V(C) \subseteq V(C_L)$ for some longest circuit C_L.* However, it remains open if $V(C)$ contains some missing vertex of every longest circuit.

Conjecture 6.6.3 (Esteva and Jensen [48]) *For every 2-connected cubic graph G, every circuit C of G has a semi-extension.*

Remark. A recent computer aided search [21] shows that Conjecture 6.6.3 holds for all cubic graphs of order at most 36.

Conjecture 6.6.6 *For every 2-connected cubic graph G, every circuit*

C of G has a weak-semi-extension. (See Definition 6.6.5 for *weak-semi-extension*.)

Conjecture 4.3.1 (Hoffmann-Ostenhof [111]) *Let C be a circuit of a 2-connected cubic graph G. Then there is a 5-even subgraph double cover of G such that C is a subgraph of one of these five even subgraphs.*

6.9 Exercises

Exercise 6.1   Let G be a smallest counterexample to the strong circuit double cover conjecture. Show that the girth of G is at least five.

Exercise 6.2   (Tarsi [221], Corollary 4.2.8 in this book) Let G be a cubic graph containing a Hamilton path. Use the method introduced in this chapter to show that G has a circuit double cover.

Exercise 6.3   Let G be a 2-edge-connected cubic graph and S be an even subgraph of G. If there is a circuit C such that $V(C) \supseteq V(S)$ and $E(C) \neq E(S)$ then show that the graph $G - (E(S) - E(C))$ is 2-edge-connected.

Definition 6.9.1   A cubic graph G is hypohamiltonian if, for every vertex $v \in V(G)$, $G - v$ contains a Hamilton circuit.

Exercise 6.4   (Fleischner and Häggkvist [72], [98]) Let G be a hypohamiltonian cubic graph and S be an even subgraph of G that is not a spanning subgraph of G. Then G has an even subgraph double cover $\mathcal{F}$ that $S \in \mathcal{F}$.

Exercise 6.5   Let G be a 2-edge-connected cubic graph of order n and C be a circuit of G. If G has an even subgraph S such that
   (1) $V(S) \subseteq V(C)$ and
   (2) the circuit C in the suppressed cubic graph $\overline{G - (E(S) - E(C))}$ has some extension-inheritable property,
then G has an even subgraph double cover $\mathcal{F}$ with $S \in \mathcal{F}$.

Exercise 6.6   (Part of Proposition B.2.27) Let w be an eulerian $(1,2)$-weight of the Petersen graph P_{10}. Then (P_{10}, w) has no faithful circuit cover if and only if $E_{w=1}$ induces a 2-factor of P_{10}.

Exercise 6.7   Let C be a circuit of a cubic graph G. If $|V(G) - V(C)| \leq 5$, then G has a circuit double cover.

Remark. Note that the CDC obtained in this exercise may not be a strong CDC containing C. Also note that this exercise is a weak version of Theorem 6.7.1, and its proof does not involve an application of Proposition 6.3.3.

Exercise 6.8 Prove that the graph illustrated in Figure 6.2 has a circuit double cover containing the given circuit C (give a direct proof without applying the computer aided result Proposition 6.3.3).

Exercise 6.9 (Esteva and Jensen [48]) For the cubic graph G and the circuit C illustrated in Figure 6.2, show that C has a semi-extension.

Exercise 6.10 For a pair (G, C), if $H = G - V(C)$ contains a Hamilton path and is of order ≤ 9, then G has a CDC containing C.

Exercise 6.11 For a pair (G, C), if $H = G - V(C)$ contains either a Hamilton path or a Y-tree and is of order ≤ 7, then G has a strong CDC.

Exercise 6.12 For a pair (G, C), if $H = G - V(C)$ is connected and of order at most 5, then (G, C) has a strong CDC.

Remark. Exercise 6.12 slightly improves an early result by Fleischner and Häggkvist [72] for $|V(G) - V(C)| \leq 4$. Exercises 6.11 and 6.12 can be further improved if the girth of G and H, the edge-connectivity of G are in detailed consideration.

Exercise 6.13 Let C_0 and C_1 be two distinct 2-factors of P_{10}. Then $P_{10} - (E(C_0) - E(C_1))$ has a bridge,

Exercise 6.14 (Esteva and Jensen [48], Theorem 6.6.8) If Conjecture 6.6.3$'$ is true for every 2-edge-connected graph, then every bridgeless graph has an even subgraph double cover containing a given connected even subgraph.

7

Spanning trees, supereulerian graphs

In the first a few chapters, we concentrated on *cubic graphs*. One may notice that 3-*edge-coloring* techniques were used extensively in the study of cubic graphs. How about non-cubic graphs? Certainly, 3-edge-coloring is no longer a suitable technique. In this chapter (and the next few chapters) we will discuss the circuit covering problems for graphs in general (no requirement of 3-regularity). The techniques presented here are adapted from the *theory of integer flow* (see Chapter 8 and Appendix C). The property of having 4-flows can be considered as an extension of 3-edge-coloring for general graphs. And techniques developed in flow theory also provide powerful tools for handling highly connected (sub)graphs.

Before formally introducing the theory of integer flow (Chapter 8), we present some fundamental results that serve as a bridge between the subjects of even subgraph covering and integer flow.

The graphs considered in this chapter are not necessarily cubic.

7.1 Jaeger Theorem: 2-even subgraph covers

⋆ Supereulerian graphs, even subgraph covers

Definition 7.1.1 A graph G is supereulerian if G contains a spanning, connected, even subgraph.

Theorem 7.1.2 (Jaeger [123], [127]) *If G is a supereulerian graph, then G has a 2-even subgraph cover.*

Proof Let F be a spanning, connected, even subgraph of G. Let T be a spanning tree of G contained in F. By Lemma A.2.15, let P be a parity subgraph of G such that

$$E(P) \subseteq E(T) \subseteq E(F).$$

Let $F' = G - E(P)$. Then $\{F, F'\}$ is a 2-even subgraph cover of G. □

⋆ Spanning trees, supereulerian graphs

Lemma 7.1.3   (Jaeger [123], [127]) *If G contains a pair of edge-disjoint spanning trees, then G is supereulerian.*

Proof Let T_1 and T_2 be a pair of edge-disjoint spanning trees of G. By Lemma A.2.15, let P_1 be a parity subgraph contained in T_1 and $F_1 = G - E(P_1)$. Note that T_2 is disjoint from P_1, and therefore T_2 is completely contained in F_1. So, the even subgraph F_1, which contains a spanning tree T_2, must be connected and spanning.   □

⋆ 4-edge-connected graphs

The following lemma is a corollary of Theorem A.1.5 (or Theorem A.1.6).

Lemma 7.1.4   (Tutte [232], Nash-Williams [184]) *Every 4-edge connected graph contains two edge-disjoint spanning trees.*

The following theorem is an immediate corollary of Lemma 7.1.3 and 7.1.4.

Theorem 7.1.5   (Jaeger [123], [127]) *Every 4-edge-connected graph is supereulerian.*

By Theorem 7.1.2 and 7.1.5, we have the following well-known theorem by Jaeger in the theory of integer flow – *the 4-flow theorem*. (Before we introduce the concept of integer flow, the theorem is presented as the *even subgraph covering problem* and we will re-present it later in terms of integer flow in Section 8.1.)

Theorem 7.1.6   (Jaeger [123], [127]) *Every 4-edge-connected graph has a 2-even subgraph cover.*

Corollary 7.1.7   *Every 4-edge-connected graph has a 3-even subgraph double cover.*

Proof By Theorem 7.1.6, let $\{F_1, F_2\}$ be an even subgraph cover of G. The set $\{F_1, F_2, F_1 \triangle F_2\}$ is a 3-even subgraph double cover of G.   □

In summary, we have shown the following families of graphs having 2-even subgraph covers (and therefore 3-even subgraph double covers):

$$\{4\text{-}edge\text{-}connected\ graphs\}$$
$$\subset \{Graphs\ with\ a\ pair\ of\ edge\text{-}disjoint\ spanning\ trees\}$$
$$\subset \{Supereulerian\ graphs\}$$
$$\subset \{Graphs\ with\ 2\text{-}even\ subgraph\ cover\}$$
$$= \{Graphs\ with\ 3\text{-}even\ subgraph\ double\ cover\}$$
$$= \{Graphs\ with\ nowhere\text{-}zero\ 4\text{-}flows\}.$$

The last family of graphs with nowhere-zero 4-flows will be discussed in Section 8.1.

By Corollary 7.1.7, bridgeless graphs without any 3-edge-cuts have circuit double covers. How about graphs with *some* 3-edge-cuts?

Theorem 7.1.8   (Catlin [25], or see [27]) *If a bridgeless graph G has at most ten 3-cuts, then either G has a 2-even subgraph cover or G is contractible to the Petersen graph.*

Theorem 7.1.8 is further generalized for graphs with more 3-cuts in [26].

7.2 Jaeger Theorem: 3-even subgraph covers

The following theorem is a well-known theorem in integer flow theory, called the 8-*flow theorem*. In this section, it is presented as an even subgraph covering problem.

Theorem 7.2.1   (Jaeger [123], [127], Kilpatrick [141]) *Every bridgeless graph has a 3-even subgraph cover.*

Note that the existence of a 2-even subgraph cover $\{F_1, F_2\}$ implies a circuit (even subgraph) double cover since $\{F_1, F_2, F_1 \triangle F_2\}$ is a 3-even subgraph double cover. However, there is no proof yet that the existence of a 3-even subgraph cover implies the existence of a circuit double cover since some edge may be covered by *all three even subgraphs*.

⋆ Smallest counterexample to the theorem

First, we show that *it is sufficient to show the theorem for a 3-edge-connected graph*. Or, more specifically, we show that *a smallest counterexample to the theorem is 3-edge-connected*.

For a smallest counterexample G to the theorem, assume that $T = \{e_1, e_2\}$ is a 2-edge-cut of G. Let $G^* = G/e_1$ be the graph obtained from G by contacting the edge e_1. Then let $\mathcal{F}^* = \{F_1^*, F_2^*, F_3^*\}$ be a 3-even subgraph cover of G^*. For each $F_i^* \in \mathcal{F}$,

(1) let F_i be the even subgraph of G induced by edges of the corresponding F_i^*, if $e_1 \notin F_i^*$,

(2) let F_i be the even subgraph of G induced by edges of the corresponding F_i^* together with the edge e_2, if $e_1 \in F_i^*$.

Then $\{F_1, F_2, F_3\}$ is a 3-even subgraph cover of G. This contradicts that G is a counterexample.

Secondly, we may also show that *it is sufficient to show the theorem for cubic graphs*. Or, more specifically, we show that *a smallest counterexample to the theorem is cubic*.

Let $v \in V(G)$ with $d(v) \geq 4$. By applying the splitting lemma (Theorem A.1.14), one may split a pair of edges $\{e_1, e_2\}$ away from v, and the resulting graph $G^* = G_{[v;\{e_1,e_2\}]}$ remains bridgeless. It is obvious that any 3-even subgraph cover of the suppressed graph $\overline{G^*}$ can be easily modified to a 3-even subgraph cover of G.

$\star$ The first proof of Theorem 7.2.1

This proof is based on the Tutte and Nash-Williams Theorem (Theorem A.1.5).

Let $G' = 2G$, which is obtained from G by replacing each edge of G with a pair of parallel edges. Thus, G' is 6-edge-connected, and by Theorem A.1.6, G' contains three edge-disjoint spanning trees T_1, T_2 and T_3. Since T_1, T_2 and T_3 are also spanning trees of G, by Lemma A.2.15, let P_i be a parity subgraph of G contained in T_i. P_1, P_2 and P_3 may not be edge-disjoint in G, but each edge of G is contained in at most two of them. So, $E(P_1) \cap E(P_2) \cap E(P_3) = \emptyset$ in G, and, therefore,

$$\{G - E(P_1), \ G - E(P_2), \ G - E(P_3)\}$$

is a family of 3 even subgraph cover of G.

$\star$ The second proof of Theorem 7.2.1

This proof is based on Theorem A.1.9 (the matching polyhedron theorem) and Theorem 7.1.6 (about 2-even subgraph cover). Corollary A.1.10 is to be applied here since it provides some special perfect matching M_1 of G such that $|M_1 \cap T| = 1$ for every 3-edge-cut T of G.

Lemma 7.2.2 *Let G be a 3-edge-connected graph. If G has an even subgraph C which intersects all 3-edge-cuts of G, then G has two even subgraphs C_1, C_2 such that $\{C_1, C_2, C\}$ is an even subgraph cover of G.*

Proof The contracted graph G/C has no 3-edge-cut since C intersects all 3-edge-cuts of G. Thus, G/C is 4-edge-connected. Furthermore, by Theorem 7.1.6, G/C has a 2-even subgraph cover $\{C_1, C_2\}$. By adding some edges of C to each of $\{C_1, C_2\}$, C_1 and C_2 can be extended as even subgraphs of G and, $E(G)$ is therefore covered by $\{C, C_1, C_2\}$. $\quad\square$

Proof of Theorem 7.2.1. By the discussion above, we may assume that G is 3-edge-connected and cubic (an r-graph). By Corollary A.1.10, G has a perfect matching M intersecting each 3-edge-cut of G in precisely one edge. Thus, the even subgraph $C = G - M$ intersects each 3-edge-cut of G in precisely two edges. Applying Lemma 7.2.2, we obtain a 3-even subgraph cover of G. $\quad\square$

7.3 Even subgraph $2k$-covers

Definition 7.3.1 A family $\mathcal{F}$ of even subgraphs of a graph G is called an **even subgraph k-cover** if each edge of G is contained in k even subgraphs of $\mathcal{F}$.

The following proposition is trivial (Exercise 7.4).

Proposition 7.3.2 *For an odd integer k, a graph G has an even subgraph k-cover if and only if G itself is an even graph.*

When $k = 2$, the problem of even subgraph k-cover is the circuit double cover conjecture. For an even integer greater than 2, by Theorems 7.3.4 and 7.3.5, the following has been proved.

Theorem 7.3.3 [50] *For each even integer k greater than two, every bridgeless graph has an even subgraph k-cover.*

⋆ 4-covers

The following theorem, which is implied by Theorem 7.2.1, is due to Bermond, Jackson and Jaeger [12].

Theorem 7.3.4 (Bermond, Jackson and Jaeger [12]) *Every bridgeless graph has a 7-even subgraph 4-cover.*

Proof   By Theorem 7.2.1, let $\mathcal{F}_1 = \{C_1, C_2, C_3\}$ be a 3-even subgraph cover of G. We claim that

$$\mathcal{F}_2 = \{C_1,\ C_2,\ C_3,\ C_1 \triangle C_2,\ C_1 \triangle C_3,\ C_2 \triangle C_3,\ C_1 \triangle C_2 \triangle C_3\}$$

is a 7-even subgraph 4-cover of G. If an edge e is contained in only one even subgraph of $\mathcal{F}_1$, say C_1, then $e \in E(C_1)$, $E(C_1 \triangle C_2)$, $E(C_1 \triangle C_3)$ and $E(C_1 \triangle C_2 \triangle C_3)$. If e is contained in two even subgraphs of $\mathcal{F}_1$, say C_1 and C_2, then $e \in E(C_1)$, $E(C_2)$, $E(C_1 \triangle C_3)$ and $E(C_2 \triangle C_3)$. If e is contained in all three even subgraphs of $\mathcal{F}_1$, then $e \in E(C_1)$, $E(C_2)$, $E(C_3)$ and $E(C_1 \triangle C_2 \triangle C_3)$. $\qquad\square$

⋆ 6-covers

Theorem 7.3.3 is the combination of Theorem 7.3.4 and the following theorem.

Theorem 7.3.5   (Fan [50]) *Every bridgeless graph has a 10-even subgraph 6-cover.*

The proof of this theorem will be presented in Section 8.3 since it is based on the 6-flow theorem (Theorem 8.3.1).

⋆ Berge–Fulkerson conjecture

Conjecture 7.3.6   (Berge and Fulkerson [80])   *Let G be a bridgeless cubic graph and let $2G$ be the graph obtained from G by replacing each edge of G with a pair of parallel edges. Then $2G$ is 6-edge-colorable.*

Note that the complement of a perfect matching in a cubic graph is a 2-factor. We have the following equivalent version of the Berge–Fulkerson conjecture.

Conjecture 7.3.6′   *Every bridgeless cubic graph has a 6-even subgraph 4-cover.*

It is pointed out by Jaeger in [131] that the following conjecture, which eliminates the "cubic" restriction in Conjecture 7.3.6′, is equivalent to Conjecture 7.3.6′ (Exercise 7.3).

Conjecture 7.3.6″   *Every bridgeless graph has a 6-even subgraph 4-cover.*

The following is a weak version of the Berge–Fulkerson conjecture (Conjecture 7.3.6).

Conjecture 7.3.7 *There is an integer K $(K \geq 2)$ such that every bridgeless cubic graph has a $3h$-even subgraph $2h$-cover for some $h \leq K$.*

The case $K = 1$ of Conjecture 7.3.7 is the 3-edge-coloring problem (Theorem 1.3.2), which is not generally true because of the Petersen graph and snarks. The case $K = 2$ is the Berge–Fulkerson conjecture (Conjecture 7.3.6″).

Theorem 7.3.4 is a partial result to both Conjectures 7.3.6″ and 7.3.7. And Theorem 7.3.5 is considered as an approach to Conjecture 7.3.7 for the case $K = 3$. However, Theorem 7.3.5 cannot be further improved to a 9-even subgraph 6-cover. Seymour [204] pointed out that the Petersen graph P_{10} does not have a $3h$-even subgraph $2h$-cover if h is an odd integer (Proposition B.2.22).

For the problem of perfect matching coverings, there are a few results and conjectures related to Conjecture 7.3.6.

Theorem A.1.9 (Edmonds [43], or see [204]) *Let G be an r-graph. Then there is an integer p and a family $\mathcal{M}$ of perfect matchings such that each edge of G is contained in precisely p members of $\mathcal{M}$. (Note that it is not necessary that the members of $\mathcal{M}$ are distinct.)*

Motivated by this result, we have the following conjectures.

Conjecture 7.3.8 (Seymour [204]) *Every r-graph G has a family $\mathcal{F}$ of perfect matchings such that each edge of G is contained in precisely two members of $\mathcal{F}$.*

Conjecture 7.3.9 *There is an integer k such that every r-graph has a family of perfect matchings that covers every edge precisely k times.*

7.4 Catlin's collapsible graphs

Contraction is one of the most useful and powerful operations in the inductive study of graph theory – if the graph resulting from a contraction preserves a given graph theory property. In this section, we introduce the concepts of *contractible configurations* and *collapsible graphs*: contractions of such subgraphs preserve some properties, including circuit double cover property.

Readers are referred to the most comprehensive survey articles [27] and [36] on this topic.

Definition 7.4.1 Let $\mathcal{P}$ be a graph theory property. A graph H is a contractible configuration of $\mathcal{P}$ if, for every supergraph G of H, G/H has the property $\mathcal{P}$ if and only if G has the property $\mathcal{P}$.

For a graph G, the set of all vertices with odd degree (odd vertices) is denoted by $O(G)$.

Definition 7.4.2 A graph H is collapsible if, for each $X \subseteq V(H)$ of even order, H has a connected, spanning subgraph H_X such that $X = O(H_X)$.

The concept of collapsible graph and its related reduction method were introduced by Catlin [24] in the study of supereulerian graphs, which provides a powerful reduction method for graphs containing a relatively dense subgraph.

Recall the definition of supereulerian graphs.

Definition 7.1.1 A graph is supereulerian if it contains a connected, spanning even subgraph.

Theorem 7.4.3 (Catlin [24]) *Collapsible graphs are contractible configurations for the supereulerian problem.*

Proof Let H be collapsible. We are to show that, *for a supergraph G of H, G is supereulerian if and only if G/H is supereulerian.*

"$\Rightarrow$" is obvious. Thus, we need to prove only "$\Leftarrow$". Let C be a spanning connected even subgraph of G/H. For each vertex $v \in V(H)$, denote the set of edges of C incident with v by $E_C(v)$. Let $X = \{v \in V(H) : |E_C(v)| \text{ is odd}\}$. Note that C is an even subgraph of G/H, and the vertex created by contracting H is of even degree in C. Thus, $|X|$ is even. Since H is a collapsible subgraph of G, let H_X be a connected spanning subgraph of H with $O(H_X) = X$. Thus, the subgraph of G induced by the edges $E(C) \cup E(H_X)$ is a connected, spanning even subgraph of G. $\qquad\square$

Theorem 7.4.4 (Catlin [25]) *Collapsible graphs are contractible configurations for the 2-even subgraph cover problem.*

Proof Let H be a collapsible graph with odd vertex set $O(H)$. Let G be a supergraph of H. We are to show that G has a 2-even subgraph cover if G/H has a 2-even subgraph cover.

Let $\{C_1, C_2\}$ be a 2-even subgraph cover of G/H. Let $O(C_i)$ be the set of odd vertices of C_i in G (note that C_i is an even subgraph of G/H

that contains no edge of H). Since H is collapsible, let R_1 be a connected spanning subgraph of H with the odd vertex set $O(R_1) = O(C_1)$. Here $C_1 \cup R_1$ is an even subgraph of G. Note that, in H, $E(H) - E(R_1)$ is not covered by $E(C_1 \cup R_1)$.

By Lemma A.2.17, let Q_2 be a T-join of R_1 with $O(Q_2) = O(C_2) \triangle O(H)$. Then $\{C_1 \cup R_1, C_2 \cup (H - E(Q_2))\}$ is a 2-even subgraph cover of G. □

⋆ Examples of collapsible graphs

The following lemmas are some examples of collapsible graphs.

Lemma 7.4.5 (Catlin [24], [27]) *A graph containing a pair of edge-disjoint spanning trees is collapsible.*

Proof Let H be a graph containing a pair of edge-disjoint spanning trees F_1, F_2 and with odd vertex set $O(H)$. Let $X \subseteq V(H)$ of even order. By Lemma A.2.17, let Q be a T-join of F_1 with $O(Q) = X \triangle O(H)$. Note that $H - E(Q)$ is a connected, spanning subgraph of H since the spanning tree $F_2 \subseteq H - E(F_1) \subseteq H - E(Q)$. Furthermore, $O(H - E(Q)) = X$ since $X = O(Q) \triangle O(H)$. □

Definition 7.4.6 A graph G is *"μ edges short of having a pair of edge-disjoint spanning trees"* if G has a supergraph G^+ that G^+ contains a pair of edge-disjoint spanning trees, and $|E(G^+) - E(G)| \le \mu$.

Lemma 7.4.7 (Catlin [24]) *A bridgeless graph G is collapsible if it is one edge short of having a pair of edge-disjoint spanning trees.*

The triangle (circuit of length 3) is an example which is one edge short of having a pair of edge-disjoint spanning trees. And the complete bipartite graph $K_{2,t}$ is an example which is two edges short of having a pair of edge-disjoint spanning trees.

Lemma 7.4.8 (Catlin, Han and Lai [28]) *Let G be a graph with at most two edges short of having two edge-disjoint spanning trees. Then either G is collapsible or it is contractible to one of $\{K_2, K_{2,t}\}$ where t is a positive integer.*

⋆ Maximal collapsible subgraph and graph reduction

The following results are some important properties about collapsible graphs.

Lemma 7.4.9   (Catlin [24]) *Let H_1 and H_2 be two collapsible subgraphs of G. If $V(H_1) \cap V(H_2) \neq \emptyset$, then $H_1 \cup H_2$ is also a collapsible subgraph of G.*

Proof Let $H = H_1 \cup H_2$. It is obvious that H/H_i is collapsible for each $i = 1, 2$. Let $X \subseteq V(H)$ of even order. Let $X_1 = [X - V(H_2)] \cup Y_1$ where either Y_1 is empty if $|X \cap V(H_2)|$ is even, or Y_1 is the single vertex of H/H_2 created by the contraction of H_2 if $|X \cap V(H_2)|$ is odd. Let H_{X_1} be a connected spanning subgraph of H/H_2 with $O(H_{X_1}) = X_1$. For each $v \in V(H_1) \cap V(H_2)$, let $E_{H_{X_1}}(v)$ be the set of edges of H_{X_1} incident with v. Let

$$Z_o = \{v \in V(H_1) \cap V(H_2) : |E_{H_{X_1}}(v)| \text{ is odd}\},$$

and

$$Y_2 = [V(H_1) \cap V(H_2)] \cap [Z_o \triangle X].$$

Let

$$X_2 = [X - V(H_1)] \cup Y_2.$$

Then let H_{X_2} be a connected spanning subgraph of H_2 with $O(H_{X_2}) = X_2$. It is easy to see that the subgraph H_X of H induced by $E(H_{X_1}) \cup E(H_{X_2})$ is spanning and connected and $O(H_X) = X$. This proves that $H = H_1 \cup H_2$ is collapsible. $\square$

Definition 7.4.10   Let $H_1, \ldots, H_t$ be the set of all *maximal* collapsible subgraphs of a graph G. The graph G^r obtained from G by contracting each H_i to a single vertex is called the **reduction** of G and G^r itself is called a **C-reduced graph** of G.

By Lemma 7.4.9, the C-reduced graph G^r of a given graph G is well defined and unique. Furthermore, we have the following corollary of Lemma 7.4.9 for C-reduced graphs.

Theorem 7.4.11   (Catlin [24]) *A graph is C-reduced if and only if it has no non-trivial collapsible subgraph.*

Proof Exercise 7.11. $\square$

The following are some properties of C-reduced graphs (we omit the proofs).

Theorem 7.4.12   (Catlin [24]) *If G is a C-reduced graph, then the girth of G is at least 4 and the minimum degree of G is at most 3.*

⋆ Contractible configurations

There are many applications of the above reduction method in the studies of problems of *circuit double covers, 4-flows, supereulerian graphs, hamiltonian line graphs,* etc. The following lemma is a generalization of Lemma 7.4.9 and Theorem 7.4.11 for some general graphic properties.

Lemma 7.4.13 *Let $\mathcal{P}$ be a graph theory property and let H be a $\mathcal{P}$-contractible configuration. Then, for any supergraph G of H, G/H is a $\mathcal{P}$-contractible configuration if and only if G is a $\mathcal{P}$-contractible configuration.*

Lemma 7.4.13 is a useful lemma for contractible configurations in inductive proofs. It is proved as an exercise (Exercise 7.12).

7.5 Exercises

⋆ 3-even subgraph covers

Exercise 7.1 Let $\{C_a, C_b, C_c\}$ be a 3-even subgraph cover of G. Let C be an even subgraph contained in the subgraph $C_a \cup C_b$. Then

$$\{C_a, C_b, C_c \triangle C\}$$

is also a 3-even subgraph cover of G.

Exercise 7.2 Let $\{C_a, C_b, C_c\}$ be a 3-even subgraph cover of G and let C be an even subgraph contained in C_c. Then

$$\{C_a \triangle C, C_b \triangle C, C_c\}$$

is also a 3-even subgraph cover of G.

Remark. Lemma 14.3.2 is another result that constructs new 3-even subgraph covers from linear combinations of an existing one.

⋆ Berge–Fulkerson conjecture

Exercise 7.3 (Jaeger [131]) Show that Conjecture 7.3.6′ and Conjecture 7.3.6″ are equivalent. That is, the statement that *"every bridgeless cubic graph has a 6-even subgraph 4-cover"* is equivalent to the same statement without the condition *"cubic."*

Exercise 7.4   For an odd integer k, a graph G has an even subgraph k-cover if and only if G itself is an even graph.

Definition 7.5.1   Let $\mathcal{F}$ be an even subgraph cover of a graph G. For each edge $e \in E(G)$, let

$$ced_{\mathcal{F}}(e) = |\{C : C \in \mathcal{F}, e \in E(C)\}|$$

which is called the **edge depth** of e (with respect to $\mathcal{F}$). The **edge depth** of the graph G (with respect to $\mathcal{F}$) is

$$ced_{\mathcal{F}}(G) = \max\{ced_{\mathcal{F}}(e) : e \in E(G)\}.$$

Exercise 7.5   Let r be a positive integer and G be a bridgeless graph. If G has an r-even subgraph cover, then, for a given edge e, G has an r-even subgraph cover $\mathcal{F}$ such that $ced_{\mathcal{F}}(e) = 1$.

Exercise 7.6   Let G be a graph and $T = \{x, y\}$ be a 2-vertex-cut of G which separates the graph G into two parts M_1 and M_2 (that is, $M_1 \cup M_2 = G$ and $M_1 \cap M_2 = T$). Let H_i be the graph obtained from M_i by adding a new edge joining x and y. Let r be an integer. If G does not have an r-even subgraph cover, then either H_1 or H_2 does not have an r-even subgraph cover.

⋆ **Collapsible graphs**

Exercise 7.7   (Catlin [24], [27]) If every edge of a graph G is contained in a circuit of length at most 3, then G is collapsible and, therefore, has a 2-even subgraph cover.

Definition 7.5.2   A graph G is locally connected if, for every $v \in V(G)$, $G[N(v)]$ is connected.

Exercise 7.8   (Kriesell [158] and also see [246]) If a graph G is locally connected, then G has a circuit double cover.

Exercise 7.9   (Cai and Corneil [22]) The line graph of every graph G with minimum degree at least 3 has a circuit double cover.

Exercise 7.10   (Cai and Corneil [22]) If a graph G has a CDC, then its line graph $L(G)$ also has a CDC.

Exercise 7.11   (Theorem 7.4.11) A graph is C-reduced if and only if it has no non-trivial collapsible subgraph.

Exercise 7.12 (Lemma 7.4.13) Let $\mathcal{P}$ be a graph theory property and let H be a $\mathcal{P}$-contractible configuration. Then, for any supergraph G of H, G/H is a $\mathcal{P}$-contractible configuration if and only if G is a $\mathcal{P}$-contractible configuration.

8

Flows and circuit covers

Readers are referred to Appendix C for definitions and fundamental properties in the integer flow theory.

In this chapter, we present some results about circuit cover problems directly resulting from and related to integer flow theory.

8.1 Jaeger Theorems: 4-flow and 8-flow

Theorem 8.1.1 (Matthews [174]) *Let r be a positive integer. A graph G admits a nowhere-zero 2^r-flow if and only if G has an r-even subgraph cover.*

Proof "$\Rightarrow$": Induction on r. By Theorem C.2.3, let (D, f_1) be a 2^{r-1}-flow of G and (D, f_2) be a 2-flow of G such that $supp(f_1) \cup supp(f_2) = E(G)$. By the inductive hypothesis, $supp(f_1)$ is covered by even subgraphs $\{C_1, \ldots, C_{r-1}\}$. Thus, $\{C_1, \ldots, C_{r-1}, C_r\}$ is an r-even subgraph cover of G where C_r is the even subgraph of G induced by $supp(f_2)$ (by Theorem C.1.8).

"$\Leftarrow$": Let $\{C_1, \ldots, C_r\}$ be an r-even subgraph cover of G and let (D, f_i) be a 2-flow of G with $supp(f_i) = E(C_i)$ for each $i \in \{1, \ldots, r\}$ (by Theorem C.1.8). Then $(D, \sum_{i=1}^{r} 2^{i-1} f_i)$ is a nowhere-zero 2^r-flow of G. $\qquad\square$

By Theorems 8.1.1, 7.1.6, and 7.2.1, we have the following well-known theorems (for the cases of $r = 2$ and 3) in the integer flow theory.

Theorem 8.1.2 (The 4-flow theorem, Jaeger [123], [127]) *Every 4-edge-connected graph admits a nowhere-zero 4-flow.*

Proof   By Theorems 7.1.6 and 8.1.1 for $r = 2$.   □

Theorem 8.1.3   (The 8-flow theorem, Jaeger [127], Kilpatrick [141])
Every bridgeless graph admits a nowhere-zero 8-flow.

Proof   By Theorem 7.2.1 and Theorem 8.1.1 for $r = 3$.   □

The following theorem is directly related to the circuit double cover
problem.

Theorem 8.1.4   (Jaeger [131]) *If a graph G has a 2^r-even subgraph
double cover then G admits a nowhere-zero 2^r-flow.*

Proof   Let D be an arbitrary orientation of $E(G)$ and $\mathcal{F}$ be a 2^r-even
subgraph double cover of G. Let m be a one-to-one mapping: $\mathcal{F} \mapsto (\mathbb{Z}_2)^r$.
For each $C \in \mathcal{F}$, let (D, f_C) be a 2-flow of G with $supp(f_C) = E(C)$.
Then (D, f) with

$$f = \sum_{C \in \mathcal{F}} m(C) f_C$$

is a group $\mathbb{Z}_2^r$-flow of G. And (D, f) is nowhere-zero since m is a one-to-
one mapping and for each pair $a, b \in \mathbb{Z}_2^r$, $a + b = 0$ if and only if $a = b$.
By Theorem C.1.4, G admits a nowhere-zero integer 2^r-flow.   □

8.2 4-flows

⋆ Even subgraph covers

As we have already discussed in Sections 7.1 and 8.1, the problems of
even subgraph covers and integer flows are very closely related to each
other, especially for the problem of nowhere-zero 4-flow.

Definition 8.2.1   A family of even subgraphs $\mathcal{F}$ of a graph G is called
an even subgraph $(1, 2)$-cover of G if each edge of G is contained in one
or two members of $\mathcal{F}$.

In the following theorem, the equivalence of (i) and some other statements for cubic graphs was originally observed by Tutte [229] and further
generalized and reformulated by Jaeger [127], [130] and Seymour [208].

Theorem 8.2.2   *Let $G = (V, E)$ be a graph. The following statements
are equivalent:*
(1) G admits a nowhere-zero 4-flow;
(2) G has a 2-even subgraph $(1, 2)$-cover;

(3) G *has a* 3-*even subgraph* $(1,2)$-*cover;*

(4) G *has a* 3-*even subgraph double cover;*

(5) G *has a* 4-*even subgraph double cover.*

Proof It is trivial that $(2) \Rightarrow (3)$, $(4) \Rightarrow (5)$, $(4) \Rightarrow (2)$ and $(5) \Rightarrow (3)$.

For $(2) \Rightarrow (4)$ and $(3) \Rightarrow (5)$: Let $\{C_1, \ldots, C_t\}$ be a t-even subgraph $(1,2)$-cover. Then $\{C_1, \ldots, C_t, C_1 \triangle \cdots \triangle C_t\}$ is a $(t+1)$-even subgraph double cover since $C_1 \triangle \cdots \triangle C_t$ is an even subgraph covering all edges of G that were covered only once by $\{C_1, \ldots, C_t\}$ (by Lemma A.2.4).

$(1) \Leftrightarrow (2)$: by Theorem 8.1.1 for $r = 1$.

$(5) \Rightarrow (4)$: Let $\mathcal{F} = \{C_0, C_1, C_2, C_3\}$ be an even subgraph double cover of G. Now we claim that $\{C_0 \triangle C_1, C_0 \triangle C_2, C_0 \triangle C_3\}$ is an even subgraph double cover of G. For an edge e contained in C_0 and C_1, we have that $e \in C_0 \triangle C_2$ and $C_0 \triangle C_3$; for an edge e contained in C_1 and C_2, $e \in C_0 \triangle C_1$ and $C_0 \triangle C_2$. $\qquad \square$

It is evident by Theorem 8.2.2 that the circuit double cover conjecture holds for graphs admitting nowhere-zero 4-flows.

⋆ Parity subgraph decompositions

Recall the definition of parity subgraphs (Definition A.2.9): *a spanning subgraph* H *of a graph* G *is called a* parity subgraph *of* G *if, for each vertex* $v \in V(G)$,

$$d_H(v) \equiv d_G(v) \pmod 2.$$

Definition 8.2.3 Let $G = (V, E)$ be a bridgeless graph.

(1) A decomposition of $E(G)$ is called a parity subgraph decomposition of G if each member of the decomposition induces a parity subgraph of G.

(2) A parity subgraph decomposition of G is trivial if it has only one member (the graph G itself).

Theorem 8.2.4 *A graph* G *admits a nowhere-zero 4-flow if and only if* G *has a non-trivial parity subgraph decomposition.*

Proof See Exercise 8.5. $\qquad \square$

Remark. If the empty graph $\emptyset$ (the graph has no edge) is a parity subgraph of a graph G, then G must be even. Hence, if a graph G is even, then G admits a nowhere-zero 2-flow (therefore, a 4-flow) and $\{G, \emptyset, \emptyset\}$ is a non-trivial parity subgraph decomposition.

For a cubic graph, a non-trivial parity subgraph decomposition must consist three perfect matchings. Hence, we have the following corollary.

Corollary 8.2.5 (Tutte [230]) *A cubic graph G admits a nowhere-zero 4-flow if and only if G is 3-edge-colorable.*

The parity subgraph decomposition of a graph is equivalent to parity 3-edge-coloring of the graph (see Definition 5.2.10).

⋆ Evenly spanning even subgraphs

Every 3-edge-colorable cubic graph G has a 2-factor such that every component is of even length (even 2-factor). This concept is generalized for 4-flows as follows.

Definition 8.2.6 An evenly spanning even subgraph S is a spanning even subgraph in G such that each component of S contains an even number of odd degree vertices of G.

It is obvious that every supereulerian graph contains an evenly spanning even subgraph. The topic of evenly spanning even subgraph was studied in [8] for face-colorings of graphs.

Theorem 8.2.7 *A graph G admits a nowhere-zero 4-flow if and only if G has an evenly spanning even subgraph.*

Proof See Exercise 8.8. □

It is evident that Theorem 8.2.7 is a generalization of the following well-known facts.

(1) *A cubic graph G is 3-edge-colorable if and only if G has a 2-factor C such that each component of C is a circuit of even length.*

(2) *If a cubic graph G contains a Hamilton circuit, then G is 3-edge-colorable.*

The following is a generalized definition of *oddness of general graphs* (oddness for cubic graphs was defined in Definition 4.2.2).

Definition 4.2.9 Let G be a bridgeless graph. For a spanning even subgraph S of G, the oddness of S, denoted by $odd(S)$, is the number of components of S that contains an odd number of odd degree vertices of

G. The oddness of G, denoted by $odd(G)$, is the minimum of $odd(S)$ for all spanning even subgraph S of G.

Obviously, a graph G is of oddness zero if and only if the graph G contains an evenly spanning even subgraphs.

⋆ Faithful cover

The following theorem (Theorem 8.2.8) is a generalization of Lemma 2.2.1. The equivalence of (1) and (2) in the following theorem for cubic graphs (Lemma 2.2.1) was first observed by Seymour in [205] for cubic graphs and was further formulated for the general case in [254]. The equivalence of (1) and (3) was presented in [83].

Theorem 8.2.8   (Goddyn [83], Zhang [254]) *Let G be a bridgeless graph and w be an eulerian $(1,2)$-weight of G. Denote $E_{w=1} = \{e \in E(G) : w(e) = 1\}$. Then the following statements are equivalent:*

(1) G admits a nowhere-zero 4-flow;

(2) (G, w) has a faithful even subgraph cover consisting of at most three even subgraphs;

(3) G has a 4-even subgraph double cover $\mathcal{F}$ such that $G[E_{w=1}] \in \mathcal{F}$.

$(2) \Rightarrow (1)$ is proved in Theorem 8.2.2. $(3) \Leftrightarrow (2)$ is a corollary of Proposition 2.1.6. Thus, we only need to prove the part $(1) \Rightarrow (2)$ of the theorem: it is a corollary of Theorem 8.2.2 and the following lemma.

Lemma 8.2.9   *Let G be a graph having an even subgraph double cover $\{C_1, C_2, C_3\}$ and let $w : E(G) \mapsto \{1, 2\}$ be an eulerian $(1,2)$-weight of G. Denote $E_{w=1} = \{e \in E(G) : w(e) = 1\}$. Then*

$$\{G[E_{w=1}] \triangle C_1, \quad G[E_{w=1}] \triangle C_2, \quad G[E_{w=1}] \triangle C_3\}$$

is a faithful even subgraph cover of (G, w).

Proof Exercise 8.9.   □

Remark. Note that Theorem 8.2.8 may not be true if the weight w has a higher value at some edge of G ($w(e) \geq 3$ for some e).

Let $\mathcal{F}$ be a faithful even subgraph cover described in Theorem 8.2.8-(2). If $\mathcal{F}$ consists of at most *two* even subgraphs, we have the following structural result.

Theorem 8.2.10 *Let G be a bridgeless graph and w be an eulerian $(1,2)$-weight of G. Denote $E_{w=i} = \{e \in E(G) : w(e) = i\}$. Then the following statements are equivalent:*

(1) (G, w) has a faithful even subgraph cover consisting of at most two even subgraphs;

(2) G has a parity subgraph decomposition $\mathcal{P} = \{P_1, P_2, P_3\}$ such that the subgraph $G[E_{w=2}]$ of G induced by $E_{w=2}$ is a member of $\mathcal{P}$;

(3) the subgraph $G[E_{w=1}]$ of G induced by $E_{w=1}$ is an evenly spanning even subgraph;

(4) G has a 3-even subgraph double cover $\mathcal{F}$ such that $G[E_{w=1}] \in \mathcal{F}$.

Proof Exercise 8.11. □

8.3 Seymour Theorem: 6-flow

In this section, we present the best approach to the 5-flow conjecture (Conjecture C.1.10) ([207], also see [131], [250]).

Theorem 8.3.1 (Seymour [207]) *Every 2-edge-connected graph admits a nowhere-zero 6-flow.*

A series of lemmas is presented first from which the theorem follows.

Let G be a graph and C be a circuit (note that C may not be contained in G, and G and C may have a non-empty intersection). Consider the following operation for constructing a new graph obtained from G and C:

Φ_k: *add the circuit C into G if $|E(C) - E(G)| \leq k$.*

Let $\mathcal{G}_k$ be the family of graphs which can be obtained from the graph K_1 via a finite series of Φ_k constructions.

Lemma 8.3.2 (Seymour [207]) *Every graph in $\mathcal{G}_{k-1}$ admits a nowhere-zero k-flow.*

Lemma 8.3.2 is implied by following modified lemma which is presented by Jaeger in [131].

Lemma 8.3.3 (Jaeger [131]) *Let $G \in \mathcal{G}_{k-1}$ and D be an orientation of G. For each mapping $m : E(G) \mapsto \mathbb{Z}_k$, there exists a mod-$k$-flow (D, f) of G such that $f(e) \not\equiv m(e) \pmod{k}$ for each edge $e \in E(G)$.*

Proof We proceed by induction on $|E(G)|$. If $|E(G)| = 0$, there is nothing to prove. Suppose that G is constructed from G' by adding a circuit C into G' with $|E(C) - E(G')| \leq k - 1$ (an operation Φ_{k-1}). Let m be a mapping from $E(G)$ into $\mathbb{Z}_k$ and D be an orientation on $E(G)$. By Theorem C.1.8, let (D, f_1) be an integer valued 2-flow of G with $supp(f_1) = E(C)$. Since $|E(G) - E(G')| \leq k - 1$, let

$$\mu \in \mathbb{Z}_k - \{f_1(e)m(e) : e \in E(G) - E(G')\}.$$

By the inductive hypothesis, there is a mod-k-flow (D, f_2) of G' such that $f_2(e) \not\equiv m(e) - \mu f_1(e) \pmod{k}$ for each $e \in E(G')$. Then $(D, f_2 + \mu f_1)$ is a mod-k-flow of G such that $f_2(e) + \mu f_1(e) \not\equiv m(e) \pmod{k}$ for each $e \in E(G)$. $\qquad\square$

By Lemma 8.3.3, Theorem 8.3.1 can be proved if, for a bridgeless graph G, one can find an even subgraph S such that $G/S \in \mathcal{G}_2$, then G admits a 3-flow (D, f) such that $supp(f) \supseteq E(G) - E(S)$ and therefore G admits a nowhere-zero 6-flow $(D, 2f + f')$ where (D, f') is a 2-flow with the support $E(S)$.

Lemma 8.3.4 (Seymour [207]) *Let G be a 3-edge-connected graph. Then G has an even subgraph S such that the contracted graph $G/S \in \mathcal{G}_2$.*

Proof We choose an even subgraph S and a subgraph $R(S)$ of G such that

(a) $R(S)$ is connected and contains S;
(b) $R(S)/S \in \mathcal{G}_2$;
(c) subject to (a) and (b), $|E(R(S))|$ is as large as possible.

Assume that $G - V(R(S))$ is not empty and let H be a component of $G - V(R(S))$. If H has some bridge, choose a bridge e so that one component H' of $H - \{e\}$ is as small as possible. If H has no bridge, then simply let $H' = H$. Since G is 3-edge-connected, there are two distinct edges xx', $yy' \in (V(H'), V(R(S)))$ where $x, y \in V(H')$ and $x', y' \in V(R(S))$. Furthermore, since H' has no bridge, H' has two edge-disjoint paths P_1, P_2 joining x and y. Then $S' = S \cup P_1 \cup P_2$ is an even subgraph of G and $R(S') \supseteq R(S) \cup P_1 \cup P_2 \cup \{xx', yy'\}$ satisfies (a) and (b) and is bigger than $R(S)$. This contradicts the choice of S and $R(S)$, and therefore completes the proof of the lemma. $\qquad\square$

Lemma 8.3.5 *If G is a smallest bridgeless graph not admitting a nowhere-zero 6-flow, then G must be 3-edge-connected.*

Proof Exercise C.1. $\qquad\square$

With Lemmas 8.3.4 and 8.3.5, we have the following structural result.

Theorem 8.3.6 (Seymour [207], an equivalent statement of the 6-flow theorem) *Every bridgeless graph G admits a 2-flow (D, f_1) and a 3-flow (D, f_2) such that $supp(f_1) \cup supp(f_2) = E(G)$.*

By applying Lemma C.2.2, Theorems 8.3.1 and 8.3.6 are equivalent.

Seymour also mentioned an alternative proof of the 6-flow theorem in [207].

Theorem 8.3.7 (Seymour [207]) *Let G be a 3-connected cubic graph. Then there exists a spanning tree T of G such that $G/[E(G) - E(T)] \in \mathcal{G}_2$.*

$\star$ Even subgraph 6-covers

With the 6-flow theorem (Theorem 8.3.1 or Theorem 8.3.6), we are ready to prove the following theorem.

Theorem 7.3.5 (Fan [50]) *Every bridgeless graph has a 10-even subgraph 6-cover.*

Proof Let D be an orientation of $E(G)$. By Theorem 8.3.6 (an equivalent statement of Theorem 8.3.1), let (D, f_0) be an integer 2-flow of G and (D, f_1) be an integer 3-flow of G such that $supp(f_1) \cup supp(f_0) = E(G)$. Let $f_2 \equiv f_1 + f_0$ and $f_3 \equiv f_1 + 2f_0 \pmod{3}$. Here each (D, f_i) ($i = 1, 2, 3$) is considered as a mod-3-flow of G with $supp_3(f_i) \cup supp_3(f_0) = E(G)$ (note that $2 + 1 = 3 \equiv 0 \pmod{3}$). Let $A_i = E(G) - supp_3(f_i)$ (for $i = 1, 2, 3$) and S_0 be the even subgraph of G induced by the support of the 2-flow f_0. We claim that

$(\star)$ $\{A_1, A_2, A_3\}$ *is a partition of* $E(S_0)$.

For an edge $e \in E(S_0)$, let $f_1(e) = a \in \{0, \pm 1, \pm 2\}$ and $f_0(e) = b \in \{1, -1\}$. Then $e \in A_x$ if and only if x is the solution of the equation

$$a + bx \equiv 0 \pmod{3}.$$

By Theorem 8.2.2, let $\mathcal{F}_i$ be a 3-even subgraph double cover of $G - A_i$ for each $i = 1, 2, 3$. Then, by $(\star)$, $\mathcal{F}_1 \cup \mathcal{F}_2 \cup \mathcal{F}_3$ is a 9-even subgraph cover that covers each edge of $G - E(S_0)$ six times and covers each edge of S_0 four times. Furthermore,

$$\{S_0 \triangle C : C \in \mathcal{F}_1 \cup \mathcal{F}_2 \cup \mathcal{F}_3\}$$

is a 9-even subgraph cover of G that covers each edge of $G - E(S_0)$ (still) six times and covers each edge of S_0 five times. Thus,

$$\{S_0\} \cup \{S_0 \triangle C : C \in \mathcal{F}_1 \cup \mathcal{F}_2 \cup \mathcal{F}_3\}$$

is a 10-even subgraph 6-cover of G. $\qquad\qquad\qquad\qquad\qquad\qquad$ □

8.4 Contractible configurations for 4-flow

Since it is known that a spanning eulerian subgraph is an evenly spanning even subgraph, one can use the reduction method described in Section 7.4 in the study of nowhere-zero 4-flows.

Theorem 8.4.1 (Catlin [25]) *Let G be a graph and H be a subgraph of G. If H is collapsible or a circuit of length 4, then G admits a nowhere-zero 4-flow if and only if G/H admits a nowhere-zero 4-flow. (That is, collapsible graphs and 4-circuits are contractible configurations for the nowhere-zero 4-flow problem.)*

Proof By Theorem 7.4.4 and Theorem 8.2.2, collapsible graphs are contractible configurations for nowhere-zero 4-flow problem. For circuit of length 4, it is solved in the following lemma. $\qquad\qquad\qquad\qquad$ □

The following lemma is a generalization of Lemma B.1.9 for non-cubic graphs.

Lemma 8.4.2 (Catlin [25]) *Let G be a graph and C_0 be a circuit of G of length at most 4. If G admits a 4-flow (D, f) such that $supp(f) \supseteq E(G) - E(C_0)$, then G admits a nowhere-zero 4-flow.*

Proof By Exercise C.14, choose a mod-4-flow (D, f) such that under the orientation D, C_0 is a directed circuit. Assume that

$$\mathbb{Z}_4 - [\{0\} \cup \{f(e) : e \in E(C_0)\}] \neq \emptyset.$$

Let $\lambda \in \mathbb{Z}_4 - [\{0\} \cup \{f(e) : e \in E(C_0)\}]$, and let (D, f') be a non-negative 2-flow of G with $supp(f') = E(C_0)$. Then $(D, f - \lambda f')$ is a nowhere-zero 4-flow of G.

Thus, we assume that $\mathbb{Z}_4 - [\{0\} \cup \{f(e) : e \in E(C_0)\}] = \emptyset$ and therefore C_0 is of length 4 and $|E(C_0) - supp(f)| = 1$. Let $C_0 = v_1 \ldots v_4 v_1$ and $e_0 = v_4 v_1$ where $f(e_0) = 0$ (that is, $supp(f) = E(G) - \{e_0\}$). Let $\{P_1, P_2, P_3\}$ be a parity subgraph decomposition of $G - \{e_0\}$ (by Theorem 8.2.4). We are to find an evenly spanning even subgraph of G which will imply that G admits a nowhere-zero 4-flow (by Theorem 8.2.7). If

$$E(P_1) \cap E(C_0) = \emptyset,$$

then, without loss of generality, let $|E(P_2) \cap E(C_0)| \leq 1$ since the length of C_0 is at most 4. Thus, $[P_1 \cup P_2] \triangle C_0$ is an evenly spanning even subgraph of G. So, we assume that $|E(P_i) \cap E(C_0)| \geq 1$ for each i. Thus, $|E(P_i) \cap E(C_0)| = 1$ for each i. Without loss of generality, let P_1 and P_2 contain the edges $v_1 v_2$ and $v_3 v_4$, respectively. If v_1 and v_4 are contained in the same component of $P_1 \cup P_2$, then $P_1 \cup P_2$ is an evenly spanning even subgraph of G. So, we assume that $v_1 v_2$ is contained in the component D of $P_1 \cup P_2$ which does not contain $v_3 v_4$. Here, $\{P_1', P_2', P_3\}$ is a parity subgraph decomposition of $G - \{e_0\}$ where $P_1' = P_1 \triangle D$ and $P_2' = P_2 \triangle D$. Now $E(P_1') \cap E(C_0) = \emptyset$ which is the case we have already done. $\square$

8.5 Bipartizing matching, flow covering

The following theorem was obtained by Tutte [229] and reformulated by Jaeger in [127].

Theorem 8.5.1 (Tutte [229]) *Let G be a cubic graph. G admits a nowhere-zero 3-flow if and only if G is bipartite.*

Proof "$\Rightarrow$": Let (D, f) be a nowhere-zero mod-3-flow of G. Define a new mod-3-flow (D', f') as follows: $f'(e) = 1$ for every edge of G, and D' is obtained from D by reversing the direction of every arc e if $f(e) = 2$. Note that every vertex v, under the orientation D', has either three incoming arcs, or three outgoing arcs. That is, $V(G)$ has a partition $\{V^+, V^-\}$ where $d_{D'}^+(v) = 3$ for every $v \in V^+$ and $d_{D'}^-(v) = 3$ for every $v \in V^-$. Therefore, G is bipartite since there is no arc between any pair of vertices of the same part.

"$\Leftarrow$": If G is a bipartite cubic graph with a bi-partition $\{A, B\}$, let D be an orientation on $E(G)$ such that the direction of each edge e is from A to B and let f be a weight on $E(G)$ with $f(e) = 1$ for every edge e of G. It is obvious that (D, f) is a nowhere-zero mod-3-flow of G. $\square$

By the proof of Theorem 8.5.1, we also have the following structural lemma.

Lemma 8.5.2 *Let G be a bipartite cubic graph. For every perfect matching M of the graph, G admits a nowhere-zero integer 3-flow (D, f) with*

$$M = \{e : \ f(e) = \pm 2\}.$$

The following concept of bipartizing matching was introduced in [67] by Fleischner.

Definition 8.5.3 Let G be a cubic graph with a dominating circuit C. A *bipartizing matching* M *of G with respect to C* is a matching that $M \subseteq E(G) - E(C)$, $V(M) \supseteq V(G) - V(C)$, and either $\overline{G - M}$ is bipartite or $G - M$ is 2-regular.

It is proved in [67] that for a dominating circuit C of a cubic graph G, a bipartizing matching with respect C always exists in G. However, does G have a pair of edge-disjoint bipartizing matchings with respect to some dominating circuit? This remains as an open problem (Conjecture 8.5.5). And this problem has a direct relation to some circuit covering problems, including the CDC conjecture (Theorem 8.5.4).

The following theorem (by Fleischner) solves Sabidussi conjecture (Conjecture 1.5.2) under an assumption of the existence of a pair of edge-disjoint bipartizing matchings. (Other partial results for the Sabidussi conjecture (Conjecture 1.5.2 or Conjecture 10.5.2) can be found in Section 10.5 and Section 10.8 as exercises.)

Theorem 8.5.4 (Fleischner [67]) *Let G be a cubic graph with a dominating circuit C. If G contains a pair of edge-disjoint bipartizing matchings M_1 and M_2 with respect to the circuit C, then G has a 5-even subgraph double cover $\mathcal{F}$ with $C \in \mathcal{F}$.*

Proof Let $G_j = \overline{G - M_j}$ for each $j = 1, 2$. By Lemma 8.5.2, the bipartite cubic graph G_j admits a nowhere-zero 3-flow (D'_j, f'_j) with $E_{f'_j = \pm 1} = E(C)$ and $E_{f'_j = \pm 2} = E(G_j) - E(C)$. Consider (D'_j, f'_j) as a 3-flow of G with $E_{f'_j = 0} = M_j$.

Let D be an arbitrary orientation of G. Change the sign of the flow value if necessary: let (D, f_j) be the integer 3-flow of G with $f_j(e) = f'_j(e)$ if e is oriented in the same direction under both D'_j and D, and $f_j(e) = -f'_j(e)$ if e is oriented in the opposite directions under both D'_j and D.

Let $(D, g^+) = (D, \frac{f_1 + f_2}{2})$ and $(D, g^-) = (D, \frac{f_1 - f_2}{2})$. It is easy to see that both (D, g^+) and (D, g^-) are 3-flows of G with the following properties:

(1) if $e \in M_j$ for some j then $e \in E_{g^+ = \pm 1}$ and $E_{g^- = \pm 1}$;

(2) if $e \in E(C)$ then either $e \in E_{g^+ = \pm 1} \cap E_{g^- = \pm 0}$ or $e \in E_{g^+ = \pm 0} \cap E_{g^- = \pm 1}$;

(3) if $e \in E(G) - M_1 - M_2 - E(C)$ then either $e \in E_{g^+ = \pm 2} \cap E_{g^- = \pm 0}$ or $e \in E_{g^+ = \pm 0} \cap E_{g^- = \pm 2}$.

By Theorem 8.2.10, let $\mathcal{F}^+$ be a 2-even subgraph cover of $supp(g^+)$ that covers $E_{g^+ = \pm 1}$ once, and $E_{g^+ = \pm 2}$ twice. And, similarly, let $\mathcal{F}^-$ be the corresponding 2-even subgraph cover of $supp(g^-)$. By the properties listed above, $\mathcal{F}^+ \cup \mathcal{F}^-$ is a 4-even subgraph cover of G that covers C once and all other edges twice. $\qquad\square$

Fleischner proposed the following conjecture about the existence of such pair bipartizing matchings.

Conjecture 8.5.5 (Fleischner [68]) *If G is a cyclically 4-edge-connected snark containing some dominating circuit. Then G must contain at least one dominating circuit C such that G has a pair of edge-disjoint bipartizing matchings with respect to C.*

Remark. Conjecture 8.5.5 may not be true if the dominating circuit C is specified. Some cyclically 4-edge-connected snark G has been discovered [110], [111], in which there is a dominating circuit C' such that the graph G *does not* have a pair of edge-disjoint bipartizing matchings with respect to C'. That explains why the dominating circuit C in Conjecture 8.5.5 cannot be specified.

An open problem similar to Conjecture 8.5.5 with respect to a specified dominating circuit is proposed in [111].

Conjecture 8.5.6 (Fleischner and Hoffmann-Ostenhof [111]) *Let G be a cyclically 4-edge-connected snark, C be a dominating circuit of G, $W = [V(G) - V(C)]$ and L be the set of all chords of C. Then the set $R = W \cup L$ has a partition $\{R_1, R_2\}$ such that each $G - R_i$ has a pair of edge-disjoint bipartizing matchings with respect to C.*

Although Conjecture 8.5.6 is not as strong as Conjecture 8.5.5, it still implies the CDC conjecture in a relatively weaker conclusion (not the strong CDC problem).

Theorem 8.5.7 (Fleischner and Hoffmann-Ostenhof [111]) *Let G be* *a cubic graph, C be a dominating circuit of G, W = [V(G) − V(C)] and L be the set of all chords of C. If the set R = W ∪ L has a partition {R_1, R_2} such that each G − R_i has a pair of edge-disjoint bipartizing matchings with respect to C, then G has an 8-even subgraph double cover.*

Proof By Theorem 8.5.4, each $(G − R_i)$ has a 5-even subgraph double cover $\mathcal{F}_i$ containing C. Hence, $[\mathcal{F}_1 − C] ∪ [\mathcal{F}_2 − C]$ is a 8-even subgraph double cover of G. □

Without mentioning the dominating circuit, Theorem 8.5.4 can be further generalized as follows in terms of flows (with the same proof as Theorem 8.5.4).

Theorem 8.5.8 [245] *Let G be a cubic graph. If G admits a pair of 3-flows (D, f_1) and (D, f_2) such that*

(1) $supp(f_1) ∪ supp(f_2) = E(G)$,

(2) $E_{f_1 = ±1} = E_{f_2 = ±1}$,

then G has a 5-even subgraph double cover $\mathcal{F}$ with $C ∈ \mathcal{F}$ where $E(C) = E_{f_1 = ±1} = E_{f_2 = ±1}$.

Remark. The 5-flow conjecture (Conjecture C.1.10) is also proved [67] under the assumption of the existence of a pair of edge-disjoint bipartizing matchings (see Exercise 8.18). Further generalizations of Exercise 8.18 related to flows or circuit covers can be found in [245].

8.6 Exercises

⋆ 4-flows

Exercise 8.1 Let $\{C_0, C_1, C_2, C_3\}$ be a 4-even subgraph double cover. Show that $\{C_1 \triangle C_2, C_1 \triangle C_3, C_2 \triangle C_3\}$ is a 3-even subgraph double cover.

Remark. Compare with the proof of Theorem 8.2.2 ((v) ⇒ (iv)): $\{C_1 \triangle C_2, C_1 \triangle C_3, C_2 \triangle C_3\}$ is the same even subgraph double cover as $\{C_0 \triangle C_1, C_0 \triangle C_2, C_0 \triangle C_3\}$ of G since $C_0 = C_1 \triangle C_2 \triangle C_3$.

Exercise 8.2 (A corollary of Theorem 8.1.2) Let G be a graph with odd-edge-connectivity at least 5. Then G admits a nowhere-zero 4-flow.

Exercise 8.3 If a graph G is not even, then every parity subgraph decomposition of G has an odd number of members.

Exercise 8.4 Let G be a graph.
(1) If G has a 3-even subgraph double cover $\{C_1, C_2, C_3\}$, then

$$\{G - E(C_1), \quad G - E(C_2), \quad G - E(C_3)\}$$

is a parity subgraph decomposition of G.
(2) If G has a parity subgraph decomposition $\{P_1, P_2, P_3\}$, then

$$\{P_1 \cup P_2, \quad P_1 \cup P_3, \quad P_2 \cup P_3\}$$

is a 3-even subgraph double cover of G.

Exercise 8.5 (Theorem 8.2.4) A graph G admits a nowhere-zero 4-flow if and only if G has a non-trivial parity subgraph decomposition.

Exercise 8.6 (Tutte [230]) (Corollary 8.2.5) A cubic graph G admits a nowhere-zero 4-flow if and only if G is 3-edge-colorable.

Exercise 8.7 An even subgraph S of a graph G is an evenly spanning even subgraph of G if and only if S is the union of two edge-disjoint parity subgraphs of G.

Exercise 8.8 (Theorem 8.2.7) A graph G admits a nowhere-zero 4-flow if and only if G has an evenly spanning even subgraph.

⋆ Faithful covers

Exercise 8.9 (Lemma 8.2.9) Let G be a graph having an even subgraph double cover $\{C_1, C_2, C_3\}$ and let $w : E(G) \mapsto \{1, 2\}$ be an eulerian $(1, 2)$-weight of G. Denote $E_{w=1} = \{e \in E(G) : w(e) = 1\}$. Then

$$\{G[E_{w=1}] \triangle C_1, \quad G[E_{w=1}] \triangle C_2, \quad G[E_{w=1}] \triangle C_3\}$$

is a faithful cover of (G, w).

Exercise 8.10 Let G be a bridgeless graph and w be an eulerian $(1, 2)$-weight of G. Suppose that $\{S_1, S_2\}$ is a faithful even subgraph cover of (G, w). Then, for each even subgraph R with $E(R) \subset E(S_1 \triangle S_2)$, $\{R \triangle S_1, R \triangle S_2\}$ is also a faithful even subgraph cover of (G, w).

Exercise 8.11 (Theorem 8.2.10) Let G be a bridgeless graph and w be an eulerian $(1, 2)$-weight of G. Denote $E_{w=i} = \{e \in E(G) : w(e) = i\}$. Then the following statements are equivalent:

(1) (G, w) has a faithful even subgraph cover consisting of two even subgraphs;

(2) G has a parity subgraph decomposition $\mathcal{P} = \{P_1, P_2, P_3\}$ such that the subgraph $G[E_{w=2}]$ of G induced by $E_{w=2}$ is a member of $\mathcal{P}$;

(3) The subgraph $G[E_{w=1}]$ of G induced by $E_{w=1}$ is an evenly spanning even subgraph;

(4) G has a 3-even subgraph double cover $\mathcal{F}$ such that $G[E_{w=1}] \in \mathcal{F}$.

Exercise 8.12 [257] The following statements are equivalent:

(1) a graph G has a 6-even subgraph double cover;

(2) $(G, 2)$ has a weight decomposition $(G_1, w_1) + (G_2, w_2) = (G, 2)$ such that both G_1 and G_2 admits nowhere-zero 4-flows;

(3) G has two subgraphs A and B such that
    (i) $E(A) \cup E(B) = E(G)$,
    (ii) $A \cap B = C$ is even and
    (iii) both A and B admit nowhere-zero 4-flows.

Exercise 8.13 [257] A graph G has a 7-even subgraph double cover if and only if G has two subgraphs A, B such that

(1) $E(A) \cup E(B) = E(G)$,

(2) the subgraph A has an even subgraph C, the subgraph B has an even subgraph D and $E(A) \cap E(B) \subseteq E(C)$, $E(A) \cap E(B) \subseteq E(D)$, and

(3) both A and B admit nowhere-zero 4-flows.

Exercise 8.14 (Celmins [31], Jaeger [131]) Let G be a bridgeless graph and $e \in E(G)$. If $G - \{e\}$ admits a nowhere-zero 4-flow, then G has a 5-even subgraph double cover. (Give a direct proof without applying Theorem 4.2.3.)

⋆ Seymour's operation

Exercise 8.15 Let T be a spanning tree of G. Show that $G/T \in \mathcal{G}_1$ and hence G admits a 2-flow (D, f) with $E(G) - E(T) \subseteq supp(f)$.

Exercise 8.16 Show that every graph with minimum degree at least 3 and embeddable on the sphere or the projective plane belongs to $\mathcal{G}_5$.

Exercise 8.17 Let G be a 2-edge-connected planar graph. Assume that G has an even subgraph S such that $G/S \in \mathcal{G}_2$ and each component of S is a facial circuit. Show that G is 4-face-colorable.

⋆ Miscellanies

Exercise 8.18   (Fleischner [67]) Let G be a cubic graph with a dominating circuit C. If G contains a pair of edge-disjoint bipartizing matchings M_1 and M_2 with respect to the circuit C, then G admits a nowhere-zero 5-flow.

Exercise 8.19   (1) Find a graph G such that G contains two edge-disjoint spanning trees but is not 4-edge-connected.

(2) Find a supereulerian graph G which does not contain two edge-disjoint spanning trees.

(3) Find a graph G such that G admits a nowhere-zero 4-flow but is not supereulerian.

9

Girth, embedding, small cover

9.1 Girth

The girth of a smallest counterexample to the circuit double cover was first studied by Goddyn [82], in which, a lower bound 7 of girth was found. Later, this bound was improved as follows: at least 8 by McGuinness [179], and at least 9 by Goddyn [83] (a girth bound of 10 was also announced in [83]). The following theorem, proved with a computer aided search, remains the best bound up to today.

Theorem 9.1.1 (Huck [115]) *The girth of a smallest counterexample to the circuit double cover conjecture is at least* 12.

It was conjectured in [132] that cyclically 4-edge-connected snarks may have bounded girths. If this conjecture were true, then the circuit double cover conjecture would be followed immediately by Theorem 9.1.1 (or an earlier result in [82] for girth 7). But this is not the case: in [146], Kochol gave a construction of cyclically 5-edge-connected snarks of arbitrarily large girths.

However, Theorem 9.1.1 (or its earlier results) remains useful in the studies of some families of embedded graphs with small genus since the girth of such graphs is bounded (see Section 9.2).

9.2 Small genus embedding

The circuit double cover conjecture is trivial if a bridgeless graph is planar: the collection of face boundaries is a double cover. How about graphs embeddable on surfaces other than a sphere? Although it is known that every bridgeless graph has a 2-cell embedding on some surface, it is not guaranteed that face boundaries are circuits (see Section 1.4).

The following early results verified the circuit double cover conjecture for graphs embeddable on some surfaces with small genus.

Theorem 9.2.1 (Zha [251], [252], [253]) *Let G be a bridgeless graph. If G has a 2-cell embedding on a surface with at most 5 crosscaps, or at most 2 handles, then G has a circuit double cover.*

If Σ is a surface and a bridgeless graph G has a 2-cell embedding on Σ, then, by the Euler formula (Theorem A.1.1),

$$\xi(\Sigma) = |V(G)| - |E(G)| + |F(G)| \qquad (9.1)$$

where $\xi(\Sigma)$ is the Euler characteristic of the surface Σ. Surfaces with at most 5 crosscaps, or at most 2 handles are of Euler characteristic $\xi \geq -3$. Applying the latest result about girth (Theorem 9.1.1), we present a generalization of Theorem 9.2.1.

Theorem 9.2.2 *Let $\mathcal{G}$ be the family of all bridgeless graphs each of which has a 2-cell embedding on some surface with Euler characteristic $\xi \geq -8$. Then every member of $\mathcal{G}$ has a circuit double cover.*

A surface with Euler characteristic $\xi \geq -8$ contains at most 10 crosscaps or at most 5 handles (or some combinations of both).

Before the proof of Theorem 9.2.2, we first present some observations about 2-cell embedding.

Lemma 9.2.3 *Let G be a bridgeless graph with a 2-cell embedding on a surface Σ. If the embedding is not a circular 2-cell embedding, then*

$$|F(G)| + 1 \leq \frac{2|E(G)|}{g}$$

where g is the girth of G.

Proof See Exercise 9.3. □

Proof of Theorem 9.2.2. Let G be a smallest counterexample to the theorem. By the splitting lemma (Theorem A.1.14) that preserves the embedding, G is cubic. Similarly to Lemma 1.2.3, G is 3-connected and essentially 4-edge-connected.

Since G is cubic,

$$3|V(G)| = 2|E(G)|.$$

By Theorem 9.1.1, the girth of G is at least 12. So, by Lemma 9.2.3,

$$\frac{2|E(G)|}{12} - 1 \geq |F(G)|.$$

Substituting these into Equation (9.1) (the Euler formula, Theorem A.1.1), we have that

$$\frac{-|E(G)|}{6} - 1 \geq \xi \geq -8.$$

So $|E(G)| \leq 42$.

Let $\mathcal{F}$ be a circuit double cover of $G - e_0$ for some edge $e_0 = x_0 y_0 \in G$ with $|\mathcal{F}|$ as large as possible (satisfying the condition of Lemma 3.4.1-(4)). Since $G - e_0$ has at most 41 edges, the total length of all members of $\mathcal{F}$ is at most 82. Note that the girth of G is at least 12. We have that $|\mathcal{F}| \leq 6$.

Let $C' \in \mathcal{F}$ be a circuit containing x_0 and $C'' \in \mathcal{F}$ be a circuit containing y_0. By Lemma 3.4.1-(4),

$$C' \neq C'', \quad C' \cap C'' = \emptyset, \tag{9.2}$$

for otherwise, x_0 and y_0 are joined by a circuit chain of length ≤ 2. Hence, we may let $\mathcal{F} = \{C_1, \ldots, C_r\}$ with $r \leq 6$ and $x_0 \in C_1 \cap C_2$, $y_0 \in C_3 \cap C_4$. By Equation (9.2),

$$\{C_1 \cup C_3, C_2 \cup C_4, C_5, \ldots, C_r\}$$

is a 4-even subgraph double cover of the cubic graph $\overline{G - e_0}$. By Theorem 1.3.2, the cubic graph $\overline{G - e_0}$ is 3-edge-colorable. By Theorem 2.2.4, G has a 5-even subgraph double cover. This contradicts that G is a counterexample and completes the proof. $\qquad\square$

Theorem 9.2.2 was recently further generalized by Mohar to the following theorem for surfaces with larger genus.

Theorem 9.2.4 (Mohar [181]) *Let $\mathcal{G}$ be the family of all bridgeless graphs each of which has a 2-cell embedding on some surface with Euler characteristic $\xi \geq -31$. Then every member of $\mathcal{G}$ has a circuit double cover.*

9.3 Small circuit double covers

The following conjectures were proposed by Bondy in [16].

Conjecture 9.3.1 (Bondy [16]) *Every 2-edge-connected simple graph G of order n has a circuit double cover $\mathcal{F}$ such that $|\mathcal{F}| \leq n - 1$.*

Conjecture 9.3.2 (Bondy [16]) *Every 2-edge-connected simple cubic graph G $(G \neq K_4)$ of order n has a circuit double cover $\mathcal{F}$ such that $|\mathcal{F}| \leq \frac{n}{2}$.*

Theorem 9.3.3 (Goddyn and Richter, see [166]) *Let $\mathcal{F}$ be a circuit double cover of a cubic graph G. Then $\mathcal{F}$ contains at most $\frac{|V(G)|}{2} + 2$ circuits.*

Proof Each circuit of $\mathcal{F}$ can be considered as the boundary of a disk. The graph G is therefore embedded on a surface Σ established by joining all of these disks at the edges of G. Since the Euler characteristic of any surface is not greater than 2, by Euler's formula (Theorem A.1.1) we have that

$$|V(G)| + |\mathcal{F}| - |E(G)| \leq 2.$$

Note that $|E(G)| = \frac{3|V(G)|}{2}$ since G is cubic. Therefore, no circuit double cover of G contains more than $\frac{|V(G)|}{2} + 2$ circuits. □

The equivalent relation (Theorem 9.3.4) between the circuit double cover conjecture and a small circuit double cover conjecture (Conjecture 9.3.2) are proved in [166].

Theorem 9.3.4 (Lai, Yu and Zhang [166]) *If a simple cubic graph G $(G \neq K_4)$ has a circuit double cover, then the graph G has a circuit double cover containing at most $\frac{|V(G)|}{2}$ circuits.*

Here, we give only an outline of the proof of Theorem 9.3.4. We can use induction on $|V(G)|$ for a slightly more general result. For a circuit double cover $\mathcal{F}$ of G, if some member C of $\mathcal{F}$ is of length at most 5, then we may delete one edge e of C and modify $\mathcal{F}$ to a circuit double cover of the resulting smaller graph G'. Let $\mathcal{F}'$ be a circuit double cover of $\overline{G'} = \overline{G - e}$ with $|\mathcal{F}'|$ as small as possible. By modifying $\mathcal{F}'$, we can have a circuit double cover of G which satisfies the theorem (or a slightly stronger result). Thus, we assume that no member of $\mathcal{F}$ is of length at most 5. The result follows immediately by counting the numbers of edges and vertices.

There is another similar problem related to the small circuit double cover problem: *find a circuit cover $\mathcal{F}$ of a 2-connected graph G with $|\mathcal{F}|$ as small as possible* (without the requirement that $\mathcal{F}$ is a double cover). A conjecture about this problem is proposed by Bondy in [16] and some partial results to this conjecture can be found in [163] and [10].

Conjecture 9.3.1 has been verified for some families of graphs ([57], [164], [171], [186], [187], [201], [202], [203]).

9.4 Exercises

Embedded graphs

Exercise 9.1 Let H be a connected even graph with multiplicity at most 2 such that, for every cut-vertex x (if it exists) of H, no component of $H - x$ is a single vertex (that is, no degree 2 vertex is incident with a pair of parallel edges). Then either H itself is a circuit or H contains two edge-disjoint circuits of lengths at least 3.

Exercise 9.2 Let G be a bridgeless graph with a 2-cell embedding on a surface Σ. Let g be the girth of G. For each face f of G in Σ, if the closure of f is homeomorphic to a closed disk, then the degree of f is at least g; if the closure of f is *not* homeomorphic to a closed disk, then the degree of f is at least $2g$.

Exercise 9.3 (Lemma 9.3) Let G be a bridgeless graph with a 2-cell embedding on a surface Σ. If the embedding is not a circular 2-cell embedding, then

$$|F(G)| + 1 \leq \frac{2|E(G)|}{g}$$

where g is the girth of G.

10

Compatible circuit decompositions

10.1 Introduction

The following is a "proof" of the CDC conjecture (Exercise 1.2).

Let $2G$ be the graph obtained from G by replacing every edge with a pair of parallel edges. Since the resulting graph $2G$ is an even graph, by Lemma A.2.2, the even graph $2G$ has a circuit decomposition $\mathcal{F}$. Is it obvious that $\mathcal{F}$ is a circuit double cover of the original graph G?

What is wrong in this "proof"? The circuit decomposition $\mathcal{F}$ of the eulerian graph $2G$ may contain some digons (circuit of length 2) which corresponds to an edge instead of a circuit in G.

Can we add some restriction to a circuit decomposition of $2G$? For example, a restriction to the decomposition $\mathcal{F}$ so that no member of $\mathcal{F}$ is a digon. This is one of the motivations of compatible circuit decompositions for eulerian graphs [59], [60].

Definition 10.1.1 Let H be an eulerian graph. For a vertex v in H, a forbidden set incident with v, denoted by $\mathcal{P}(v)$, is a partition of $E(v)$ (the set of edges incident with v). A member of $\mathcal{P}(v)$ is called a forbidden part (incident with v). The collection $\mathcal{P} = \bigcup_{v \in V(H)} \mathcal{P}(v)$ is called a forbidden system of H.

A circuit decomposition of H is a set of edge-disjoint circuits (of H) whose union is H. We say that $(H, \mathcal{P})$ has a compatible circuit decomposition (CCD, in abbreviation) if H has a circuit decomposition $\mathcal{F}$ such that $|E(C) \cap P| \leq 1$ for every circuit $C \in \mathcal{F}$ and every forbidden part $P \in \mathcal{P}$.

A forbidden system $\mathcal{P}$ is admissible if $|P \cap T| \leq \frac{|T|}{2}$ for every forbidden part $P \in \mathcal{P}$ and every edge-cut T of H. If $\mathcal{P}$ is an admissible forbidden system of H, we simply say that $(H, \mathcal{P})$ is admissible.

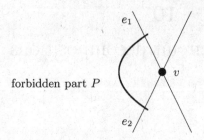

Figure 10.1 *A forbidden part* $P = \{e_1, e_2\}$

Similar to the faithful circuit covering problem, admissibility is obviously a necessary condition for compatible circuit decomposition. (But, it is not sufficient for the CCD problem, see Section 10.3.)

Definition 10.1.2   A forbidden part is trivial if it consists of a single edge, and non-trivial otherwise. A vertex v in $(H, \mathcal{P})$ is trivial (with respect to $\mathcal{P}$) if every forbidden part incident with v is trivial.

Clearly, if every vertex of $(H, \mathcal{P})$ is trivial, then any circuit decomposition of H is a CCD of $(H, \mathcal{P})$.

Definition 10.1.3   Let $\mathcal{F} = \{C_1, C_2, \ldots, C_m\}$ be a set of even subgraphs of H. $\mathcal{F}$ is called a compatible even subgraph decomposition of $(H, \mathcal{P})$ if $|C_i \cap P| \leq 1$ for every C_i, $1 \leq i \leq m$, and every $P \in \mathcal{P}$.

By this definition, if $\mathcal{F} = \{C_1, \ldots, C_m\}$ is a compatible even subgraph decomposition of $(H, \mathcal{P})$, then the collection of circuit decompositions of all C_i's is a CCD of $(H, \mathcal{P})$. Therefore, $(H, \mathcal{P})$ *has a CCD if and only if it has a compatible even subgraph decomposition.*

In some figures of this chapter, edges of a forbidden part are indicated by a crossing bold curve (see Figure 10.1).

10.2 Relation with faithful circuit cover

Fleischner [59], [60] observed that there are close connections between the compatible circuit decomposition problem and the faithful circuit cover problems.

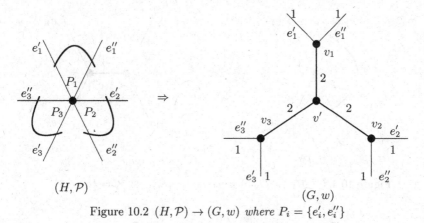

Figure 10.2 $(H, \mathcal{P}) \to (G, w)$ where $P_i = \{e'_i, e''_i\}$

⋆ From CCD to faithful cover

Operation 1. (Figure 10.2) Let H be an eulerian graph with an admissible forbidden system $\mathcal{P}$. For each vertex v, let $\mathcal{P}(v) = \{P_1, P_2, \ldots, P_k\}$. We split v into k vertices $v_1, v_2, \ldots, v_k$ such that v_i is incident with the edges in P_i and then add a new vertex v' joining each v_i with a new edge of weight $|P_i|$, $1 \leq i \leq k$. Let G be the new graph obtained by applying this operation to every vertex v of H, and complete the weight function by assigning weight 1 to every old edge (edge of H).

If we denote this weight function by w, then $(H, \mathcal{P})$ has a compatible circuit decomposition if and only if (G, w) has a faithful circuit cover (Lemma 10.2.1).

Operation 2. (For degree 4 vertex, Figure 10.3) For a degree 4, non-trivial vertex v with $E(v) = \{e_1, e_2, e_3, e_4\}$ and a forbidden part $P_1 = \{e_1, e_2\}$, one may split v into two vertices v' and v'' with e_1, e_2 incident with v' and e_3, e_4 incident with v'' and add a weight 2 edge between v' and v''. All original edges are assigned with weight $w = 1$.

Similar to the discussion for Operation 1, the weighted graph (G, w) obtained by Operations 1 and 2 has the property that $(H, \mathcal{P})$ has a CCD if and only if (G, w) has a faithful circuit cover.

⋆ From faithful cover to CCD

Operation 3. (Figure 10.4) If G is a graph with an admissible, eulerian weight w, let H be the even graph obtained from G by replacing each

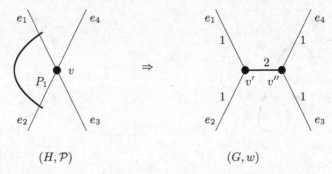

Figure 10.3 $(H, \mathcal{P}) \to (G, w)$ where $P_1 = \{e_1, e_2\}$, $P_2 = \{e_3, e_4\}$

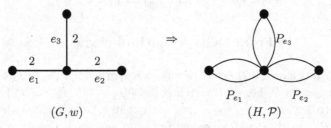

Figure 10.4 $(G, w) \to (H, \mathcal{P})$ where $|P_{e_i}| = w(e_i)$

edge e by a set P_e of $w(e)$ parallel edges (thus deleting e if $w(e) = 0$ and leaving e unaltered if $w(e) = 1$). Consider each P_e as a forbidden part (of H) incident with either end of e (in G). Then the set $\mathcal{P}$ of all these forbidden parts is an admissible forbidden system of H.

Evidently, (G, w) has a faithful circuit cover if and only if $(H, \mathcal{P})$ has a CCD (Lemma 10.2.1).

The following lemma summarizes the relation between CCD and faithful cover.

Lemma 10.2.1 (Fleischner [59], [60]) *(1) (Operations 1 and 2) Let H be an eulerian graph and $\mathcal{P}$ be an admissible forbidden system of H. Then $(H, \mathcal{P})$ has a CCD if and only if the corresponding weighted graph (G, w), constructed by Operations 1 and 2, has a faithful circuit cover.*

(2) (Operation 3) Let G be a bridgeless graph with an admissible eulerian weight $w : E(G) \mapsto \mathbb{Z}^+$. Then the weighted graph (G, w) has a faithful circuit cover if and only if the corresponding $(H, \mathcal{P})$, constructed by Operation 3, has a CCD.

Since we are concentrating on the circuit double cover conjecture in

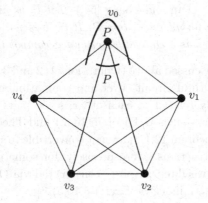

Figure 10.5 K_5 *with* $\mathcal{P}_5 = \{\{v_{i-1}v_i, v_iv_{i+1}\}, \{v_{i-2}v_i, v_iv_{i+2}\} : i \in \mathbb{Z}_5\}$

this book, we will mostly discuss the forbidden system $\mathcal{P}$ with $|P| \leq 2$ in this chapter. Generalizations for arbitrary forbidden systems will be summarized in this chapter, and will be further investigated in later chapters (Chapter 12, etc.).

10.3 Counterexamples and graph minor related results

It is clear that admissibility is necessary for $(H, \mathcal{P})$ to have a CCD. But, it is not sufficient. This can be seen from the following example. Let K_5 be the complete graph on five vertices $\{v_i : 0 \leq i \leq 4\}$. The forbidden set incident with v_i is defined by $\mathcal{P}(v_i) = \{\{v_iv_{i-1}, v_iv_{i+1}\}, \{v_iv_{i-2}, v_iv_{i+2}\}\}$, where $0 \leq i \leq 4$ and the subscripts are read modulo 5 (see Figure 10.5). Set $\mathcal{P}_5 = \bigcup_{i=0}^{i=4} \mathcal{P}(v_i)$. Then $(K_5, \mathcal{P}_5)$ is admissible, but has no CCD (see Exercise 10.3).

The following are major results about the compatible circuit decomposition problem.

Theorem 10.3.1 (Fleischner [59]) *Let H be an eulerian graph and $\mathcal{P}$ be an admissible forbidden system on H. If $|P| \leq 2$ for each $P \in \mathcal{P}$ and H is planar, then $(H, \mathcal{P})$ has a CCD.*

Theorem 10.3.2 (Fleischner and Frank [70]) *Let H be an eulerian graph and $\mathcal{P}$ be an admissible forbidden system on H. If H is planar, then $(H, \mathcal{P})$ has a CCD.*

Theorem 10.3.3 (Fan and Zhang [54]) *Let H be an eulerian graph with an admissible forbidden system $\mathcal{P}$. If H does not contain K_5 as a minor, then $(H, \mathcal{P})$ has a compatible circuit decomposition.*

Though, as we discussed above (Operations 1, 2 and 3, Lemma 10.2.1), the problems of faithful circuit cover and compatible circuit decomposition are so closely related to each other, some pairs of results (with minor-free conditions, such as Theorem 10.3.1 and Theorem 2.2.3, Theorem 10.3.3 and Theorem 12.1.2) are not convertible from one to another by the above constructions. More precisely, for some planar graph H, the corresponding weighted graph G constructed via Operations 1 and 2 may contain a P_{10}-minor (see Exercise 10.5).

10.4 Planar graphs

In this section, we present the proof of a pioneer result (Theorem 10.3.1) by Fleischner.

Theorem 10.3.1 (Fleischner [59]) *Let H be an eulerian graph and $\mathcal{P}$ be an admissible forbidden system on H. If $|P| \leq 2$ for each $P \in \mathcal{P}$ and H is planar, then $(H, \mathcal{P})$ has a CCD.*

Proof Let $(H, \mathcal{P})$ be a counterexample to the theorem such that
(1) $|E(H)|$ is as small as possible,
(2) subject to (1), the number of non-trivial forbidden parts of $(H, \mathcal{P})$ is as small as possible.
It is obvious that the minimum degree of H is at least 4. And, since $(H, \mathcal{P})$ is a counterexample, there must be some non-trivial vertex, for otherwise, any circuit decomposition of the eulerian graph H is a CCD.

For each non-trivial forbidden part $P \in \mathcal{P}$ at a vertex v, a forbidden system $\mathcal{P}_P$ obtained from $\mathcal{P}$ by replacing P with two trivial forbidden parts is called *a discharge of $\mathcal{P}$ at P*.

I. We claim that H *is 4-regular.*
Let $P_0 = \{e', e''\}$ be a non-trivial forbidden part of H at the vertex v_0 (that is, $P_0 \subseteq E(v_0)$). Let $\mathcal{P}_{P_0}$ be the forbidden system obtained from $\mathcal{P}$ by discharging P_0 (that is, replacing P_0 with two trivial forbidden parts $\{e'\}$ and $\{e''\}$). By the choice of $(H, \mathcal{P})$, the new planar eulerian graph $(H, \mathcal{P}_{P_0})$, which has a smaller number of non-trivial forbidden parts, has a CCD $\mathcal{F}^*$. If no member of $\mathcal{F}^*$ contains both e' and e'', then $\mathcal{F}^*$ is

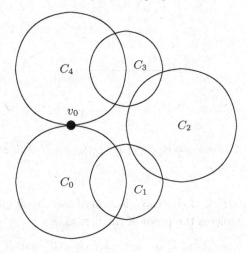

Figure 10.6 *Circuit chain: $C_0, \ldots, C_4$ with common end v_0*

also a CCD of $(H, \mathcal{P})$. This contradicts that $(H, \mathcal{P})$ is a counterexample. Hence, we have that $C_0 \in \mathcal{F}^*$ contains both e' and e''.

Let $\mathcal{C}^* = \{C_0, \ldots, C_t\} \subseteq \mathcal{F}^*$ be a shortest circuit chain that

$$C_i \cap C_j - \{v_0\} \neq \emptyset \ \text{ if and only if } i = j \pm 1$$

and

$$v_0 \in V(C_0) \cap V(C_t).$$

(See Figure 10.6.)

Let H' be the subgraph of H induced by edges of $\cup_{i=0}^{t} E(C_i)$, and $\mathcal{P}'$ be the restriction of $\mathcal{P}$ on the graph H', that is,

$$\mathcal{P}' = \{P \cap E(H') : P \in \mathcal{P}\}.$$

Note that H' does not have a cut-vertex since it is induced by a circuit chain with a common end v_0 ($v_0 \in V(C_0) \cap V(C_t)$).

We claim that $\mathcal{P}'$ is admissible. Suppose that there exists a forbidden part $P \in \mathcal{P}'$ and a cut T of H' with $|P' \cap T| > \frac{|T|}{2}$. Note that $|P| = 2$. Then $P = T$ and T must be $E(x)$ for some vertex $x \in V(H')$ and therefore, x is a cut-vertex of H'. This contradicts that H' has no cut-vertex.

If H' is a proper subgraph of H, then let $\mathcal{F}'$ be a CCD of $(H', \mathcal{P}')$. Note that $\mathcal{F}' \cup (\mathcal{F}^* - \mathcal{C}^*)$ is a CCD of $(H, \mathcal{P})$. A contradiction again.

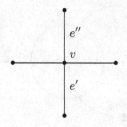

Figure 10.7 *A crossing vertex v of the embedded H with a forbidden part $P = \{e', e''\}$*

So, we have that $H' = H$. By the selection of the circuit chain C^*, H' is regular. This completes the proof of the first claim.

II. In the proof of Claim I, we not only proved that H is 4-regular, but also proved that, *for each non-trivial vertex v_0 and a non-trivial forbidden part P_0 at v_0, $(H, \mathcal{P}_{P_0})$ has a CCD which is a circuit chain $C_0, \ldots, C_t$ with the common end at v_0.*

III. At a vertex v_0 with a non-trivial forbidden part P_0, let $\{C_0, \ldots, C_t\}$ be a CCD (a circuit chain) of $(H, \mathcal{P}_{P_0})$ with t as large as possible. *We claimed that*

$$t \geq 2.$$

From $(H, \mathcal{P})$, construct a cubic graph G with an eulerian weight $w : E(G) \mapsto \{1, 2\}$ as follows. For each non-trivial vertex v, apply Operation 2 (defined in Section 10.1). Let v_0' and v_0'' be the pair of vertices obtained from v_0 via Operation 2, and e_0 be the new weight 2 edge between v_0' and v_0''.

By Lemma 10.2.1, we have the following:

$(H, \mathcal{P}_{P_0})$ *has a CCD if and only if $(G - e_0, w)$ has a faithful circuit cover; and*

$(H, \mathcal{P})$ *has a CCD if and only if (G, w) has a faithful circuit cover.*

Thus, the claim is proved by applying Lemma 3.4.1-(4) to the eulerian $(1, 2)$-weighted graph (G, w).

IV. Let H be embedded on a plane. A non-trivial vertex v of $(H, \mathcal{P})$ is called a **$\mathcal{P}$-crossing vertex** if there is a $P_1 = \{e', e''\} \in \mathcal{P}(v)$ such that e' and e'' are not consecutive in a facial circuit (see Figure 10.7).

V. We claim that $(H, \mathcal{P})$ *has no $\mathcal{P}$-crossing vertex.* Assume that v_0 is a $\mathcal{P}$-crossing vertex of $(H, \mathcal{P})$. Let $E(v_0) = \{e_1, e_2, e_3, e_4\}$ be arranged on the plane in the cyclic order as e_1, e_2, e_3, e_4, e_1 and let $P_0 = \{e_1, e_3\}$. Let $\mathcal{F}'$ be a CCD of $(H, \mathcal{P}_{P_0})$ (where $\mathcal{P}_{P_0}$ is obtained from $\mathcal{P}$ by discharging P_0) and $\mathcal{F}$ is chosen so that $|\mathcal{F}|$ is as large as possible. By II and III, $\mathcal{F} = \{C_0, \ldots, C_t\}$ is a circuit chain closed at v with $t \geq 2$. Without loss of generality, let $e_1, e_3 \in E(C_0)$ and $e_2, e_4 \in E(C_t)$. Thus, C_0 and C_t "cross" each other on the plane at the vertex v_0. Since both C_0 and C_t are circuits, they must "cross" each other again on the plane at another vertex. Thus, by the definition of circuit chain, $t = 1$ since $|V(C_0) \cap V(C_t)| \geq 2$. This contradicts III and proves our claim.

VI. By V, $(H, \mathcal{P})$ has no $\mathcal{P}$-crossing vertex. Applying Operation 2 (defined in Section 10.1), we obtain a planar graph G with an eulerian weight $w : E(G) \mapsto \{1, 2\}$. Since G is planar, by Theorem 2.2.3 or Theorem 3.3.1, (G, w) has a faithful circuit cover, which corresponds to a CCD of $(H, \mathcal{P})$. This contradicts that $(H, \mathcal{P})$ is a counterexample and completes the proof. $\qquad\square$

10.5 Dominating circuit and Sabidussi Conjecture

In previous sections, we discussed the relation between the problem of faithful circuit cover and the problem of compatible circuit decomposition. In this section, we further investigate another circuit decomposition problem which is a special case of the strong circuit double cover conjecture (Conjecture 1.5.1).

Definition 10.5.1 Let H be an eulerian graph. Let T be an Euler tour with the edge sequence $\{e_0, \ldots, e_{m-1}\}$ and $\mathcal{C}$ be a circuit decomposition of H. T and $\mathcal{C}$ are compatible to each other if, for each $i \in Z_m$ and each $C \in \mathcal{C}$,

$$|\{e_i, e_{i+1}\}| \cap E(C)| \leq 1.$$

Conjecture 10.5.2 (Sabidussi, see [60], or [121], and Conjecture 2.4 in [2]) *Let H be an eulerian graph of minimum degree at least four and T be an Euler tour of H. Then H has a circuit decomposition compatible to T.*

An equivalent version of the above conjecture is the following conjecture (see Lemma 10.5.5 for the proof).

Conjecture 1.5.2 (Sabidussi and Fleischner [60]) *Let G be a cubic graph* *such that G has a dominating circuit C. Then G has a circuit double cover F such that C is a member of F.*

Note that the existence of a dominating circuit in a cyclically 4-edge-connected graph was proposed as a conjecture by Fleischner, Ash and Jackson.

Conjecture 10.5.3 (Fleischner [61], Ash and Jackson [9]; or see Conjecture 25 in [63]) *Every cyclically 4-edge-connected cubic graph contains a dominating circuit.*

Proposition 10.5.4 (Fleischner [61], [73]) *Conjectures 1.5.2 and 10.5.3 together imply the circuit double cover conjecture.*

Lemma 10.5.5 (Fleischner [60]) *Conjectures 10.5.2 and 1.5.2 are equivalent.*

Proof. **I.** If the maximum degree of the eulerian graph H is at most 6, Conjectures 10.5.2 and 1.5.2 are equivalent by applying Operation 1 to degree 6 vertices, and Operation 2 to all degree 4 vertices (Lemma 10.2.1).

II. If v is a vertex of degree $d \geq 8$ in the eulerian graph H, then, along the Euler tour, split v to be a degree 4 vertex and a degree $(d-4)$ vertex. Repeat this procedure for every such vertex or resulting vertex, and apply I. □

10.6 Construction of contra pairs

In this section, we only consider 4-regular graphs and forbidden systems $\mathcal{P}$ with $|P| \leq 2$ for every $P \in \mathcal{P}$.

Similar to Section 2.3, we discuss the constructions of some larger contra pair from smaller ones for the CCD problem.

Definition 10.6.1 Let H be an eulerian graph and $\mathcal{P}$ be an admissible forbidden system of H. Then $(H, \mathcal{P})$ is a contra pair if $(H, \mathcal{P})$ has no compatible circuit decomposition.

By applying Operation 2 (on page 119) to non-trivial vertices, we convert all operations discussed in Section 2.3 for the CCD problem.

The following two construction methods correspond to the $\oplus_2$-sums at a pair of weight 1 edges or a pair of weight 2 edges for the faithful cover problem.

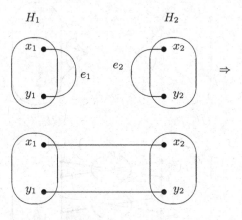

Figure 10.8 $(H_1, \mathcal{P}_1) \oplus_{2e} (H_2, \mathcal{P}_2)$

Definition 10.6.2 ($\oplus_{2e}$-sum at an edge) Let H_i be an eulerian graph associated with a forbidden system $\mathcal{P}_i$ and an edge $e_i = x_i y_i \in E(H_i)$ ($i = 1, 2$). (See Figure 10.8.)

Construct $(H, \mathcal{P}) = (H_1, \mathcal{P}_1) \oplus_{2e} (H_2, \mathcal{P}_2)$ as follows: H is the graph obtained from H_1 and H_2 by replacing a pair of edges $e_i = x_i y_i \in E(H_i)$ with new edges $x_1 x_2$ and $y_1 y_2$, and $\mathcal{P}$ is the union $\mathcal{P}_1 \cup \mathcal{P}_2$.

Definition 10.6.3 ($\oplus_{2v}$-sum at a vertex) Let H_i be an eulerian graph associated with a forbidden system $\mathcal{P}_i$ ($i = 1, 2$), and $x_i \in V(H_i)$ with

$$\mathcal{P}_1(x_1) = \{\{x_1 u_1, x_1 u_2\}, \{x_1 u_3, x_1 u_4\}\},$$

and

$$\mathcal{P}_2(x_2) = \{\{x_2 v_1, x_2 v_2\}, \{x_2 v_3, x_2 v_4\}\}.$$

(See Figure 10.9.)

Construct $(H, \mathcal{P}) = (H_1, \mathcal{P}_1) \oplus_{2v} (H_2, \mathcal{P}_2)$ as follows: H is the graph obtained from H_1 and H_2 by deleting x_1, x_2 and adding new vertices y_1, y_2 and new edges $\{y_1 u_1, y_1 u_2, y_1 v_1, y_1 v_2, y_2 u_3, y_2 u_4, y_2 v_3, y_2 v_4\}$, and $\mathcal{P}$ is obtained from $\mathcal{P}_1 \cup \mathcal{P}_2 - \mathcal{P}_1(x_1) - \mathcal{P}_2(x_2)$ by adding new forbidden parts:

$$\{y_1 u_1, y_1 u_2\}, \{y_1 v_1, y_1 v_2\}, \{y_2 u_3, y_2 u_4\}, \{y_2 v_3, y_2 v_4\}.$$

The following construction method corresponds to the $\oplus_3$-sum at a pair of vertices for the faithful cover problem.

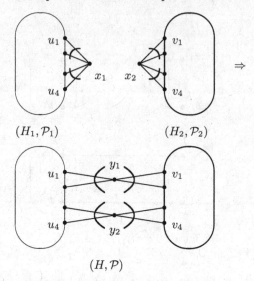

$(H, \mathcal{P})$

Figure 10.9 $(H_1, \mathcal{P}_1) \oplus_{2v} (H_2, \mathcal{P}_2)$

Definition 10.6.4 ($\oplus_3$-sum) Let H_i be an eulerian graph associated with a forbidden system $\mathcal{P}_i$ ($i = 1, 2$), and $x_i \in V(H_i)$ with

$$\mathcal{P}_1(x_1) = \{\{x_1u_1, x_1u_2\}, \{x_1u_3, x_1u_4\}\},$$

and

$$\mathcal{P}_2(x_2) = \{\{x_2v_1, x_2v_2\}, \{x_2v_3, x_2v_4\}\}.$$

(See Figure 10.10.)

Construct $(H, \mathcal{P}) = (H_1, \mathcal{P}_1) \oplus_3 (H_2, \mathcal{P}_2)$ as follows: H is the graph obtained from H_1 and H_2 by deleting x_1, x_2 and adding a new vertex y and new edges $\{u_1v_1, u_2v_2, yu_3, yu_4, yv_3, yv_4\}$, and $\mathcal{P}$ is obtained from $\mathcal{P}_1 \cup \mathcal{P}_2 - \mathcal{P}_1(x_1) - \mathcal{P}_2(x_2)$ by adding two new forbidden parts:

$$\{yu_3, yu_4\}, \{yv_3, yv_4\}.$$

The following construction method corresponds to the $\oplus_{IJF}$-product for the faithful cover problem, which was discovered by Fleischner (see [121]) and is equivalent to the Isaacs–Fleischner–Jackson product defined in Section 2.3.

Definition 10.6.5 ($\otimes_{IFJ}$-product or Isaacs–Fleischner–Jackson product) Let $(H_M, \mathcal{P}_M)$ and $(H_F, \mathcal{P}_F)$ be a pair of eulerian graphs associated with forbidden systems. (See Figure 10.11.)

Construct $(H, \mathcal{P}) = (H_F, \mathcal{P}_F) \otimes_{IFJ} (H_M, \mathcal{P}_M)$ as follows: let $f_1 =$

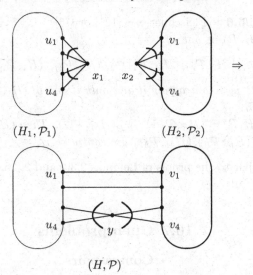

$(H_1, \mathcal{P}_1)$ $(H_2, \mathcal{P}_2)$ $\Rightarrow$

$(H, \mathcal{P})$

Figure 10.10 $(H_1, \mathcal{P}_1) \oplus_3 (H_2, \mathcal{P}_2)$

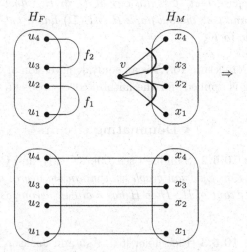

Figure 10.11 $(H_F, \mathcal{P}_F) \otimes_{IFJ} (H_M, \mathcal{P}_M)$

$u_1 u_2, f_2 = u_3 u_4 \in E(H_F)$, and $v \in V(H_M)$ with $\mathcal{P}(v) = \{\{vx_1, vx_2\}, \{vx_3, vx_4\}\}$. The graph H is obtained from H_F, H_M by deleting the vertex v, edges f_1, f_2 and adding edges $u_i x_i$ ($i = 1, 2, 3, 4$), and the forbidden system $\mathcal{P} = \mathcal{P}_F \cup \mathcal{P}_M - \mathcal{P}(v)$.

Lemma 10.6.6   (Fleischner, see [121] or [75])
(1) Let $(H,\mathcal{P})$ be one of

$$(H_1,\mathcal{P}_1) \oplus_{2e} (H_2,\mathcal{P}_2), \quad (H_1,\mathcal{P}_1) \oplus_{2v} (H_2,\mathcal{P}_2), \quad (H_1,\mathcal{P}_1) \oplus_3 (H_2,\mathcal{P}_2).$$

Then $(H,\mathcal{P})$ is a contra pair if and only if one of $(H_1,\mathcal{P}_1),(H_2,\mathcal{P}_2)$ is a contra pair.

(2) Let $(H,\mathcal{P}) = (H_F,\mathcal{P}_F) \otimes_{IFJ} (H_M,\mathcal{P}_M)$. Then $(H,\mathcal{P})$ is a contra pair if both $(H_F,\mathcal{P}_F),(H_M,\mathcal{P}_M)$ are contra pairs.

Proof   Similar to the proofs of Lemmas 2.3.2 and 2.3.6.   □

10.7 Open problems

⋆ Contra pairs

Conjecture 10.7.1   (Goddyn [87] (revised)) *Let H be a 4-regular graph with an edge-decomposition consisting of a pair of Hamilton circuits C_1 and C_2. Suppose that, for any circuit D in H whose edges alternate between C_1 and C_2, the subgraph $H - E(D)$ has a cut-vertex. Then H is isomorphic to K_5.*

Remark.   Conjecture 10.7.1 is an equivalent version of Conjecture 3.5.3 (Operation 2 is applied) for the faithful cover problem.

⋆ Dominating circuits

Conjecture 10.5.2 (Sabidussi see [60], or [121], and Conjecture 2.4 in [2]) *Let H be an Eulerian graph with minimum degree at least four and T be an Euler tour of H. Then H has a circuit decomposition compatible to T.*

Conjecture 10.5.3 (Fleischner [61], Ash and Jackson [9]; or see Conjecture 25 in [63]) *Every cyclically 4-edge-connected cubic graph contains a dominating circuit.*

The following is another open problem related to dominating circuits proposed by Fleischner.

Conjecture 10.7.2   (Fleischner [61] p. 237) *Every cyclically 4-edge-connected, non-3-edge-colorable cubic graph (snark) contains a dominating circuit.*

Remark. It is proved in [147] that Conjectures 10.5.3 and 10.7.2 are equivalent. Some related further studies about dominating circuits in cyclically 4-edge-connected cubic graphs can be found in [73].

⋆ Miscellanies

Definition 10.7.3 A pair of circuit decompositions $\mathcal{C}_1, \mathcal{C}_2$ of an eulerian graph H are compatible to each other if, for every vertex $v \in V(H)$ and every $C_1 \in \mathcal{C}_1$, $C_2 \in \mathcal{C}_2$,

$$|E(v) \cap E(C_1) \cap E(C_2)| \leq 1.$$

Conjecture 10.7.4 (Fleischner, Hilton and Jackson [74]) *Every 2-connected Eulerian graph with minimum degree at least 4 has a pair of mutually compatible circuit decompositions.*

Remark. Exercises 10.8, 10.9 and 11.2 are partial results of Conjecture 10.7.4.

10.8 Exercises

⋆ Operations

Exercise 10.1 (Operation 1, see Figure 10.2.) Let H be an eulerian graph with an admissible forbidden system $\mathcal{P}$. For each vertex $v \in V(H)$, let $\mathcal{P}(v) = \{P_1, P_2, \ldots, P_k\}$. We split v into k vertices $v_1, v_2, \ldots, v_k$ such that v_i is incident with the edges in P_i and then add a new vertex v' joining each v_i with a new edge of weight $|P_i|$, $1 \leq i \leq k$. Let G be the new graph obtained by applying this operation to every vertex v of H, and complete the weight function by assigning to every old edge (edge of H) weight 1. We denote this weight function by w. Show that $(H, \mathcal{P})$ has a CCD if and only if (G, w) has a faithful circuit cover.

Exercise 10.2 (Operation 3, see Figure 10.4.) If G is a graph with an admissible, eulerian weight w, let H be the eulerian graph obtained from G by replacing each edge e by a set P_e of $w(e)$ parallel edges (thus deleting e if $w(e) = 0$ and leaving e unaltered if $w(e) = 1$). Consider each P_e as a forbidden part (of H) incident with either end of e in H. Then show that the set $\mathcal{P}$ of all these forbidden parts is an admissible forbidden system of H, and show that (G, w) has a faithful circuit cover if and only if $(H, \mathcal{P})$ has a CCD.

⋆ Counterexamples, minors

Exercise 10.3 Let $\{v_1, \ldots, v_5\}$ be the set of vertices of the complete graph K_5. A forbidden system of K_5 is defined as:

$$\mathcal{P}_5 = \{\{v_i v_{i-j}, v_i v_{i+j}\} : i = 1, \ldots, 5, \; j = 1, 2, \quad (\text{mod } 5)\}.$$

(See Figure 10.5.) Prove that $(K_5, \mathcal{P}_5)$ does not have a CCD.

Exercise 10.4 Let $\mathcal{X}$ be a collection of graphs. The family of graphs without X-minor, for every $X \in \mathcal{X}$, is denoted by $\Lambda_{\overline{\mathcal{X}}}$.

If each eulerian graph of $\Lambda_{\overline{\mathcal{X}}}$ has a circuit decomposition compatible to each admissible forbidden system, then show that each bridgeless graph of $\Lambda_{\overline{\mathcal{X}}}$ has a faithful circuit cover with respect to each admissible eulerian weight.

Exercise 10.5 Find a family of planar, 4-regular graphs associated with forbidden systems such that the corresponding weighted graphs constructed by Operation 2 have P_{10}-minors.

⋆ Admissibility

Exercise 10.6 Let H be an eulerian graph with minimum degree ≥ 4 and $\mathcal{P}$ be a forbidden system of H with $|P| \leq 2$ for every $P \in \mathcal{P}$. Then $\mathcal{P}$ is not admissible if and only if there is a cut-vertex v of H separating H into Q_1, Q_2 (that is, $Q_1 \cup Q_2 = H$ and $Q_1 \cap Q_2 = v$) and a forbidden part $P \in \mathcal{P}(v)$, $P = E(H_i) \cap E(v)$, for some $i \in \{1, 2\}$.

⋆ Compatible or faithful

Exercise 10.7 Let H be a 4-regular graph and $\mathcal{P}$ be an admissible forbidden system of H. Let (G, w) be obtained from $(H, \mathcal{P})$ by Operation 2. If each component of $E_{w=1}$ contains an even number of edges, then $(H, \mathcal{P})$ has a compatible circuit decomposition.

Exercise 10.8 (Jackson [121]) Let H be an eulerian graph. If $d(v) \equiv 0$ (mod 4) and H has a circuit decomposition $\mathcal{F}_e$ such that each circuit of $\mathcal{F}_e$ is of even length, then H has another circuit decomposition $\mathcal{F}_c$ compatible to $\mathcal{F}_e$. (See Definition 10.7.3.)

Exercise 10.9 If an eulerian graph H contains no K_5-minor and with the minimum degree at least 4, then, for any circuit decomposition $\mathcal{F}_1$ of H, there is another circuit decomposition $\mathcal{F}_2$ compatible to $\mathcal{F}_1$.

Remark. Exercises 10.8 and 10.9 are partial results of Conjecture 10.7.4

Exercise 10.10   Conjecture 10.5.2 is true if H is $4k$-regular.

Exercise 10.11   Conjecture 10.5.2 is true if H is bipartite and $6k$-regular.

⋆ Embeddings

Exercise 10.12   [258] Every 2-connected graph containing no K_5-minor has a circular 2-cell embedding on some 2-manifold surface.

11

Other circuit decompositions

11.1 Restricted circuit decompositions

By looking at the incorrect proof presented in Section 10.1, one may suggest some different restrictions so that some circuit decompositions of eulerian graphs may imply the CDC conjecture.

Example 1 Let H be the even graph obtained from a bridgeless cubic graph G by adding a path of length 2 between every pair $\{x, y\}$ if $xy \in E(G)$. If H has a circuit decomposition $\mathcal{F}$ such that every member of $\mathcal{F}$ is of even length, then $\mathcal{F}$ corresponds to a circuit double cover of the original graph G.

Example 2 Let H be the even graph described in the previous example. If H has a circuit decomposition $\mathcal{F}'$ such that no member of $\mathcal{F}$ is of length 3, then $\mathcal{F}'$ corresponds to a circuit double cover of the original graph G.

Those two examples motivated some related studies of special circuit decompositions of eulerian graphs.

Theorem 11.1.1 (Seymour [206]) *If H is a planar even graph such that every block contains an even number of edges, then H has a circuit decomposition consisting of even circuits.*

Theorem 11.1.2 (Zhang [255]) *If H is a K_5-minor-free even graph such that every block contains an even number of edges, then H has a circuit decomposition consisting of even circuits.*

Theorem 11.1.2 is an application of Theorem 10.3.3. A proof of this theorem different from the original paper [255] can be found in [259] which uses a removable circuit theorem in [87] and, therefore, it is an application of both Theorems 10.3.3 and 12.1.2.

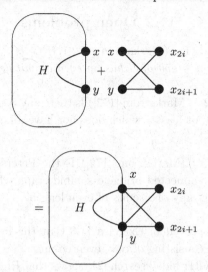

Figure 11.1 *Construction of a $\Delta(2i+1)$-graph from a $\Delta(2i-1)$-graph*

Note that K_5 is not the only 3-connected eulerian graph without even circuit decomposition. An infinite family of 3-connected eulerian graphs without even circuit decomposition was constructed in [121], an infinite family of 4-connected such graphs was constructed in [194], and a 4-regular and 4-connected such graph was discovered in [172].

Theorem 11.1.3 (Markström [172]) *If G is a 2-connected cubic graph with oddness ≤ 2, then the line graph of G has an even circuit decomposition.*

Theorem 11.1.4 (Heinrich, Liu and Zhang [105]) *If H is a P_{10}-minor-free even graph such that the number of edges in every block is a multiple of 3, then H has a triangle-free circuit decomposition except for a well-characterized family of Δ-graphs.*

A Δ-graph is defined recursively as follows (see Figure 11.1). The family $\Delta(1)$ contains only the triangle $x_1x_2x_3$. Suppose we have constructed all graphs in $\Delta(2i-1)$. A graph in $\Delta(2i+1)$ is constructed from a graph H in $\Delta(2i-1)$ by choosing an edge $e = xy$ (called a base edge) of H and adding two new vertices x_{2i} and x_{2i+1}, attaching the 4-circuit $xx_{2i+1}yx_{2i}x$. In successively choosing base edges no three may ever form a triangle and a base edge may be chosen more than once.

Theorem 11.1.4 is proved by applying Theorems 3.3.1 and 12.1.2.

11.2 Open problems

Conjecture 11.2.1 (Markström [172]) *If G is a 2-connected cubic graph then the line graph of G has an even circuit decomposition.*

Problem 11.2.2 (Markström [172]) Is there any 3-connected 4-regular graphs with girth at least 4 which does not have an even circuit decomposition?

Problem 11.2.3 (Markström [172]) Is the Petersen graph the only cyclically 4-edge connected bridgeless cubic graph which does not have a CDC consisting only of circuits of even length?

Remark. It is proved in Exercise 11.3 that the Petersen graph does not have a CDC consisting of only even circuits.

A recent computer aided search [21] shows that Problem 11.2.3 is true for all cubic graphs of order at most 36.

11.3 Exercises

Exercise 11.1 Let G be a 3-edge-colorable cubic graph. Show that the line graph of G has an even circuit decomposition (without using Theorem 11.1.3).

Exercise 11.2 Conjecture 10.7.4 is true for line graphs $H = L(G)$ if G is a 2-connected cubic graph with oddness ≤ 2.

Exercise 11.3 (Markström [172]) Show that the Petersen graph does not have an even circuit double cover.

Reductions of weights, coverages

For an eulerian weighted graph (G, w), if (G, w) is a contra pair, can we find another admissible eulerian weight w^* of G such that (G, w^*) remains as a contra pair while $w^*(G) < w(G)$ and $0 \leq w^*(e) \leq w(e)$ for every edge e? If the answer to this question is "yes," then we should concentrate on eulerian $(1, 2)$-weights in the study of contra pairs.

It is obvious that every bridgeless graph has a circuit cover. However, we do not know yet how "small" the maximum coverage would be. If one is able to find another circuit cover that reduces the coverage while the parity of coverage is retained, then one is able to reduce the coverage recursively down to 1 or 2, and the CDC conjecture is followed.

These two problems are both related to reductions: reduction of weight in a contra pair, and reduction of coverage of an existing cover. The first problem has a complete answer, and is studied in Section 12.1. The second problem, as we can see already, remains as an approach to the CDC conjecture (Section 12.2).

Note that reduction of total coverage without preserving the parity of coverage is the shortest cycle cover problem, which is discussed separately in Chapter 14.

12.1 Weight reduction for contra pairs

The following weight reduction theorem was originally proved by Seymour [205] for planar graphs, and generalized in [4] for general graphs.

Theorem 12.1.1 (Alspach, Goddyn and Zhang [4]) *Let G be a bridgeless graph associated with an admissible eulerian weight $w : E(G) \mapsto \mathbb{Z}^+$.*

If (G, w) is a contra pair, then G has an admissible eulerian weight $w^* : E(G) \mapsto \{0, 1, 2\}$ such that
(1)

$$0 \le w^*(e) \le w(e)$$

for every $e \in E(G)$ and
(2) the admissible eulerian weighted graph (G, w^*) remains as a contra pair.

⋆ Application of Theorem 12.1.1

The following theorem is a generalization of Theorem 3.3.1 that the range of eulerian weights is extended to all non-negative integers.

Theorem 12.1.2 (Alspach, Goddyn and Zhang [4]) *A bridgeless graph G has a faithful circuit cover with respect to every admissible eulerian weight $w : E(G) \mapsto \mathbb{Z}^*$ if and only if G does not contain a subdivision of the Petersen graph.*

Proof Note that zero is an element in $\mathbb{Z}^*$. If G contains a Petersen subdivision H, then define an eulerian weight $w : E(G) \mapsto \{0, 1, 2\}$ by assigning 0 to all edges in $E(G) - E(H)$, and an eulerian $(1, 2)$-weight to $E(\overline{H})$ corresponding to (P_{10}, w_{10}) (see Figure B.8). By Proposition B.2.26, (G, w) has no faithful cover. This completes the proof of the "only if" part of the theorem.

For the "if" part, by Theorem 3.3.1, it is sufficient to show that, as a minimal counterexample, the maximum weight w is upper bounded by 2. Thus, it is an immediate corollary of Theorem 12.1.1. □

⋆ Outline of the proof of Theorem 12.1.1

Let (G, w) be a smallest counterexample to the theorem. We may find an edge $e_0 = xy$ with $w(e_0) \ge 2$ such that (G, w^*) has a faithful even subgraph cover $\mathcal{F}^*$ where $w^* : E(G) \mapsto \mathbb{Z}^+$ is defined as follows

$$w^*(e) = \begin{cases} w(e) & \text{if } e \ne e_0 \\ w(e) - 2 & \text{if } e = e_0. \end{cases}$$

The remaining part of the proof is to find some local adjustments of the faithful cover $\mathcal{F}^*$ so that the resulting faithful cover $\mathcal{F}^{**}$ contradicts some special "maximality" or "minimality" choices of $\mathcal{F}^*$ (such as (α), (β) and (γ) described on pages 148, 149).

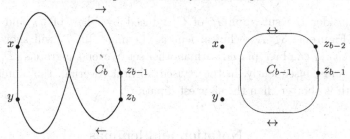

Figure 12.1 *A passing circuit C_b, and a free even subgraph C_{b-1}*

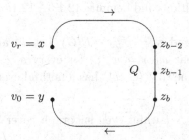

Figure 12.2 *Admissible dipath Q in A from x to y*

Among all members of $\mathcal{F}^*$, those containing the edge e_0 are *circuits* (and are called **passing members of $\mathcal{F}^*$**), those not containing e_0 are *connected even subgraphs* (and are called **free members of $\mathcal{F}^*$**). (See Figure 12.1.)

Construct an auxiliary directed graph A with the vertex set $V(G)$ and $u \to v$ if u and v are contained in a passing member $C \in \mathcal{F}^*$ in the ordering $y \ldots u \ldots v \ldots xy$, or both are contained in a free member.

It will be proved that there is a directed path Q (called an **admissible dipath**) in A from x to y since w is admissible in G (Lemma 12.1.6). By applying Lemma 12.1.6, every member of $\mathcal{F}^*$ must be involved in Q, otherwise, it is "removable."

Among all possible faithful even subgraph covers $\mathcal{F}^*$, choose one such that a corresponding admissible dipath Q is *as short as possible*. (See Figure 12.2.) The final contradiction is to find another faithful cover, for which an admissible dipath is *shorter* than Q.

In the dipath Q, there is a segment $z_{b-2}z_{b-1}z_b$ in Q such that the arc $z_{b-1}z_b$ comes from a passing member C_b, and the arc $z_{b-2}z_{b-1}$ comes from a free member C_{b-1} (that is, z_{b-1}, z_b are along the passing circuit C_b in the ordering $y \ldots z_{b-1} \ldots z_b \ldots xy$, and both $z_{b-1}, z_{b-2} \in V(C_{b-1})$). (See Figures 12.1 and 12.2.)

Consider the subgraph H of G induced by edges of C_b and C_{b-1} (see Figure 12.3). A technical lemma (Lemma 12.1.7) will show that $\{C_b, C_{b-1}\}$ can be replaced with another set of even subgraphs $\{D_0', D_1', \ldots\}$ and consequently, in the revised faithful covering, the admissible dipath is shorter than the shortest dipath Q.

⋆ Notation and lemmas

Admissible dipath

In the following Definitions and Lemmas 12.1.3 – 12.1.6, we use the same notation as follows.

Let G be a graph and $e_0 = x_0 y_0 \in E(G)$. Let w^* and w be eulerian weights of G such that $w(e) = w^*(e)$ for every $e \neq e_0$ and $w(e_0) = w^*(e_0) + 2$. And assume that (G, w^*) has a faithful even subgraph cover $\mathcal{F}^*$.

Definition 12.1.3   A faithful even subgraph cover $\mathcal{F}^*$ of (G, w^*) is e_0-special if

(1) each member not containing e_0 is a connected even subgraph and is called a **free member** or **free even subgraph**,

(2) each member containing e_0 is a circuit and is called a **passing member** or **passing circuit**,

(3) $E(C') \cap E(C'') \neq \emptyset$, for each pair of free members C', C'' with $V(C') \cap V(C'') \neq \emptyset$ (that is, each free member is "maximal" in some sense).

Definition 12.1.4   For an e_0-special faithful cover $\mathcal{F}^*$ of (G, w^*), construct an auxiliary directed graph $A(G, w^*, e_0 = x_0 y_0; \; \mathcal{F}^*)$ with the vertex set $V(G)$ as follows: for each passing member (circuit) C, the path of $C - e_0$ is replaced with a transitive tournament from y_0 to x_0 (with $C - e_0$ as the directed Hamilton path from y_0 to x_0); for each free member (connected even subgraph) C', it is replaced with a complete directed graph (two opposite arcs joining each pair of vertices).

Definition 12.1.5   An arc of $A(G, w^*, e_0 = x_0 y_0; \mathcal{F}^*)$ induced by a member C of $\mathcal{F}^*$ is called a **C-arc**. Furthermore, an arc induced by a free member (or passing member) of $\mathcal{F}^*$ is called a **free arc** (or **passing arc**, respectively).

Lemma 12.1.6   (Seymour [205]) *The auxiliary graph $A(G, w^*, e_0 = x_0 y_0; \; \mathcal{F}^*)$ has a directed path from x_0 to y_0 if and only if w is admissible in G.*

The directed path in $A(G, w^*, e_0 = x_0y_0; \mathcal{F}^*)$ from x_0 to y_0 is called an admissible dipath.

Recall that a weight w is admissible in G if, for every edge e and every edge-cut T containing e,

$$w(e) \geq \frac{w(T)}{2}.$$

An edge-cut T is critical with respect to w if there is an edge $e \in T$ such that

$$w(e) = \frac{w(T)}{2}.$$

Proof Let $r = w^*(e_0)$. There are precisely r passing circuits.

"$\Rightarrow$": Suppose that w is not admissible in G. Then there is an edge-cut $T = (X, Y)$ such that $e_0 \in T$ and $w(e_0) > \frac{1}{2}w(T)$. Since w^* is admissible, the cut T is critical with respect to w^*:

$$w(e_0) = r + 2 \quad \text{while} \quad w(T - e_0) = w^*(T - e_0) = r.$$

Since $w^*(e_0) = r = w^*(T - e_0)$ is the number of passing circuits, every passing circuit must pass through $T - e_0$ precisely once, and no free member passes through T. Let $T = (X, Y)$ with $x_0 \in X$ and $y_0 \in Y$. In the auxiliary graph $A(G, w^*, e_0 = x_0y_0; \mathcal{F}^*)$, each arc joining X and Y must be in the same direction: from Y to X, since every passing circuit passes through $T - e_0 = (X, Y) - e_0$ only once. This contradicts the existence of a directed path from x_0 to y_0 in the auxiliary graph.

"$\Leftarrow$": Suppose that there is no directed path from x_0 to y_0 in the auxiliary graph $A(G, w^*, e_0 = x_0y_0; \mathcal{F}^*)$. Let X be the subset of all vertices that are reachable from x_0 in $A(G, w^*, e_0 = x_0y_0; \mathcal{F}^*)$ (and let $Y = V(G) - X$ and $T = (X, Y)$). By the construction of $A(G, w^*, e_0 = x_0y_0; \mathcal{F}^*)$, every arc joining X and Y is a passing-arc in the same direction: from Y to X. Furthermore, $T - e_0$ contains precisely one edge of each passing member, for otherwise, there would be an arc from X to Y in the auxiliary graph. That is, $w^*(T - e_0)$ is the number of the passing members, which is r and $r = w^*(e_0)$. Thus, T is critical for w^*. This contradicts that w is admissible since $w(e_0) = w^*(e_0) + 2 = r + 2$. $\square$

A technical lemma for adjustment of faithful cover

As described in the outline, the following lemma will be used for local adjustment of a faithful even subgraph cover. In the application of this lemma, $e_0 = v_r v_0$ where $v_r = x_0$, $v_0 = y_0$, and D_0 is a passing circuit while D_1 is a free even subgraph.

Lemma 12.1.7 *Let H be a bridgeless graph and w_H be an eulerian $(1, 2)$-weight of H with a faithful even subgraph cover $\mathcal{F} = \{D_0, D_1\}$ and the following conditions (see Figure 12.3):*

(1) D_0 is a circuit, D_1 is a connected even subgraph;

(2) $D_0 = v_0 \ldots v_r v_0$ contains two vertices v_{a_1}, v_{a_2} with $0 \leq a_1 < a_2 \leq r$; let $P'' = v_{a_2} \ldots v_r$ be the segment of D_0 not containing v_0, and $P' = v_0 \ldots v_{a_2}$ be the segment of D_0 not containing v_r;

(3) $v_{a_1} \in D_0 \cap D_1$;

(4) $V(D_1) \cap V(P'') \neq \emptyset$;

(5) Case 1, if $E_{w_H=2} \neq \emptyset$ then $E_{w_H=2} \subseteq E(P'')$;

   Case 2, if $E_{w_H=2} = \emptyset$ then let u be a given vertex in D_1.

Then (H, w_H) has another e_0-special faithful even subgraph cover $\mathcal{F}' = \{D'_0, \ldots, D'_q\}$ such that

(a) either

$$|\mathcal{F}'| \geq 3,$$

(b) or

$$|\mathcal{F}'| = 2, \qquad e_0 = v_r v_0 \in E(D'_0),$$
$$\text{in Case 1} \quad v_{a_2} \in V(D'_1) \text{ and } E_{w_H=2} \subseteq E(D'_1),$$
$$\text{in Case 2} \quad v_{a_2}, u \in V(D'_1).$$

Proof Let (H, w_H) be a counterexample with $|E(\overline{H})|$ as small as possible.

I. If $d(v_{a_2}) > 2$ then the faithful cover $\mathcal{F}$ itself satisfies the conclusion (b) of the lemma. Hence, it is obvious that any vertex other than v_{a_2} and u (in Case 2) must be of degree ≥ 3 in this smallest counterexample H.

And H is *2-connected*.

Since this lemma is not a minor-closed result, one may apply vertex splitting at some high degree vertex as long as the conditions described in lemma are preserved. Thus, we have the following two claims.

II. We claim that D_1 *is a circuit*.

Since D_1 is a connected even subgraph, one may split any high degree vertex z of D_1 by preserving an Euler tour of D_1. The resulting graph H' satisfies all conditions of the lemma, and the suppressed graph $\overline{H'}$

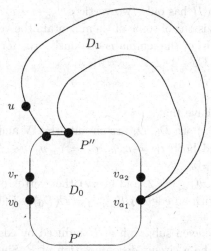

Figure 12.3 (H, w_H) *and an even subgraph cover* $\{D_0, D_1\}$

is smaller than $\overline{H}$. Thus, (H', w_H) has another faithful circuit cover $\mathcal{F}_1$ satisfying the conclusions of the lemma. With no confusion, $\mathcal{F}_1$ can be further considered as a faithful *even subgraph* cover of (H, w_H) satisfying the conclusions of the lemma. This contradicts the assumption that (H, w_H) is a counterexample.

Hence, *no vertex of H is of degree > 4.*

III. We claim that *(i) for Case 1 $(E_{w_H=2} \neq \emptyset)$,*

$$d(v_{a_1}) = 4 \ and \ d(z) \leq 3 \ for \ every \ z \in V(H) - \{v_{a_1}\};$$

(ii) for Case 2 $(E_{w_H=2} = \emptyset)$,

$$d(v_{a_1}) = d(v_r) = 4 \ and \ d(z) \leq 3 \ for \ every \ z \in V(H) - \{v_{a_1}, v_r\}.$$

Suppose that $z \in V(D_1)$ is a degree 4 vertex of H other than v_{a_1}, v_r. One may split the vertex z by preserving the circuits D_0, D_1, the suppressed resulting graph is smaller than $\overline{H}$. By the same argument as in II, we obtain a faithful even subgraph cover of (H, w_H) satisfying the conclusions of the lemma. This is a contradiction.

Note that, by conditions (2) and (5), v_{a_1} is not incident with any edge of $E_{w_H=2}$. Hence, $d(v_{a_1}) = 4$. It is similar for v_r in Case 2.

IV. We claim that Case 2 does not occur.

Suppose that Case 2 does occur (that is, $E_{w_H=2} = \emptyset$). Then, by III, the suppressed graph $\overline{H}$ has only two vertices v_{a_1} and v_r with four parallel edges (the subdivisions of some of them contain the vertices u and v_{a_2}). The conclusion (b) of the lemma is obviously true for this case. Hence, we only consider Case 1 ($E_{w_H=2} \neq \emptyset$).

V. Summary

Therefore, we have that

(a) $E_{w_H=2} \neq \emptyset$, both D_0, D_1 are circuits (by IV and II),

(b) $a_1 = 0$ and both $e_0 = v_r v_0$, $e_a = v_{a_1} v_{a_2}$ ($\in E_{w_H=1}$) are edges incident with $v_0 = v_{a_1}$,

(c) $d(v_{a_1}) = 4$, $d(v_{a_2}) = 2$, and every other vertex of H is of degree 3 and is incident with an edge of $E_{w_H=2} = E(D_0) \cap E(D_1)$.

VI. Let R be the even subgraph of H induced by edges of $E_{w_H=1}$ and let $\{Q_1, \ldots, Q_s\}$ be a circuit decomposition of R. Since v_{a_1} is the only degree 4 vertex in R, by V, we have that $s \geq 2$. Let $e_0 \in Q_1$, and $E(v_{a_1}) \subseteq E(Q_1) \cup E(Q_2)$. There are two cases: $e_a = v_0 v_{a_2} \in E(Q_1)$ or $e_a \in E(Q_2)$.

VII. Color $E(H)$ as follows, $c : E(H) \mapsto \{Red, Blue, Purple\}$:

$$c(e) = \begin{cases} Red & \text{if } e \in E(D_0) \cap E_{w_H=1} = E(D_0) - E(D_1) \\ Blue & \text{if } e \in E(D_1) \cap E_{w_H=1} = E(D_1) - E(D_0) \\ Purple & \text{if } e \in E(D_0) \cap E(D_1) = E_{w_H=2}. \end{cases}$$

Note that both e_0 and e_a are colored with red.

Subcase 1-1. $e_a \in Q_1$. Modify the coloring by interchanging the red-blue colors along the circuit Q_2. Then, under the new coloring, the set of red-purple bi-colored edges induces an even subgraph $D_0 \triangle Q_2$, the set of blue-purple bi-colored edges induces another even subgraph $D_1 \triangle Q_2$, and v_0 is incident with four red edges ($\in D_0 \triangle Q_2$). With a non-trivial circuit decomposition of $D_0 \triangle Q_2$, we obtain an e_0-special faithful even subgraph cover with at least three members. This proves the conclusion (a).

Subcase 1-2. $e_a \in Q_2$. Similarly, modify the coloring by interchanging the red-blue colors along the circuit Q_2. Then, under the new coloring, the set of red-purple bi-colored edges induces an even subgraph $D_0 \triangle Q_2$, the set blue-purple bi-colored edges induces another even subgraph $D_1 \triangle Q_2$. Now, e_a is colored with blue. Hence, conclusion (b) is proved where $\{D_0 \triangle Q_2, D_1 \triangle Q_2\} = \{D'_0, D'_1\}$. $\qquad \square$

⋆ Proof of Theorem 12.1.1

Let (G, w) be a counterexample to the theorem with

$$\sum_{e \in E(G)} (w(e) + 1)^2 \qquad (12.1)$$

as small as possible.

Part one

I. As a smallest counterexample (with respect to (12.1)), $w(e) > 0$ for every $e \in E(G)$.

II. We claim that $|E_{w \geq 2}| \geq 2$.

Suppose that $e = xy$ is the only edge with weight ≥ 2. Apply Menger's Theorem (Theorem A.1.3) to the graph $G - e$, there are k edge-disjoint paths $P_1, \ldots, P_k$ joining x and y where $w(e) = k$. Hence,

$$\{P_\mu + e: \mu = 1, \ldots, k\} \cup \{G - e - \cup_{\mu=1}^{k} E(P_\mu)\}$$

is a faithful even subgraph cover of (G, w).

III. If we are able to find an eulerian weight w_0 of G such that
    (i) $0 \leq w_0(e) \leq w(e)$ and the total weight $w_0(G)$ is smaller,
    (ii) w_0 is admissible,
    (iii) (G, w_0) has no faithful circuit cover,
then we can apply the theorem to the smaller weight w_0 and the theorem is proved.

IV. Let

$$p = \min\{w(e): \ w(e) \geq 2\}, \quad e_0 = x_0 y_0 \in E(G) \text{ with } w(e_0) = p. \quad (12.2)$$

V. We claim that (G, w^*) *has a faithful cover* where $w^* : E(G) \mapsto \mathbb{Z}^+$ is defined as follows

$$w^*(e) = \begin{cases} w(e) & \text{if } e \neq e_0 \\ w(e) - 2 & \text{if } e = e_0. \end{cases}$$

Suppose that (G, w^*) *does not have a faithful cover.*

We consider the weighted graph (G^S, w^S): G^S is obtained from G by adding a new edge e_0' parallel with e_0. (See Figure 12.4.) And $w^S : E(G^S) \mapsto \mathbb{Z}^\star$:

$$w^S(e) = \begin{cases} w(e) & \text{if } e \neq e_0, e_0' \\ 1 & \text{if } e = e_0' \\ w(e) - 1 & \text{if } e = e_0. \end{cases}$$

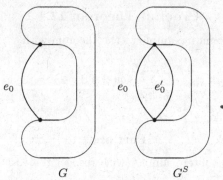

Figure 12.4 G^S *is obtained from* G *by adding a new edge* e'_0 *parallel to* e_0

It is obvious that w^S is an admissible eulerian weight of G^S since w is admissible in G (Exercise 12.2). Note that w^S has a smaller value of (12.1) than w does. Therefore, there are two cases:

(1) either (G^S, w^S) has a faithful cover,

(2) or (G^S, w^S) is a contra pair (the theorem can be applied here).

If (1) occurs, either (G, w) or (G, w^*) has a faithful cover. A contradiction. So, only (2) happens.

By applying the theorem to the smaller (G^S, w^S), there is an admissible eulerian weight $w_1^S : E(G^S) \mapsto \{0, 1, 2\}$ such that (G^S, w_1^S) is a contra pair where

$$0 \leq w_1^S(e) \leq \min\{2, w^S(e)\}$$

for every edge $e \in E(G^S)$. If $w_1^S(e'_0)$ or $w_1^S(e_0) = 0$, then w_1^S can be considered as a weight of G and therefore the theorem is proved. Hence, we assume that

$$w_1^S(e'_0) = 1 \quad \text{and} \quad 2 \geq w_1^S(e_0) \geq 1.$$

Further define two new eulerian weights $(w_1^S)^{-D}$, $(w_1^S)^M$ of G as follows:

$$(w_1^S)^M(e) = \begin{cases} w_1^S(e) & \text{if } e \neq e_0 \\ w_1^S(e) + 1 & \text{if } e = e_0, \end{cases}$$

$$(w_1^S)^{-D}(e) = \begin{cases} w_1^S(e) & \text{if } e \neq e_0 \\ w_1^S(e) - 1 & \text{if } e = e_0 \end{cases}$$

(where $(w_1^S)^M$ is a *merging weight* induced from w_1^S by merging e_0, e'_0

together, and, $(w_1^S)^{-D}$ is a *digon-deletion weight* induced from w_1^S by deleting a digon $\{e_0, e_0'\}$).

Since (G^S, w_1^S) is a contra pair, it is obvious that neither $(G, (w_1^S)^{-D})$ nor $(G, (w_1^S)^M)$ has a faithful cover (by Exercise 12.1). If $(w_1^S)^{-D}$ is admissible in G, then the weight $(w_1^S)^{-D}$ is what the theorem requested. So, $(w_1^S)^{-D}$ is not admissible in G. By Exercise 12.2, $(w_1^S)^M$ must be admissible since w_1^S is admissible. By III, $(w_1^S)^M = w$ for otherwise, one may further apply the theorem to the smaller one. But, $w(e_0) = 3$ while all other edges are of weight 1 or 2. This contradicts the definition of p (12.2) and the choice of e_0 unless no edge is of weight 2. This contradicts II.

VI. Summary of part one. The main conclusion of this part is the claim in V: (G, w^*) *has a faithful even subgraph cover.*

Part two

VII. Let (G, w^*) be defined in V. By V, let $\mathcal{F}$ be any e_0-special faithful even subgraph cover of (G, w^*) as defined in Definition 12.1.3. We claim the following:

(1) there are precisely $(p-2)$ passing circuits (p is defined in (12.2) of IV);

(2) $w(e) = 1$ or $= p$, for every $e \in E(G)$;

(3) for every e with $w(e) = p$ and $e \neq e_0$, e is contained in precisely two free members of $\mathcal{F}$ and all passing circuits;

(4) for every $C \in \mathcal{F}$, there must be a C-arc in every admissible dipath of the auxiliary graph $A(G, w^*, e_0 = x_0 y_0; \ \mathcal{F})$;

(5) for free members $C_i, C_j \in \mathcal{F}$,

$$C_i \cap C_j \neq \emptyset \text{ if and only if } i = j \pm 1.$$

VII-i. The claim (1) is straightforward by the definition of passing circuit and (12.2).

VII-ii. The proof of claim (4).

Suppose that there is a member $C \in \mathcal{F}$ and there is an admissible dipath Q in the auxiliary graph $A(G, w^*, e_0 = x_0 y_0; \ \mathcal{F})$ (Definition 12.1.4) such that Q contains no C-arc. Then let $w_2 : E(G) \mapsto \mathbb{Z}^+$:

$$w_2(e) = \begin{cases} w(e) & \text{if } e \notin E(C) \\ w(e) - 1 & \text{if } e \in E(C) \end{cases}$$

where w_2 is an *admissible* eulerian weight (by Lemma 12.1.6). Note that (G, w_2) has no faithful cover since (G, w) does not. Thus, we may apply the theorem to the smaller one (G, w_2): G has an admissible eulerian weight satisfying the theorem.

VII-iii. Proofs of claims (2) and (3).

Choose $Q_{\mathcal{F}} = z_0 z_1 \ldots z_{s-1} z_s$ to be a shortest admissible dipath of $A(G, w^*, e_0 = x_0 y_0; \ \mathcal{F})$ from $x_0 = z_0$ to $y_0 = z_s$.

By the construction of $A(G, w^*, e_0 = x_0 y_0; \ \mathcal{F})$ and the choice of $Q_{\mathcal{F}}$, no vertex of G is contained in more than two free members of $\mathcal{F}$, for otherwise the dipath can be shortened. Thus, every edge is contained in at most p members of $\mathcal{F}$: at most two free members, and at most $p - 2$ passing circuits. Hence, $w^*(e) \leq p$ for every $e \neq e_0$. By the choice of e_0 (Equation (12.2)), $w(e) = p$ or 1, and this proves claims (2) and (3).

An immediate corollary (observation from VII-(4)) is that no member of $\mathcal{F}$ is "removable." And (5) is trivial by Definition 12.1.3.

VIII. Among all e_0-special faithful even subgraph covers of (G, w^*), we choose $\mathcal{F}_0$ such that

(α) $|\mathcal{F}_0|$ is as large as possible,

(β) subject to (α), a corresponding *admissible dipath* $Q_{\mathcal{F}_0}$ of $A(G, w^*, e_0 = x_0 y_0; \ \mathcal{F}_0)$ *is as short as possible.*

Let $Q_{\mathcal{F}_0} = z_0 \ldots z_s$ ($z_0 = x_0, z_s = y_0$).

Let $z_{j-1} z_j$ be a C_j-arc. The admissible dipath corresponds to a sequence of even subgraphs, $\{C_1, \ldots, C_s\}$, in which, by VII and the choice of the dipath, each free member appears precisely once, and each passing member appears at least once.

IX. Choose a smallest $b \in \{1, \ldots, s\}$ such that $z_{b-1} z_b$ is a passing arc. By the construction of $A(G, w^*, e_0 = x_0 y_0; \ \mathcal{F}_0)$ and the definition of $Q_{\mathcal{F}_0}$ (Definition 12.1.4), $b > 1$ since the first arc of $Q_{\mathcal{F}_0}$ must be a free arc. Let C_{b-1} be the free member of $\mathcal{F}_0$ that the arc $z_{b-2} z_{b-1}$ is a C_{b-1}-arc, and C_b be the passing member of $\mathcal{F}_0$ that the arc $z_{b-1} z_b$ is a C_b-arc.

Construct a subgraph H of G induced by $E(C_b) \cup E(C_{b-1})$ (we are to apply Lemma 12.1.7 to the graph H). (See Figure 12.3.)

Let $C_b = v_0 \ldots v_r v_0$ where $v_0 = y_0, v_r = x_0$. With no confusion, we may choose the smallest $a_1, a_2 \in \{0, \ldots, r\}$ such that v_{a_1} is z_{b-1} and v_{a_2} is z_b ($a_1 < a_2$). Let

$$w_H \leftarrow w_{\{C_{b-1}, C_b\}} \text{ (the eulerian } (1, 2)\text{-weight induced by } \{C_b, C_{b-1}\}),$$
$$D_0 \leftarrow C_b, \quad D_1 \leftarrow C_{b-1},$$

$u \leftarrow z_{b-2}$.

Add one more induction requirement:

(γ) *subject to* (α) *and* (β), *choose* $H = G[E(C_b) \cup E(C_{b-1})]$ *with the least number of edges.*

X. Let us verify all conditions of Lemma 12.1.7 one by one. There is no need to verify conditions (1), (2) and (3) since they are all given in definitions (and descriptions). We only need to verify conditions (4) and (5).

Along the segment $P' = v_0 C_b v_{a_2}$, we first claim that

$$E(P') \cap E(C_{b-1}) = \emptyset. \tag{12.3}$$

For otherwise, by VII-(2), $w(e) = p$ for each $e = v_h v_{h+1} \in E(P') \cap E(C_{b-1})$, and by VII-(3), e is contained in another free member C_j ($j \neq b - 1, b$). For either $j < b - 1$ or $j > b$, the dipath $Q_{\mathcal{F}_0}$ can be shortened by taking a short cut $z_0 \ldots z_{j-1} v_h z_b \ldots z_s$ if $j < b - 1$, or a short cut $z_0 \ldots z_{b-2} v_h z_j \ldots z_s$ if $j > b$ (contradicts (β)).

Along the segment $P'' = v_{a_2} C_b v_r$, we further claim that

$$V(P'') \cap V(C_{b-1}) \neq \emptyset.$$

If $b - 1 = 1$, then $v_r = x_0 \in V(C_{b-1}) \cap V(C_b)$. If $b - 1 > 1$, then, by Definition 12.1.3, let $e' \in E(C_{b-1}) \cap E(C_{b-2})$. By VII-(3), e' is also contained in the passing member C_b. Hence, by the above claim (Equation (12.3)), $e' \in E(P'') \cap E(C_b)$.

Final step. Now, we are ready to apply Lemma 12.1.7 to (H, w_H): (H, w_H) has a faithful cover $\mathcal{F}'$ (different from $\{C_b, C_{b-1}\}$). Let $\mathcal{F}'_0 = \mathcal{F}_0 - \{C_b, C_{b-1}\} + \mathcal{F}'$.

If Lemma 12.1.7-(a) occurs, then $|\mathcal{F}'| \geq 3$ and it contradicts the choice of $\mathcal{F}_0$. If $\mathcal{F}'_0$ remains e_0-special (according to Definition 12.1.3), then it contradicts (α). If some member of $\mathcal{F}'$ can be merged into some other free member $C_{b'}$ as a larger connected even subgraph, then, by VII-(5), the free member $C_{b'}$ must be C_{b-2} or C_{b+1}, not any others. Hence, $\mathcal{F}'_0$ is not smaller than $\mathcal{F}_0$, but it contradicts (γ).

If Lemma 12.1.7-(b) occurs but not (a), then $\mathcal{F}' = \{D'_0, D'_1\}$ with $e_0 \in E(D'_0)$ and $v_{a_2} \in V(D'_1)$. The dipath $Q_{\mathcal{F}'_0}$ will be reduced as follows.

For case 1, the new free member D'_1 contains the vertex $v_{a_2} = z_b$, and a weight 2 edge e' of (H, w_H). Note that, by VII-(3), the edge e' is a weight p edge in (G, w). Therefore, $e' \in E(C_{b-2}) \cap E(C_{b-1})$. With a similar argument in X, the dipath $Q_{\mathcal{F}_0'}$ is shorter than $Q_{\mathcal{F}_0}$ since the C_b-arc $v_{a_1} v_{a_2}$ can be skipped.

The argument is similar for case 2.

Both contradict the choice (β) for $\mathcal{F}_0$ and $Q_{\mathcal{F}_0}$. □

12.2 Coverage reduction with fixed parity

It is easy to see that *every bridgeless graph has a circuit cover* (or by the results in Section 7.2). Is it possible that we can *find a circuit double cover by recursively reducing the coverage of a given circuit cover?* This section will address the possibility of this approach.

Recall Definition 7.5.1. Let $\mathcal{F}$ be an even subgraph cover of a graph G. For each edge $e \in E(G)$, let

$$ced_{\mathcal{F}}(e) = |\{C : C \in \mathcal{F}, e \in E(C)\}|$$

which is called the **edge depth** of e (with respect to $\mathcal{F}$). The **edge depth** of the graph G (with respect to $\mathcal{F}$) is

$$ced_{\mathcal{F}}(G) = \max\{ced_{\mathcal{F}}(e) : e \in E(G)\}.$$

Theorem 12.1.1 enables us to reduce the weight and identify a minimal contra pair from a highly weighted contra pair. The following open problem, which is another type of reduction problem, will lead us directly to the solution of the circuit double cover conjecture.

Conjecture 2.4.3 [259] *Let G be a bridgeless graph and $\mathcal{F}_1$ be circuit cover of G. If $ced_{\mathcal{F}_1}(G) = 4$, then there is another circuit cover $\mathcal{F}_2$ of G such that*

$$0 < ced_{\mathcal{F}_2}(e) \leq ced_{\mathcal{F}_1}(e), \quad ced_{\mathcal{F}_2}(e) \equiv ced_{\mathcal{F}_1}(e) \pmod 2$$

for every $e \in E(G)$, and there is an edge $e_0 \in E(G)$ such that

$$ced_{\mathcal{F}_2}(e_0) < ced_{\mathcal{F}_1}(e_0).$$

Proposition 12.2.1 *Conjecture 2.4.3 implies the circuit double cover conjecture.*

Proof Let $\mathcal{F}$ be a circuit cover of G such that

(1) $ced_{\mathcal{F}}(e) \leq 4$ and $ced_{\mathcal{F}}(e) \equiv 0 \pmod 2$ for every edge $e \in E(G)$,

(2) subject to (1), the total length of $\mathcal{F}$ is as small as possible.

By Theorem 7.3.4, such a circuit cover exists. If every edge is covered twice, then we have a circuit double cover for G. So, assume that an edge e_1 is covered four times, say, covered by C_1, C_2, C_3, C_4 of $\mathcal{F}$.

Let $G' = \overline{G[C_1 \cup \cdots \cup C_4]}$. If Conjecture 2.4.3 is true, one is able to find another circuit cover C' of G' satisfying the conjecture. Let $\mathcal{F}'$ be the cover obtained from $\mathcal{F}$ by replacing $\{C_1, \ldots, C_4\}$ with C'.

Since

$$0 < cedc'(e) \le cedc(e), \quad cedc'(e) \equiv cedc(e) \pmod 2$$

for every $e \in E(G')$, and there is an edge $e_0 \in E(G')$ that

$$cedc'(e_0) < cedc(e_0),$$

this contradicts the choice of $\mathcal{F}$ that the total length of $\mathcal{F}$ is minimum. $\square$

Remark. The maximum coverage $ced_{\mathcal{F}_1}(G)$ in Conjecture 2.4.3 cannot be reduced to 3: the Petersen graph has a circuit cover $\mathcal{F}$ with $ced_{\mathcal{F}}(P_{10}) = 3$ (Proposition B.2.29), but the coverage of $\mathcal{F}$ cannot be reduced by maintaining the parity (Propositions B.2.26).

12.3 Exercises

Definition 12.3.1 Let G be a bridgeless graph and $e_0 \in E(G)$. And let G^S be the graph obtained from G by adding a new edge e_0' parallel to e_0 (G^S is a split of G at e_0). Let $\phi : E(G^S) \mapsto \mathbb{Z}^*$ be an eulerian weight of G^S with

$$\phi(e_0) \ge 1, \quad \phi(e_0') = 1.$$

Define two eulerian weights ϕ^M and ϕ^{-D} of G as follows:

$$\phi^M(e) = \begin{cases} \phi(e) & \text{if } e \ne e_0 \\ \phi(e)+1 & \text{if } e = e_0; \end{cases}$$

$$\phi^{-D}(e) = \begin{cases} \phi(e) & \text{if } e \ne e_0 \\ \phi(e)-1 & \text{if } e = e_0 \end{cases}$$

(where ϕ^M is a *merging weight induced from* ϕ and ϕ^{-D} is a *digon-deletion weight induced from* ϕ).

Exercise 12.1 (G^S, ϕ) has a faithful cover if and only if at least one of (G, ϕ^M), (G, ϕ^{-D}) has a faithful cover.

Exercise 12.2 ϕ is admissible in G^S if and only if at least one of ϕ^M, ϕ^{-D} is admissible in G.

13
Orientable cover

Attempts to prove the CDC conjecture have led to various conjectured strengthenings, such as the *faithful circuit cover problem* (Problem 2.1.4), *strong circuit double cover problem* (Conjecture 1.5.1), *even covering problems* (Conjectures 2.4.1 and 2.4.2), 5-*even subgraph double cover problem* (Conjecture 4.1.4), etc. Verification of any of those stronger problems will imply the CDC conjecture.

In this chapter, we present another type of variation of the double cover problem: *directed circuit double covering*. These are, in general, much stronger than the CDC problem. And some of them have already been completely characterized.

Historically, the paper by Tutte [229] on *orientable circuit double cover* is the earliest published article related to the CDC problem.

13.1 Orientable double cover

Definition 13.1.1 Let $G = (V, E)$ be a graph and D be an orientation of $E(G)$. A directed even subgraph H of the directed graph $D(G)$ is a subgraph of $D(G)$ such that for each vertex v of H, the indegree of v equals the outdegree of v.

Definition 13.1.2 (1) Let $\mathcal{F} = \{C_1, \ldots, C_r\}$ be an even subgraph double cover of a graph G. The set $\mathcal{F}$ is an orientable even subgraph double cover if there is an orientation D_μ on $E(C_\mu)$, for each $\mu = 1, \ldots, r$, such that

(i) $D_\mu(C_\mu)$ is a directed even subgraph, and

(ii) for each edge e contained in two even subgraphs C_α and C_β ($\alpha, \beta \in \{1, \ldots, r\}$), the directions of $D_\alpha(C_\alpha)$ and $D_\beta(C_\beta)$ are opposite on e.

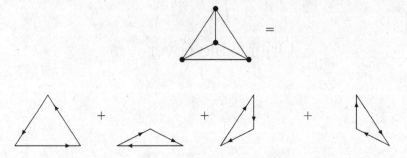

Figure 13.1 *An orientable 4-even subgraph double cover of* K_4

(2) An orientable k-even subgraph double cover $\mathcal{F}$ is an orientable even subgraph double cover consisting of k members. (See Figure 13.1.)

We discuss the relations of the following properties of graphs. Let G be a bridgeless graph and k be an integer.

(P1) *The graph G admits a nowhere-zero k-flow.*

(P2) *The graph G has an orientable k-even subgraph double cover.*

(P3) *There exists an orientable surface (2-manifold) Σ on which the graph G is k-face-colorable.*

The following theorem is a combination of some earlier results obtained by Tutte [229], and later reformulated and generalized by Jaeger and Younger (see [130] and [250]).

Theorem 13.1.3   *Let G be a bridgeless graph and k be a positive integer.*

(1) Generally,

$$(P3) \Rightarrow (P2) \Rightarrow (P1).$$

(2) If G is planar and the surface Σ is a sphere, then

$$(P3) \Leftrightarrow (P2) \Leftrightarrow (P1).$$

(3) If G is cubic, then

$$(P3) \Leftrightarrow (P2) \Rightarrow (P1).$$

(4) If $k \in \{2, 3, 4\}$

$$(P3) \Rightarrow (P2) \Leftrightarrow (P1).$$

In this chapter, we pay most attention to the relation between properties (P1) and (P2). For the relations between (P3) and others, and properties of planar graphs, readers are referred to Theorem 2.5.3 in [259].

It is always true that (P2) $\Rightarrow$ (P1) (see Lemma 13.1.5). But there are only a few verified cases for (P1) $\Rightarrow$ (P2). For (P2) $\Leftrightarrow$ (P1), it is trivial if $k = 2$. And the cases of $k = 3$ or 4 will be presented in Theorems 13.1.6 and 13.1.7.

Some conjectures stronger than *Tutte's 5-flow conjecture* (Conjecture C.1.10) and *the CDC conjecture* have been proposed. With Theorem 13.1.3, the following conjecture implies the 5-flow conjecture and the CDC conjecture (see Lemma 13.1.5).

Conjecture 13.1.4 (Archdeacon [8] and Jaeger [131]) *Every bridgeless graph has an orientable 5-even subgraph double cover.*

Together with Tutte's 5-flow conjecture, Conjecture 13.1.4 implies that (P1) $\Leftrightarrow$ (P2) for all bridgeless graphs.

Lemma 13.1.5 *If a graph G has an orientable k-even subgraph double cover, then G admits a nowhere-zero k-flow.*

Proof Let $\{C_1, \ldots, C_k\}$ be an orientable k-even subgraph double cover, and D_μ be the orientation of C_μ ($\mu = 1, \ldots, k$). Let (D_μ, f_μ) be a nonnegative 2-flow of G with $supp(f_\mu) = E(C_\mu)$ ($\mu = 1, \ldots, k$).

Let D be an arbitrary orientation of G. Let (D, f'_μ) be a 2-flow on C_μ such that $f'_\mu(e) = f_\mu(e)$ if D and D_μ agree with each other on the edge e, and $f'_\mu(e) = -f_\mu(e)$ if D and D_μ are opposite on the edge e. Then,

$$\left(D, \sum_{\mu=1}^{k} \mu \, f'_\mu \right)$$

is a nowhere-zero k-flow of G. $\square$

⋆ Orientable 3-even subgraph double covers

The following theorem was originally proved by Tutte [229] for cubic, bipartite graphs and reformulated and generalized by Jaeger (Theorem 13.2.8).

Theorem 13.1.6 (Tutte [229]) *A graph G admits a nowhere-zero 3-flow if and only if G has an orientable 3-even subgraph double cover.*

Proof "$\Leftarrow$" is proved in Lemma 13.1.5.

"$\Rightarrow$": Assume that G admits a nowhere-zero 3-flow. By Corollary C.2.5, let D be an orientation of G such that G is covered by two directed even subgraphs C_1, C_2 of $D(G)$. Let $\widetilde{C_2}$ be the directed even subgraph obtained from C_2 by reversing the direction of every arc. Let C_3 be the directed even subgraph with the edge set $E(C_1 \bigtriangleup C_2)$ and the direction opposite to either C_1 or $\widetilde{C_2}$. It is obvious that each edge e of G is covered by precisely two directed even subgraphs of $\{C_1, \widetilde{C_2}, C_3\}$ with opposite directions along e. $\qquad\square$

$\star$ Orientable 4-even subgraph double covers

Tutte proved the following theorem in [229] for cubic graphs and later this was generalized by Jaeger (see [127] or see [130]) and Archdeacon [8].

Theorem 13.1.7 (Tutte [229], Jaeger [127], Archdeacon [8]) *A graph G admits a nowhere-zero 4-flow if and only if G has an orientable 4-even subgraph double cover.*

Proof "$\Leftarrow$": The "if" part is not only proved in Lemma 13.1.5, but also in Theorem 8.2.2 since an orientable 4-even subgraph double cover is also a 4-even subgraph double cover. So, we need only to prove the "only if" part.

"$\Rightarrow$": By Theorem 8.2.2 again, let $\{C_1, C_2, C_3\}$ be a 3-even subgraph double cover of G. Let D be an arbitrary orientation of G. By Theorems C.1.8, let (D, f_j) (for $j = 1, 2, 3$) be a 2-flow of G with $supp(f_j) = E(C_j)$. Furthermore, let

$$\phi_0 = \frac{f_1 + f_2 + f_3}{2}, \quad \phi_1 = \frac{f_1 - f_2 - f_3}{2},$$

$$\phi_2 = \frac{-f_1 + f_2 - f_3}{2}, \quad \phi_3 = \frac{-f_1 - f_2 + f_3}{2}.$$

Since $\{C_1, C_2, C_3\}$ is an even subgraph double cover of G, each (D, ϕ_μ) $(\mu = 0, 1, 2, 3)$ is a 2-flow of G such that *each edge is contained in precisely two supports of* $\{\phi_0, \phi_1, \phi_2, \phi_3\}$, and *if* $e \in supp(\phi_i) \cap supp(\phi_j)$ *then* $\phi_i(e) = -\phi_j(e)$. For $i = 0, 1, 2, 3$, let D_i be the orientation obtained from D by reversing the orientation of D on each edge e with $\phi_i(e) = -1$, and let $S_i = supp(\phi_i)$. Then $\{D_0(S_0), D_1(S_1), D_2(S_2), D_3(S_3)\}$ is an orientable 4-even subgraph double cover of G since on each edge e the two directed even subgraphs containing e have opposite directions. $\qquad\square$

13.2 Circular double covers and modulo orientations

⋆ Circular double covers

Definition 13.2.1 An even subgraph double cover $\mathcal{F} = \{C_0, \ldots, C_{r-1}\}$ of a graph G is called a circular r-even subgraph double cover of G if $E(C_i) \cap E(C_j) \neq \emptyset$ if and only if $|j - i| \equiv 1 \pmod{r}$.

The concept of circular even subgraph double cover was introduced by Jaeger ([129], [131]). (It is also called C_r-flow in [131]).

Lemma 13.2.2 *Let G be a graph and t $(t \geq 2)$ be an integer. If G has a circular $(t+2)$-even subgraph double cover, then G has a circular t-even subgraph double cover.*

Proof Exercise 13.1. □

Theorem 13.2.3 (Jaeger [129], also see [131]) *Let t be an integer. Every $(4t)$-edge-connected graph has a circular $(2t+1)$-even subgraph double cover.*

Proof By Theorem A.1.6, the graph G contains $2t$ edge-disjoint spanning trees $T_1, \ldots, T_{2t}$. By Lemma A.2.15, let P_i be a parity subgraph of G contained in T_i for each $i = 1, \ldots, 2t$. By Lemmas A.2.12 and A.2.10, let $P_0 = G - \cup_{i=1}^{2t} E(P_i)$ which is also a parity subgraph of G. Then

$$\{P_i \cup P_{i+1} : i = 0, \ldots, 2t, \pmod{2t+1}\}$$

is a circular even subgraph double cover of G. □

Note that Theorem 8.1.2 is a special case $(t = 1)$ of Theorem 13.2.3 (by Theorem 8.2.2-(iv))

Definition 13.2.4 Let $\mathcal{F} = \{C_0, \ldots, C_{r-1}\}$ be a circular r-even subgraph double cover of a graph G. $\mathcal{F}$ is called a circular orientable r-even subgraph double cover if there is an orientation D_i on $E(C_i)$, for each $i = 0, \ldots, r-1$, such that
(1) $D_i(C_i)$ is a directed even subgraph,
(2) for each edge e contained in two even subgraphs C_i and C_{i+1} (mod r), the directions of $D_i(C_i)$ and $D_{i+1}(C_{i+1})$ are opposite on e.

Conjecture 13.2.5 (Jaeger [129], also see [131]) *Let t be an integer. Every $4t$-edge-connected graph has a circular orientable $(2t+1)$-even subgraph double cover.*

Conjecture 13.2.5 will be further discussed in terms of modulo orientations. (See Theorem 13.2.8-(1) and (4).)

⋆ Modulo orientations

Definition 13.2.6 Let G be a graph. An orientation D of G is called a mod-h-orientation if

$$d^+_{D(G)}(v) \equiv d^-_{D(G)}(v) \pmod{h}.$$

A graph G is mod-h-orientable if G has a mod-h-orientation.

The concept of mod-h-orientation was introduced by Jaeger [131], and is a generalization of the 3-flow problem that *G has a mod-3-orientation if and only if G admits a nowhere-zero 3-flow* (see Exercise C.30).

For an even integer h, mod-h-orientation is a trivial problem.

Proposition 13.2.7 *Let G be a graph and t be an integer. The following statements are equivalent:*
(1) G is mod-$(2t)$-orientable;
(2) G is an even subgraph.

Proof Exercise 13.3. □

By Proposition 13.2.7, we are only interested in mod-h-orientations with h being odd. In the following theorem, the equivalence of the statements (1), (2), (3) and (4) is due to Jaeger (see [131]), the equivalence of (1) and (5) is due to Lai [161].

Theorem 13.2.8 *Let G be a graph and t be an integer. The following statements are equivalent:*
(1) G is mod-$(2t + 1)$-orientable;
(2) G admits a mod-$(2t + 1)$-flow (D_1, f_1) such that $f_1(e) = \pm 1$ for each edge $e \in E(G)$;
(3) G admits an integer flow (D_2, f_2) such that $f_2(e) \in \{\pm t, \pm(t + 1)\}$ for each edge $e \in E(G)$;
(4) G has a circular orientable $(2t + 1)$-even subgraph double cover;
(5) G is a graph obtained from a subdivision of a $(2t + 1)$-regular bipartite graph and some circuits by vertex-identification;
(6) G admits a nowhere-zero circular $\frac{2t+1}{t}$-flow.

Proof (1) $\Rightarrow$ (2): Let D_0 be a mod-$(2t + 1)$-orientation of G and let $f_0 : E(G) \mapsto \{1\}$. Then (D_0, f_0) is a nowhere-zero mod-$(2t + 1)$-flow of G.

(2) $\Rightarrow$ (1): By Exercise C.12, let (D'_1, f'_1) be a mod-$(2t + 1)$-flow obtained from (D_1, f_1) by reversing the direction of each $e \in E_{f_1 = -1}$ and changing the flow value to be "1." Then D'_1 is a mod-$(2t + 1)$-orientation.

(2) $\Rightarrow$ (3): Let (D, f) be a mod-$(2t+1)$-flow with $f(e) = \pm 1$ for every edge e. Then (D, tf) is a mod-$(2t + 1)$-flow with $tf(e) = \pm t$ for every edge e. Note that $-t \equiv t + 1 \pmod{2t + 1}$. An integer flow is obtained by applying Theorem C.2.1.

(3) $\Rightarrow$ (2): Let (D, f) be a mod-$(2t+1)$-flow with $f(e) \in \{\pm t, \pm(t+1)\}$ for every edge e. Then $(D, 2f)$ is a mod-$(2t + 1)$-flow with $2f(e) = \pm 1$ $\pmod{2t + 1}$ for every edge e.

(4) $\Rightarrow$ (2): Let $\mathcal{F} = \{C_0, \ldots, C_{2t}\}$ be a circular $(2t+1)$-even subgraph double cover of a graph G and let D_i be an orientation on C_i, for each $i = 0, \ldots, 2t$, such that $D_i(C_i)$ is a directed even subgraph and, for each edge e contained in two even subgraphs C_i and C_{i+1} (mod $2t + 1$), the directions of $D_i(C_i)$ and $D_{i+1}(C_{i+1})$ are opposite on e. Let (D_i, f_i) be a positive 2-flow of C_i (for each $i = 0, \ldots, 2t$). Let D be an arbitrary orientation of G and let (D, f_i') be a 2-flow of G obtained from (D_i, f_i) by changing the sign of $f(e)$ if D_i and D have opposite directions on the edge e. Then (D, f) with

$$f = \sum_{i=0}^{2t} i f_i$$

is a nowhere-zero mod=$(2t+1)$-flow of G with $f(e) \equiv 1$ or -1 (mod $2t+$ 1).

(2) $\Rightarrow$ (4): By Theorem C.2.1, let (D_1', f_1') be an integer $(2t + 1)$- flow of G obtained from (D_1, f_1) such that $f_1'(e) \equiv f_1(e) \pmod{2t + 1}$. Let (D, f) be a positive flow obtained from (D_1', f_1') be reversing the directions of all edges $e \in E_{f_1' < 0}$ and changing the signs of their flow values (by Exercise C.12). Note that the ranges of these flows f_1', f are

$$f_1' \subseteq \{\pm 1, \pm 2t\}, \ f \subseteq \{1, 2t\}.$$

By Theorem C.2.4, let $\mathcal{F} = \{C_1, \ldots, C_{2t}\}$ be the set of directed even subgraphs of $D(G)$ such that each edge $e \in E(G)$ is contained in precisely $f(e)$ directed even subgraphs of $\mathcal{F}$. Note that $E(C_i) \cap E(C_j) = E_{f=2t}$ for each $i \neq j$.

For each $i \in \{1, \ldots, 2t - 1\}$, let $C_{i,i+1}$ be the directed even subgraph of G with $E(C_{i,i+1}) = E(C_i \triangle C_{i+1})$ and the direction of an edge $e \in E(C_{i,i+1})$ is the same as it is in C_{i+1} if $e \in E(C_{i+1})$ and is the opposite of what it is in C_i if $e \in E(C_i)$. Also, let C_{2t}' be the directed even subgraph with $E(C_{2t}') = E(C_{2t})$, but in the opposite direction. It is not hard to see that

$$\{C_{i,i+1} : i = 1, \ldots, 2t - 1\} \cup \{C_1, C_{2t}'\}$$

is a circular orientable $(2t + 1)$-even subgraph double cover of G.

(5) $\Rightarrow$ (2): Let H be a subdivision of a $(2t+1)$-regular bipartite graph $\overline{H}$ with a bi-partition $\{A, B\}$ and G is obtained from H and an even subgraph C by a series of vertex identification operations. Let D be an orientation of $\overline{H}$ and C such that C is a directed even subgraph, and each edge of $\overline{H}$ is oriented from A to B. Let $f : E(\overline{H}) \cup E(C) \mapsto \{1\}$. Obviously, (D, f) is a mod-$(2t+1)$-flow of $\overline{H} \cup C$ with $f(e) = 1$ for each $e \in E(\overline{H}) \cup E(C)$. Without causing any confusion, we may also consider (D, f) as a mod-$(2t+1)$-flow of $H \cup C$, as well as of G, with $f(e) = 1$ for each $e \in E(G)$ since G is obtained from $H \cup C$ by vertex identifications.

(1) $\Rightarrow$ (5): Let G be a graph and define

$$\sigma(G) = \sum_{v \in V(G)} |d(v) - (2t + 1)|.$$

For a $(2t + 1)$-regular graph G (with $\sigma(G) = 0$) having a mod-$(2t + 1)$-orientation D, it can be easily seen that G is a bipartite graph with a bi-partition $\{A, B\}$, where

$$A = \{v \in V(G) : d^-(v) = 0\}$$

and

$$B = \{v \in V(G) : d^+(v) = 0\}.$$

Thus, we assume that there is an integer N such that "(1) $\Rightarrow$ (5)" for graphs with $\sigma < N$.

Now let G be a graph with $\sigma(G) = N$ and having a mod-$(2t + 1)$-orientation D. Let $v \in V(G)$ with $d(v) \neq 2t + 1$.

If $d^+(v) > 0, d^-(v) > 0$, then let u_1v, vu_2 be two arcs of G (under the orientation D). The new graph $G' = G_{[v,\{u_1v,vu_2\}]}$ obtained from G by splitting edges u_1v, vu_2 away from v certainly has a smaller $\sigma(G')$ and by inductive hypothesis, $\overline{G'}$ (therefore G) is obtained from a subdivision of some $(2t+1)$-regular bipartite graph and an even subgraph by vertex identifications.

Now we assume that either $d^+(v) = 0$ or $d^-(v) = 0$ for each vertex $v \in V(G)$. Let $v \in V(G)$ with $d^+(v) = 0$. Obviously, $d^-(v) \equiv 0$ (mod $2t + 1$). Let $u_1v, u_2v, \ldots, u_{2t+1}v$ be arcs of $D(G)$. Then the graph $G'' = G_{[v,\{u_1v,u_2v,\ldots,u_{2t+1}v\}]}$ obtained from G by splitting $u_1v, u_2v, \ldots, u_{2t+1}v$ away from v has a smaller $\sigma(G'')$. By the inductive hypothesis,

G'' (therefore G) is obtained from a subdivision of some $(2t+1)$-regular bipartite graph and an even subgraph by vertex identifications.

(3) $\Leftrightarrow$ (6): This is the definition of *circular flow* (see Definition C.2.9). □

Corollary 13.2.9 *Let G be a graph and t ($t \geq 2$) be an integer. If the graph G has a mod-$(2t+1)$-orientation, then G has a mod-$(2t-1)$-orientation.*

Proof This is a corollary of Theorem 13.2.8-(6) and Lemma C.2.11.

By Theorem 13.2.8-(4), Lemma 13.2.2 is an alternative proof without using the concept of circular flow. □

13.3 Open problems

Conjecture 13.1.4 (Archdeacon [8] and Jaeger [131]) *Every bridgeless graph has an orientable 5-even subgraph double cover.*

Conjecture 13.2.5 (Jaeger [129], also see [131]) *Let t be an integer. Every 4t-edge-connected graph has a circular orientable $(2t+1)$-even subgraph double cover.*

Conjecture 13.3.1 (Jaeger [130]) *Every 2-connected graph has a circular 2-cell embedding on some orientable surface.*

Remark. Note that Theorems 13.1.6 and 13.1.7 show some partial results to Conjectures 13.3.1 and 13.1.4.

The following is another conjecture closely related to Conjecture 13.3.1.

Conjecture 13.3.2 (Robertson and Zha [47]) *If G is a 2-connected graph and has a circular 2-cell embedding on some non-orientable surface, then G has a circular 2-cell embedding on some orientable surface.*

Remark. Conjecture 13.3.2 is verified in [47] for cubic graphs embeddable on the projective plane. An immediate corollary of this result is that every projective planar graph has an orientable circuit double cover (by the splitting lemma (Theorem A.1.14)).

13.4 Exercises

Exercise 13.1   (Lemma 13.2.2) Let G be a graph and t ($t \geq 2$) be an integer. If G has a circular $(t + 2)$-even subgraph double cover, then G has a circular t-even subgraph double cover.

Exercise 13.2   (Jaeger [131]) If every 9-edge-connected graph is mod-5-orientable, then every bridgeless graph admits nowhere-zero 5-flow.

Exercise 13.3   (Proposition 13.2.7) Let G be a graph and t be an integer. The following statements are equivalent:
  (1) G is mod-$(2t)$-orientable;
  (2) G is even.

Exercise 13.4   (Corollary 13.2.9) Let G be a graph and t ($t \geq 2$) be an integer. If the graph G has a mod-$(2t + 1)$-orientation, then G has a mod-$(2t - 1)$-orientation.

14
Shortest cycle covers

14.1 Shortest cover and double cover

Definition 14.1.1 Let G be a bridgeless graph. An even subgraph cover of G with the minimum total length is called a **shortest cycle cover** of G.

One of the major open problems in this area is the following conjecture.

Conjecture 14.1.2 (Alon and Tarsi [1]; Jaeger, see [178]) *Every bridgeless graph with m edges has an even subgraph cover with total length at most $\frac{21}{15}m$.*

By Proposition B.2.30, the Petersen graph reaches the bound $\frac{21}{15}m$ of Conjecture 14.1.2. Note that we always use 15 as the denominator of the fractional coefficients (without reducing) since the Petersen graph has 15 edges and it is also easier for comparison of results.

This topic is included in this book because of its close relation with the circuit double cover conjecture (Theorem 14.1.4): *Conjecture 14.1.2 implies the circuit double cover conjecture.*

Recall Definition 7.5.1. Let $\mathcal{F}$ be an even subgraph (circuit) cover of a graph G and $e \in E(G)$. The **edge-depth** of $\mathcal{F}$ at the edge e is the number of members of $\mathcal{F}$ containing e and is denoted by $ced_{\mathcal{F}}(e)$. The **edge-depth** of $\mathcal{F}$ is $\max\{ced_{\mathcal{F}}(e) : e \in E(G)\}$ and is denoted by $ced_{\mathcal{F}}(G)$.

Notation 14.1.3 The total length of an even subgraph (circuit) cover $\mathcal{F}$ is denoted by $\mathcal{L}(\mathcal{F})$. The total length of a shortest cycle cover of a graph G is denoted by $SCC(G)$.

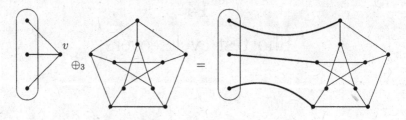

Figure 14.1 $\oplus_3$-*sum of P_{10} at a vertex v of G*

The following theorem, due to Jamshy and Tarsi [133], is the main theorem in this section.

Theorem 14.1.4   (Jamshy and Tarsi [133]) *If every bridgeless graph with m edges has a shortest cycle cover with the total length at most $\frac{21m}{15}$, then every bridgeless graph has a circuit double cover.*

Or, equivalently, *Conjecture 14.1.2 implies the circuit double cover conjecture.*

The key idea in the proof of Theorem 14.1.4 is based on the following two facts.

(1) (Lemma A.2.5) *A graph has a circuit cover with the edge-depth at most 2 if and only if the graph has a circuit double cover.*
(2) (Proposition B.2.30-(1) and (3)) *Every shortest circuit cover of the Petersen graph is of length 21 and is of edge-depth 2.*

Proof   We need to consider only cubic graphs, since a minimum counterexample to the circuit double cover conjecture is cubic (Lemma 1.2.2).

I. The construction of $G \otimes_3 P_{10}$. Let G^s be the graph obtained from G by inserting a degree-2 vertex into each edge. Let $G \otimes_3 P_{10}$ be the graph obtained from G^s by taking a $\oplus_3$-sum with a copy of the Petersen graph at every degree 3 vertex of G^s. (See Figure 14.1 and an example of $K_4 \otimes_3 P_{10}$ in Figure 14.2. The $\oplus_3$-sum is defined in Section B.1.3.)

II. The edge-depth of an SCC of $G \otimes_3 P_{10}$. In Figure 14.2, the Petersen graph added to G^s at the vertex v via a $\oplus_3$-sum is denoted by P_v (v is a degree 3 vertex of G^s). Each degree 2 vertex of G^s is labeled by e_μ if it is inserted into an edge e_μ of G.

Since $V(G)$ serves as the index set for copies of the Petersen graph, $E(G \otimes_3 P_{10}) = \cup_{v \in V(G)} E(P_v)$ and the graph $G \otimes_3 P_{10}$ has $15|V(G)|$

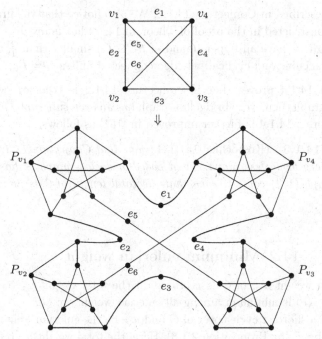

Figure 14.2 *Example:* $K_4 \otimes_3 P_{10}$

edges. With the assumption of Conjecture 14.1.2, the graph $G \otimes_3 P_{10}$ has a shortest cycle cover $\mathcal{F}$ with the total length

$$\mathcal{L}(\mathcal{F}) \leq \frac{21}{15}|E(G \otimes_3 P_{10})| = 21 \, |V(G)|. \tag{14.1}$$

For each $x \in V(G)$, the set of all members of $\mathcal{F}$ restricted in P_x is a set of circuits and paths covering P_x and is denoted by $\mathcal{F}|P_x$. By Proposition B.2.30-(1), the total length of $\mathcal{F}|P_x$ is at least 21. That is

$$\mathcal{L}(\mathcal{F}) \geq 21|V(G)|, \tag{14.2}$$

where $|V(G)|$ is the number of copies P_x of the Petersen graph. Thus, equalities hold for both (14.1) and (14.2), and therefore, the total length of the circuits and the paths in $\mathcal{F}|P_x = \{C \cap P_x : C \in \mathcal{F}\}$ is precisely 21. By Proposition B.2.30-(3), the edge-depth of $\mathcal{F}$ is 2. Therefore, $G \otimes_3 P_{10}$ has a circuit double cover, and so does G^s (by the construction of $G \otimes_3 P_{10}$). $\qquad\square$

Note that the total length of a shortest cycle cover of $G \otimes_3 P_{10}$ reaches the bound $\frac{21}{15}m$. That is, there is an infinite family of graphs that reach the bond described in Conjecture 14.1.2. We also notice that the graph $G \otimes_3 P_{10}$ constructed in the proof of Theorem 14.1.4 has many degree 2 vertices. Can we have an upper bound of $SCC(G)$ smaller than $\frac{21}{15}m$ if a graph G is cubic and cyclically 4-edge-connected unless $G = P_{10}$?

Theorem 14.1.4 proves that if Conjecture 14.1.2 is true for every bridgeless graph then every bridgeless graph has an even subgraph $(1, 2)$-cover. Theorem 14.1.4 is further improved in [151] as follows.

Theorem 14.1.5 (Kostochka [151]) *Assume that Conjecture 14.1.2 is true for every bridgeless graph. Then every bridgeless graph G has an even subgraph $(1, 2)$-cover $\mathcal{C}$ such that the total length of $\mathcal{C}$ is at most $\frac{21|E(G)|}{15}$.*

14.2 Minimum eulerian weight

An eulerian weight w_m of G is *minimum* if the total weight $w_m(G) = \sum_{e \in E(G)} w_m(e)$ is minimum among all eulerian weights of G.

Let $\mathcal{F}$ be a shortest cycle cover of G and $w_{\mathcal{F}}$ be the eulerian weight of G induced by $\mathcal{F}$ (by Proposition 2.1.8). Here, the total weight $w_{\mathcal{F}}(G) = SCC(G)$. Thus, it is obvious that

$$SCC(G) \geq w_m(G) \tag{14.3}$$

and equality holds if G has a faithful circuit cover with respect to the minimum weight w_m. Hence, $w_m(G)$ serves as a *lower bound* of $SCC(G)$.

Here, we discuss several problems.

Problem 14.2.1 **(A)** *Is there any graph such that equality of (14.3) holds?*
(B) *How small is $w_m(G)$?*
(C) *Is there any general upper bound of $SCC(G)$ for all bridgeless graphs G?*

⋆ Faithful coverable graphs

For Problem 14.2.1-(A), as we noticed already, the equality of (14.3) holds if and only if G has a faithful circuit cover with respect to w_m. (But unfortunately, some graph G may not have a faithful circuit cover with respect to its minimum weight w_m. For example, the Petersen graph P_{10} (Propositions B.2.26 and B.2.30).)

By Theorems 3.3.1 and 8.2.8, we have the following results.

Theorem 14.2.2 *$SCC(G)$ reaches the lower bound described in the Inequality (14.3) for the following families of graphs G:*

(1) if G is planar (Bermond, Jackson and Jaeger [12]; Guan and Fleischner [95]);

(2) if G admits a nowhere-zero 4-flow (Jackson [120], Zhang [254]);

(3) if G does not contain a Petersen minor (Alspach, Goddyn and Zhang [4]).

⋆ Smallest parity subgraphs

For Problem 14.2.1-(B), by Exercise 2.3, it is easy to see that w_m is an eulerian $(1, 2)$-weight of G with $E_{w=2}$ inducing a smallest parity subgraph P of G, and, therefore,

$$w_m(G) = |E(G)| + |E(P)|. \tag{14.4}$$

By Lemma A.2.19,

$$|E(P)| \leq \frac{1}{\lambda_o}|E(G)|$$

where λ_o is the odd-edge-connectivity of G. Thus, we have the following corollary. (See Exercise 14.1 for part of the proof.)

Corollary 14.2.3 *Let G be a bridgeless graph. Then $SCC(G) \leq \frac{4|E(G)|}{3}$ for graphs in the following families:*

(1) G is planar,

(2) G admits a nowhere-zero 4-flow,

(3) G contains no subdivision of the Petersen graph.

Note that the 4-flow theorem (Theorem 8.1.2) can be re-stated as follows: *a graph with odd-edge-connectivity 5 admits a nowhere-zero 4-flow.* (See Exercise 8.2.) Hence, for graphs with higher connectivity, the bound of $SCC(G)$ can be further reduced.

Theorem 14.2.4 *Let G be a bridgeless graph. Then*

$$SCC(G) \leq \frac{2h+2}{2h+1}|E(G)|$$

if

(1) the edge-connectivity of G is $2h$ for $h \geq 2$ (Jackson [120])

(2) the odd-edge-connectivity of G is $2h + 1$ for $h \geq 2$ ([213]).

Theorem 14.2.4 is proved by applying Lemma A.2.19, Theorems 8.1.2 and 8.2.8 (see Exercise 14.2).

Inspired by Equation (14.4) and Lemma A.2.15 that every spanning tree contains a parity subgraph, Itai and Rodeh [119] conjectured that $SCC(G) \le |E(G)| + |V(G)| - 1$. This conjecture was verified by Fan [52]. Note that this solved conjecture is for all bridgeless graphs G, not only those listed in Theorem 14.2.2 (with faithful covers for all eulerian weights).

⋆ Chinese postman problem

Estimation of the total weight of a minimum eulerian weight w_m is a classical optimization problem: *the Chinese postman problem* (see [44] or see Section 8.1 in [259] for more detail).

Computational complexity.

(1) (Edmonds and Johnson [44]) *The Chinese postman problem (and, equivalently, finding a minimum eulerian weight w_m) is a polynomially solvable problem.*

(2) (Thomassen [227]) *The shortest cycle cover problem is an NP-complete problem.*

Problem 14.2.1-(C) will be summarized in Section 14.3.

14.3 3-even subgraph covers

Up to now, almost all results for the shortest cycle cover problem have been 3-even subgraph covers.

Notation. Let $SCC_h(G)$ be the minimum total length of an h-even subgraph cover of a graph G.

14.3.1 Basis of cycle space

By Theorem 7.2.1, *every bridgeless graph has a 3-even subgraph cover.* Let $\mathcal{F} = \{C_1, C_2, C_3\}$ be a 3-even subgraph cover of G. One may easily find some other 3-even subgraph cover $\mathcal{F}'$ consisting of symmetric differences of members of $\mathcal{F}$. Among those 3-even subgraph covers, which one has a shorter total length? This has been studied extensively for estimations of upper bounds of $SCC(G)$.

From the view of cycle space, let $\mathcal{S}_\mathcal{F}$ be the subspace of the cycle space of G generated by $\mathcal{F} = \{C_1, C_2, C_3\}$. In this 3-dimensional linear space (on the field $\mathbb{Z}_2$), $\{C_1, C_2, C_3\}$ is a basis of $\mathcal{S}_\mathcal{F}$ and, we will show that every basis of $\mathcal{S}_\mathcal{F}$ is also a 3-even subgraph cover of G (Lemma 14.3.2).

Notation 14.3.1 The symmetric difference of two even subgraphs C_1 and C_2 is the same as their sum on the field $\mathbb{Z}_2$. That is, $C_1 \triangle C_2 = C_1 + C_2$. Let $\{C_1, \ldots, C_r\}$ be a set of even subgraphs. In general, a linear combination $\epsilon_1 C_1 + \cdots + \epsilon_r C_r$ ($\epsilon_i \in \mathbb{Z}_2$) is the same as $\triangle\{C_i : \epsilon_i = 1\}$ and is denoted by $C_{\vec{v}}$ where $\vec{v} = \langle \epsilon_1, \ldots, \epsilon_r \rangle$ is a vector in $\mathbb{Z}_2^r$.

Lemma 14.3.2 (Bermond, Jackson and Jaeger [12]) *Let* $\{C_1, \ldots, C_k\}$ *be a k-even subgraph cover of a bridgeless graph G. Let* $\{\vec{v}_1, \ldots, \vec{v}_k\}$ *be a set of k vectors in* $\mathbb{Z}_2^k$. *Then*

$$\{C_{\vec{v}_1}, \ldots, C_{\vec{v}_k}\}$$

is an even subgraph cover of G if $B = \{\vec{v}_1, \ldots, \vec{v}_k\}$ *is a basis of* $\mathbb{Z}_2^k$.

Proof For any $e \in E(G)$, without loss of generality, let $e \in E(C_1)$. It is sufficient to show that $e \in C_{\vec{v}_i}$ for some $\vec{v}_i \in B$. Since B is a basis of $\mathbb{Z}_2^k$, the vector $\langle 1, 0, \ldots, \rangle$ is a linear combination of some vectors of B, say, $\langle 1, 0, \ldots, 0 \rangle = \vec{v}_{i_1} + \cdots + \vec{v}_{i_h}$. Thus,

$$C_1 = C_{\vec{v}_{i_1}} \triangle \cdots \triangle C_{\vec{v}_{i_h}},$$

and therefore, the edge e is contained in some of $\{C_{\vec{v}_{i_1}}, \ldots, C_{\vec{v}_{i_h}}\}$. $\square$

14.3.2 3-even subgraph covers

For a given 3-even subgraph cover $\mathcal{F} = \{C_1, C_2, C_3\}$, let $c_\mathcal{F} : E(G) \mapsto \mathbb{Z}_2^3 - \{\langle 0, 0, 0 \rangle\}$ be a mapping such that, for each edge $e \in E(G)$, $c_\mathcal{F}(e) = \langle \epsilon_1, \epsilon_2, \epsilon_3 \rangle$ where $\epsilon_i = 1$ if $e \in E(C_i)$, or $\epsilon_i = 0$ otherwise. Let

$$E_{\langle \epsilon_1, \epsilon_2, \epsilon_3 \rangle} = \{e \in E(G) : c_\mathcal{F}(e) = \langle \epsilon_1, \epsilon_2, \epsilon_3 \rangle\}.$$

For cubic graphs G, the mapping $c_\mathcal{F}$ becomes a proper 7-edge-coloring of G such that, for each vertex $v \in V(G)$,

$$\sum_{e \in E(v)} c_\mathcal{F}(e) = \langle 0, 0, 0 \rangle.$$

For every vector $\vec{v} = \langle \epsilon_1, \epsilon_2, \epsilon_3 \rangle \in \mathbb{Z}_2^3$, as defined in Notation 14.3.1, denote

$$C_{\vec{v}} = \triangle\{C_{\epsilon_i} : \epsilon_i = 1, \ i = 1, 2, 3\}.$$

In particular,

$$C_{\langle 1,0,0 \rangle} = C_1, \quad C_{\langle 0,1,0 \rangle} = C_2, \quad C_{\langle 0,0,1 \rangle} = C_3.$$

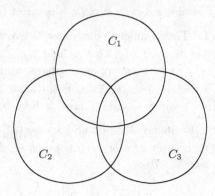

Figure 14.3 *Edge set partition according to a 3-even subgraph cover*
$\{C_1, C_2, C_3\}$

It is easy to see that

$$\{C_{\vec{v}} :\ \vec{v} \in \mathbb{Z}_2^3 - \{\langle 0, 0, 0 \rangle\}\}$$

is a 7-even subgraph 4-cover of G (Theorem 7.3.4).

By Lemma 14.3.2, each basis of $\mathbb{Z}_2^3$ corresponds to a 3-even subgraph cover of G. Note that $\mathbb{Z}_2^3$ has 28 bases, each of which corresponds to a 3-even subgraph cover of G.

(*) *Which of these has the shortest total length?*
(*) *Or what is the average of some of them?*

These are the main approaches of the next subsections and motivate the study of Fano flow (Fano plane and Fano flow will be formally introduced in Section 14.3.6).

For an average estimation, an extreme case is that each $E_{\langle \epsilon_1, \epsilon_2, \epsilon_3 \rangle}$ is of the same size ($\frac{1}{7}|E(G)|$) and, therefore, every 3-even subgraph cover generated in this way is of the same total length $\frac{12}{7}|E(G)|$ (by Theorem 7.3.4). Hence, a *strategy* for finding a shorter cover is to start with a 3-even subgraph cover $\mathcal{F}$ such that *the sizes of* $E_{\langle \epsilon_1, \epsilon_2, \epsilon_3 \rangle}$ *are not evenly distributed* (an *unbalanced* 3-even subgraph cover). By applying Lemma 14.3.2, one would be able to find a 3-even subgraph cover (a linear combination of members of $\mathcal{F}$) that covers some larger subsets $E_{\langle \epsilon_1, \epsilon_2, \epsilon_3 \rangle}$ less than other smaller subsets. This is the main idea behind most results on this subject.

14.3.3 ($\geq$ 4)-even subgraph covers

It is very unfortunate that the approaches used for 3-even subgraph cover will not reach the best bound $\frac{21}{15}|E(G)|$ because every shortest cycle cover of the Petersen graph consists of 4 even subgraphs (Proposition B.2.30-(4)). It was pointed out by Fan and Raspaud [53] that *there are infinitely many bridgeless graphs G that have no 3-even subgraph cover with total length less than $\frac{22}{15}|E(G)|$.*

Conjecture 14.3.3 (Tarsi [222]) *Every bridgeless cubic graph with m edges has a 3-even subgraph cover with total length at most $\frac{22}{15}m$.*

It was proved by Fan and Raspaud [53] that Conjecture 14.3.3 is true if the Berge–Fulkerson conjecture (Conjecture 7.3.6) is true (Theorem 14.3.5).

Furthermore, it remains unknown whether the subject of 3-even subgraph covers has any close relation with the circuit double cover conjecture.

14.3.4 Upper bounds of SCC_3

The following theorem (Theorem 14.3.4) is one of the most classical results in this area.

Theorem 14.3.4 (Bermond, Jackson and Jaeger [12], Alon and Tarsi [1]) *Let G be a bridgeless graph with m edges. Then G has a 3-even subgraph cover of total length at most $\frac{25m}{15}$.*

Proof Refer to Theorem 8.5.1 in [259]. $\qquad\qquad\square$

This theorem has two technically very different proofs (Bermond, Jackson and Jaeger [12], and independently, Alon and Tarsi [1]). In the first proof, the first *unbalanced* 3-even subgraph cover $\{C_1, C_2, C_3\}$ is found by applying the matching polyhedron theorem (Theorem A.1.9) and the 8-flow theorem (Theorem 7.2.1 or Theorem 8.1.3, and see Exercise 14.6) that $|E(C_1)| = |E_{\langle 1,0,0\rangle} \cup E_{\langle 1,1,0\rangle} \cup E_{\langle 1,0,1\rangle} \cup E_{\langle 1,1,1\rangle}| \geq \frac{2}{3}|E(G)|$, while in the second proof, the first *unbalanced* 3-even subgraph cover $\{C_1, C_2, C_3\}$ is found by applying the 6-flow theorem (Theorem 8.3.1) that $|E_{\langle 1,0,0\rangle}| \geq \frac{1}{3}|E(C_1)|$.

Theorem 14.3.4, as the most well-known classical result of SCC_3 [1], [12], has been further improved for some families of graphs. The following is a list of some improvements of this theorem for cubic graphs, and graphs without restriction of regularity.

For cubic graphs

References	SCC_3	$\leq$	
Bermond *et al.* [12], Alon *et al.* [1]	$\frac{5m}{3}$	$=$	$\frac{25m}{15}$
Jackson [122]	$\frac{64m}{39}$	$\approx$	$\frac{24.615m}{15}$
Fan [51]	$\frac{44m}{27}$	$\approx$	$\frac{24.444m}{15}$
Král *et al.* [156]	$\frac{34m}{21}$	$\approx$	$\frac{24.286m}{15}$

For general graphs with $\Delta \geq 3$

References	SCC_3	$\leq$	
Bermond *et al.* [12], Alon *et al.* [1]	$\frac{5m}{3}$	$=$	$\frac{25m}{15}$
Kaiser *et al.* [140]	$\frac{44m}{27}$	$\approx$	$\frac{24.444m}{15}$

14.3.5 Relations with other major conjectures

⋆ Berge–Fulkerson conjecture

With the assumption of the Berge–Fulkerson conjecture (Conjecture 7.3.6), Fan and Raspaud [53] were able to find some unbalanced 3-even subgraph cover with total length $\geq 2|E(G)|$. This big 3-even subgraph cover leads us to construct a smaller even subgraph cover (Theorem 14.3.5). It was further conjectured by Fan and Raspaud [53] that such a 3-even subgraph cover does exist for every bridgeless graph (Conjecture 14.4.5). Note that the bound that $SCC_3(G) \leq \frac{22}{15}m$ is sharp since the Petersen graph reaches the bound (Proposition B.2.30).

Theorem 14.3.5   (Fan and Raspaud [53]) *If the Berge–Fulkerson conjecture (Conjecture 7.3.6) is true for a cubic graph G with m edges, then the graph G has a 3-even subgraph cover of total length at most $\frac{22m}{15}$.*

Proof of Theorem 14.3.5 will be presented in Section 14.3.8.

⋆ Tutte's 5-flow conjecture

Tutte conjectured (Conjecture C.1.10) that every bridgeless graph admits a nowhere-zero 5-flow. If this conjecture is true, could we get a better upper bound of shortest cycle covers for bridgeless graphs?

Theorem 14.3.6   (Jamshy, Raspaud and Tarsi [134]) *Every graph G admitting a nowhere-zero 5-flow has a 3-even subgraph cover of total length at most $\frac{24m}{15}$ where $|E(G)| = m$.*

Proof Refer to Theorem 8.7.1 in [259]. □

The approach in the proof of Theorem 14.3.6 is somehow different from other major results. Instead of finding an unbalanced 3-even subgraph cover, the main idea behind this approach is to find four 2-flows such that their supports have an unbalanced distribution.

⋆ Tutte's 3-flow conjecture

Under the assumption of the 3-flow conjecture (Conjecture C.1.12), Fan [51] proved another upper bound for SCC_3.

Theorem 14.3.7 (Fan [51]) *If every 4-edge-connected graph admits a nowhere-zero 3-flow, then every bridgeless graph G has a 3-even subgraph cover with total length at most $\frac{92m}{57}$ where $|E(G)| = m$.*

⋆ Summary

The following table is a summary of those results under the assumptions of some major graph theory conjectures.

SCC_3 under major conjectures

References	If this conjecture is true, then	SCC_3 ≤	
Fan *et al.* [53]	Berge–Fulkerson conjecture	$\frac{22m}{15}$	
Jamshy *et al.* [134]	5-flow conjecture	$\frac{24m}{15}$	
Fan [51]	3-flow conjecture	$\frac{92m}{57}$	≈ $\frac{24.21053m}{15}$

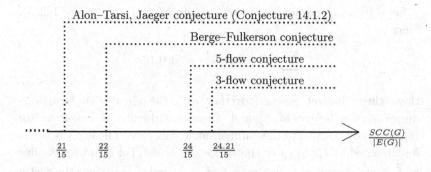

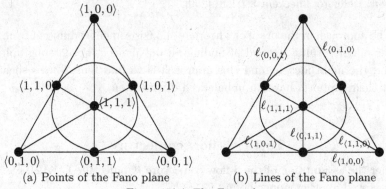

(a) Points of the Fano plane   (b) Lines of the Fano plane

Figure 14.4 *The Fano plane*

14.3.6 Fano plane and Fano flows

The discussion of SCC_3 in Section 14.3.2 motivates the use of Fano flow in the study. This subsection is a brief presentation of the *Fano plane*, *Fano flow* and some of their applications for 3-even subgraph covers. Many observations in Section 14.3.2 are now presented in terms of the notion of Fano flow.

⋆ (A) Elements of the Fano plane

Definition 14.3.8   The Fano plane consists of a set P of seven points and a set L of seven lines. Points correspond to non-zero vectors of $\mathbb{Z}_2^3$ and three points form a line if and only if they sum to zero, see Figure 14.4.

Each point $\vec{p}$ of the Fano plane is denoted by a $\mathbb{Z}_2^3$-vector $\langle \epsilon_1, \epsilon_2, \epsilon_3 \rangle$ or by $\vec{p}_{\langle \epsilon_1, \epsilon_2, \epsilon_3 \rangle}$. (See Figure 14.4-(a).)

Let $\vec{\ell}$ be a line consisting of three points $\{\vec{p}_{\langle \epsilon_{i,1}, \epsilon_{i,2}, \epsilon_{i,3} \rangle} : i = 1, 2, 3\}$. Since

$$\sum_{i=1}^{3} \langle \epsilon_{i,1}, \epsilon_{i,2}, \epsilon_{i,3} \rangle = \langle 0, 0, 0 \rangle,$$

these three distinct points (together with the identity $\vec{0}$) form a 2-dimensional subspace of $\mathbb{Z}_2^3$ and, therefore, there is a *unique* vector $\langle \delta_1, \delta_2, \delta_3 \rangle$ orthogonal to the 2-dimensional subspace. The line $\vec{\ell}$ is therefore denoted by $\vec{\ell}_{\langle \delta_1, \delta_2, \delta_3 \rangle}$ or the vector $\langle \delta_1, \delta_2, \delta_3 \rangle$. For example, the line $\vec{\ell}_{\langle 0,0,1 \rangle}$ contains three points $\vec{p}_{\langle 1,0,0 \rangle}$, $\vec{p}_{\langle 1,1,0 \rangle}$, and $\vec{p}_{\langle 0,1,0 \rangle}$ since the vector

$\langle 0, 0, 1 \rangle$ is orthogonal to $\langle 1, 0, 0 \rangle$, $\langle 1, 1, 0 \rangle$, and $\langle 0, 1, 0 \rangle$ (see Figure 14.4-(b)). That is,

($\star$) *in the Fano plane, a vector $\vec{\ell}$ is a line containing a point $\vec{p}$ if and only if the vector $\vec{\ell}$ is orthogonal to the vector $\vec{p}$.*

Note, a Fano line may have two different forms: (1) as a subset of three points $\{\vec{p}_1, \vec{p}_2, \vec{p}_3\}$, and (2) as a $\mathbb{Z}_2^3$-vector orthogonal to each $\vec{p}_i$.

Definition 14.3.9 A set $M = \{\vec{\ell}_1, \vec{\ell}_2, \vec{\ell}_3\}$ of three lines is called a basis of the Fano plane if lines $\{\vec{\ell}_1, \vec{\ell}_2, \vec{\ell}_3\}$ have no common point.

With some elementary linear algebra properties, we have the following propositions.

Proposition 14.3.10 *For each pair of points $\vec{p}_1, \vec{p}_2$ (or pair of lines $\vec{\ell}_1, \vec{\ell}_2$, or pair of bases M_1, M_2) of the Fano plane, there is an automorphism ϕ of the Fano plane such that $\phi(\vec{p}_1) = \vec{p}_2$ (or $\phi(\vec{\ell}_1) = \vec{\ell}_2$, or $\phi(M_1) = M_2$, respectively).*

Proposition 14.3.11 *Fano lines $\{\vec{\ell}_1, \vec{\ell}_2, \vec{\ell}_3\}$ form a basis (no common point) if and only if their sum $\sum_{i=1}^{3} \vec{\ell}_i \neq \langle 0, 0, 0 \rangle$.*

Hence, a basis M of the Fano plane is defined the same as a basis of the linear space $\mathbb{Z}_2^3$ (consisting of 3 linearly independent vectors when Fano lines are considered vectors).

$\star$ (B) Fano flow $\leftrightarrow$ 3-even subgraph cover

Definition 14.3.12 Let G be a bridgeless graph. A nowhere-zero group $\mathbb{Z}_2^3$-flow $f = \langle f_1, f_2, f_3 \rangle$ of G is called a **Fano flow** of G.

Without using the notion of flow theory, one may equivalently define Fano flow as an edge-coloring (as in Section 14.3.2). Let $f : E(G) \mapsto \mathbb{Z}_2^3 - \{\langle 0, 0, 0 \rangle\}$ such that $\sum_{e \in E(v)} f(e) = \langle 0, 0, 0 \rangle$ *for every vertex $v \in V(G)$.* (For cubic graphs, Fano flow is also sometimes called **Fano coloring** [175], [177], [157].)

By the 8-flow theorem (Theorem 7.2.1 or Theorem 8.1.3), every bridgeless graph has a Fano flow. Furthermore, with the same argument as in Theorem 8.1.1, we note that there is a one-to-one correspondence between Fano flows and 3-even subgraph covers. That is, by Theorem C.1.8, we have the following proposition.

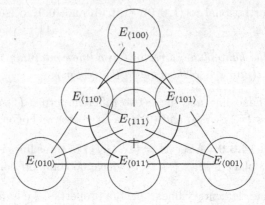

Figure 14.5 *The partition of $E(G)$ by a 3-even subgraph cover and the Fano plane*

Proposition 14.3.13   *Every Fano flow* $f = \langle f_1, f_2, f_3 \rangle$ *of* G *corresponds to a 3-even subgraph cover* $\{C_1, C_2, C_3\}$ *such that*

$$E(C_1) = \{e \in E(G): \ f(e) = \langle 1, \epsilon_2, \epsilon_3 \rangle, \ \epsilon_2, \epsilon_3 \in \mathbb{Z}_2\} = supp(f_1),$$

$$E(C_2) = \{e \in E(G): \ f(e) = \langle \epsilon_1, 1, \epsilon_3 \rangle, \ \epsilon_1, \epsilon_3 \in \mathbb{Z}_2\} = supp(f_2),$$

and

$$E(C_3) = \{e \in E(G): \ f(e) = \langle \epsilon_1, \epsilon_2, 1 \rangle, \ \epsilon_1, \epsilon_2 \in \mathbb{Z}_2\} = supp(f_3).$$

In the remaining part of this section, we always let $f = \langle f_1, f_2, f_3 \rangle$ be a Fano flow of a bridgeless cubic graph G, and C_i be the support of each f_i (here $\{C_1, C_2, C_3\}$ is a 3-even subgraph cover of G).

Because of symmetric properties (algebraic properties) of the Fano plane, Fano flow is a useful tool in the study of 3-even subgraph covers ([175], [177], [178]). Most results in this subsection were originally from [178].

⋆ (C) Fano points ↔ edge set partition

For a Fano flow f of G,

$$\{f^{-1}(\vec{p}): \ \vec{p} \in P\}$$

is a partition of $E(G)$ (see Figure 14.5), and each $f^{-1}(\vec{p})$ is denoted by $E_{\vec{p}}$ (note that P is the point set of the Fano plane).

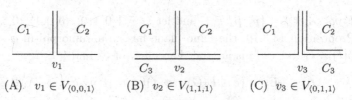

(A) $v_1 \in V_{\langle 0,0,1 \rangle}$ (B) $v_2 \in V_{\langle 1,1,1 \rangle}$ (C) $v_3 \in V_{\langle 0,1,1 \rangle}$

Figure 14.6 *Three types of vertices*

⋆ (D) Fano lines ↔ vertex set partition

Furthermore, by the definition of Fano lines, the flow f yields a corresponding mapping of the vertex set of G: $V(G) \mapsto L$ (note that L is the line set of the Fano plane).

Definition 14.3.14 Let G be a cubic graph with a Fano flow f : $E(G) \mapsto P$ (where P is the point set of the Fano plane). Define f : $V(G) \mapsto L$ (where L is the line set of the Fano plane): for each $v \in V(G)$ with $E(v) = \{e_1, e_2, e_3\}$ and $f(e_i) = \vec{p}_i$,

$$f(v) = \vec{\ell}$$

where $\vec{\ell}$ is the line consisting of the points $\vec{p}_1, \vec{p}_2, \vec{p}_3$.

(Note that, by ($\star$), the vector $\vec{\ell}$ ($= f(v)$) is orthogonal to each vector $\vec{p}_i$.) So, the vertex set $V(G)$ has a partition

$$\{f^{-1}(\vec{\ell}) : \ \vec{\ell} \in L\},$$

and each $f^{-1}(\vec{\ell})$ is denoted by $V_{\vec{\ell}}$.

Figure 14.6 illustrates three types of vertices, at which the even subgraph cover $\{C_1, C_2, C_3\}$ passes in different types (each C_i is defined as the support of f_i, see Proposition 14.3.13).

⋆ (E) Fano line ↔ even subgraph

Proposition 14.3.15 *Let G be a cubic graph G with a Fano flow f. Then, for every Fano line $\vec{\ell}$, the edge subset*

$$\cup_{\vec{p} \in P - \vec{\ell}} E_{\vec{p}}$$

induces an even subgraph of G.

Note that $P - \vec{\ell}$ is the complement of the line $\vec{\ell}$ in the Fano plane and consists of 4 points.

Proof Let $\vec{\ell} = \{\vec{p}_1, \vec{p}_2, \vec{p}_3\}$, and let $\vec{\ell}_1 = \{\langle 0,1,0\rangle, \langle 0,0,1\rangle, \langle 0,1,1\rangle\}$. By Proposition 14.3.10, the Fano plane has an automorphism ϕ such that $\phi(\vec{\ell}) = \vec{\ell}_1$. So, $\phi(f)$ is another Fano flow of G, and

$$\cup_{\vec{p}\in P - \vec{\ell}}E_{\vec{p}} = \{e \in E(G) : \phi(f)(e) = \langle 1, \delta_2, \delta_3\rangle, \ \delta_2, \delta_3 \in \mathbb{Z}_2\}$$

is an even subgraph of G (by Proposition 14.3.13). $\qquad\qquad\square$

Notation 14.3.16 Let f be a Fano flow of a bridgeless cubic graph G, and let $\vec{\ell}$ be a Fano line. The even subgraph $\cup_{\vec{p}\in P - \vec{\ell}}E_{\vec{p}}$ is denoted by $C_{\vec{\ell}}$.

By Proposition 14.3.15 and Definition 14.3.14, we have the following relation between $V_{\vec{\ell}}$ and $C_{\vec{\ell}}$ for each Fano line $\vec{\ell}$.

Lemma 14.3.17 (Máčajová, Raspaud, Tarsi and Zhu [178]) *Let G be a cubic graph with a Fano flow f. Then, for each Fano line $\vec{\ell}$ and its corresponding even subgraph $C_{\vec{\ell}}$, we have that*

$$V(C_{\vec{\ell}}) = V(G) - V_{\vec{\ell}}.$$

Furthermore, $V_{\vec{\ell}} = \emptyset$ if and only if the corresponding even subgraph $C_{\vec{\ell}}$ is a 2-factor.

Proof By Proposition 14.3.15, the edge set $E(G)$ has a partition

$$\{E(C_{\vec{\ell}}), \ \cup_{\vec{p}\in\vec{\ell}}E_{\vec{p}}\}.$$

For a vertex $v \in V_{\vec{\ell}}$, we have that $E(v) \subseteq \cup_{\vec{p}\in\vec{\ell}}E_{\vec{p}}$ (by Definition 14.3.14). Hence, a vertex $v \in V(C_{\vec{\ell}})$ if and only if $v \notin V_{\vec{\ell}}$. $\qquad\square$

$\star$ (F) Fano basis $\leftrightarrow$ 3-even subgraph cover

Proposition 14.3.18 *Let f be a Fano flow of a cubic graph G. A line subset $\{\vec{\ell}_1, \vec{\ell}_2, \vec{\ell}_3\}$ is a basis of the Fano plane if and only if $\{C_{\vec{\ell}_1}, C_{\vec{\ell}_2}, C_{\vec{\ell}_3}\}$ is a 3-even subgraph cover of G.*

Proof By the definition of basis of the Fano plane (Definition 14.3.9), lines $\vec{\ell}_1, \vec{\ell}_2, \vec{\ell}_3$ have no common point, the union of their complements covers all points of the Fano plane. That is, by Proposition 14.3.15, $\{C_{\vec{\ell}_1}, C_{\vec{\ell}_2}, C_{\vec{\ell}_3}\}$ is an even subgraph cover of G.

The proof of the other direction is similar. $\qquad\qquad\square$

The following table is a summary of the corresponding relation between a graph G (with a Fano flow f) and the Fano plane.

Graph G $\implies$	Fano plane	References
edge subset $E_{\vec{p}}$	point $\vec{p}$	page 176
vertex subset $V_{\vec{\ell}}$	line $\vec{\ell}$	Definition 14.3.14
even subgraph $C_{\vec{\ell}}$	edge-complement of a line $\vec{\ell}$	Proposition 14.3.15
3-even subgraph cover	a basis	Proposition 14.3.18

For an estimation of SCC_3, we have a corollary of Lemma 14.3.17.

Lemma 14.3.19 (Máčajová, Raspaud, Tarsi and Zhu [178]) *Let G be a cubic graph with a Fano flow f. For each line basis $B = \{\vec{\ell}_1, \vec{\ell}_2, \vec{\ell}_3\}$, the corresponding 3-even subgraph $\{C_{\vec{\ell}_1}, C_{\vec{\ell}_2}, C_{\vec{\ell}_3}\}$ has the total length*

$$3|V(G)| - |\cup_{\vec{\ell}\in B} V_{\vec{\ell}}| = 2|V(G)| + |\cup_{\vec{\ell}\notin B} V_{\vec{\ell}}|.$$

Proof Since G is cubic, for an even subgraph $C_{\vec{\ell}}$, we have that $|E(C_{\vec{\ell}})| = |V(C_{\vec{\ell}})|$. By Lemma 14.3.17,

$$|E(C_{\vec{\ell}})| = |V(G)| - |V_{\vec{\ell}}|.$$

Hence,

$$\sum_{\vec{\ell}\in B} |E(C_{\vec{\ell}})| = 3|V(G)| - |\cup_{\vec{\ell}\in B} V_{\vec{\ell}}|.$$

Note that, by Definition 14.3.14, $\{V_{\vec{\ell}}: \vec{\ell} \text{ is a Fano line}\}$ is a partition of $V(G)$. Therefore,

$$|\cup_{\vec{\ell}\in B} V_{\vec{\ell}}| = |V(G)| - |\cup_{\vec{\ell}\notin B} V_{\vec{\ell}}|.$$

$\square$

Lemma 14.3.19 motivates a strategy of finding a shorter 3-even subgraph cover: *find an Fano flow with an* unbalanced *vertex set partition* $\{V_{\vec{\ell}}: \vec{\ell} \in L\}$ (see Definition 14.3.14) *and select a basis B with* $|\cup_{\vec{\ell}\in B} V_{\vec{\ell}}|$ *is as large as possible* (see Lemma 14.3.19 and see Remarks on page 181 for a brief discussion). Both Theorems 14.3.21 and 14.3.22 are proved following this strategy. An extreme case of such un-balancedness is that some vertex subset $V_{\vec{\ell}}$ is empty.

14.3.7 Incomplete Fano flows, F_μ-flows

⋆ Line-incomplete, F_μ-flows

Definition 14.3.20 If $f : V(G) \mapsto L$ (the line set of the Fano plane) is not a surjection (see Definition 14.3.14), then the Fano flow f is line-incomplete. That is, $V_{\vec{\ell}} = \emptyset$ for some line $\vec{\ell}$. And the Fano flow f is a μ-Fano flow (or F_μ-flow) if the number of empty subsets $V_{\vec{\ell}}$ is at least $(7 - \mu)$.

By Lemma 14.3.17, if f is an F_μ-flow, then $\{C_{\vec{\ell}}: \vec{\ell} \in L\}$ contains at least $(7 - \mu)$ 2-factors.

It is evident that *every bridgeless graph has an F_6-flow* by the proof of the 8-flow theorem (Theorem 7.2.1 or Theorem 8.1.3). It was further conjectured that every bridgeless cubic graph admits an F_4-flow (Conjecture 14.3.24). The following results are upper bounds of SCC_3 if a cubic graph G admits an F_μ-flow, for $\mu = 4, 5$.

Theorem 14.3.21 (Máčajová, Raspaud, Tarsi and Zhu [178]) *If a bridgeless cubic graph G admits an F_5-flow then $SCC_3(G) \leq \frac{24}{15}|E(G)|$.*

Theorem 14.3.22 (Máčajová, Raspaud, Tarsi and Zhu [178]) *If a bridgeless cubic graph G admits an F_4-flow then $SCC_3(G) < \frac{14}{9}|E(G)| \approx \frac{23.33}{15}|E(G)|$.*

Proofs of both Theorems 14.3.21 and 14.3.22 are in Section 14.3.8.

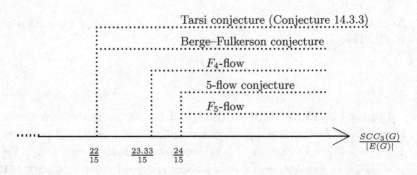

However, the existence of such line-incomplete Fano flows remains an unsolved open problem.

Conjecture 14.3.23 (Máčajová and Škoviera [175]) *Every bridgeless cubic graph admits an F_5-flow.*

Conjecture 14.3.24 (Fan and Raspaud [53]) *Every bridgeless cubic graph admits an F_4-flow.*

For Conjectures 14.3.23 and 14.3.24, Kaiser and Raspaud [138] proved that every oddness 2 cubic graph admits an F_5-flow. This result was further strengthened for F_4-flows.

Theorem 14.3.25 (Máčajová and Škoviera [176]) *Every bridgeless cubic graph of oddness at most two admits an F_4-flow.*

$\star$ Point-incomplete

A counterpart of line-incomplete Fano flow is the concept of point-incomplete Fano flow. A Fano flow f is point-incomplete (or degenerated, as in [178]) if the edge subset $E_{\vec{p}} = f^{-1}(\vec{p}) = \emptyset$ for some Fano point $\vec{p}$. It is easy to prove the following proposition.

Proposition 14.3.26 [178] *Let f be a Fano flow of G. If $f^{-1}(\vec{p}) = \emptyset$ for some Fano point $\vec{p}$, then G has a 2-even subgraph cover. Furthermore, $SCC_2(G) \leq \frac{4}{3}|E(G)|$.*

Proof Without loss of generality (by Proposition 14.3.10), let

$$f^{-1}(\langle 0, 0, 1 \rangle) = \emptyset.$$

Then $\{C_1, C_2\}$ is a 2-even subgraph cover of G. The upper bound $SCC_2(G) \leq \frac{4}{3}|E(G)|$ is an immediate corollary of Corollary 14.2.3-(2) (Exercise 14.1). $\square$

$\star$ F_4-flows, F_5-flows and 2-factors

Remark. For an F_4-flow f of a cubic graph G, let $\{\vec{\ell}_1, \vec{\ell}_2, \vec{\ell}_3\}$ be a subset of Fano lines such that each $V_{\vec{\ell}_i} = \emptyset$. That is, by Lemma 14.3.17, the corresponding $C_{\vec{\ell}_i}$ is a 2-factor. There are two cases:

(1) the lines $\{\vec{\ell}_1, \vec{\ell}_2, \vec{\ell}_3\}$ are linearly independent;
(2) the lines $\{\vec{\ell}_1, \vec{\ell}_2, \vec{\ell}_3\}$ are not linearly independent.

For Case (1), $\{\vec{\ell}_1, \vec{\ell}_2, \vec{\ell}_3\}$ form a basis, and therefore, we obtain a 3-even subgraph cover $\{C_{\vec{\ell}_1}, C_{\vec{\ell}_2}, C_{\vec{\ell}_3}\}$ which consists of three 2-factors.

For Case (2), lines $\{\vec{\ell}_1, \vec{\ell}_2, \vec{\ell}_3\}$ do not form a basis. Therefore, by Proposition 14.3.11, there is a common point $\vec{p}$ contained in all of them. This point corresponds to an empty $E_{\vec{p}}$, and thus, f is a point-incomplete flow and we are able to apply Proposition 14.3.26 for an estimation of SCC.

Remark. For an F_5-flow f of a cubic graph G, let $\{\vec{\ell}_1, \vec{\ell}_2\}$ be a subset of Fano lines such that each $V_{\vec{\ell}_i} = \emptyset$. It is evident (in linear algebra) that there is a basis $\{\vec{\ell}_1, \vec{\ell}_2, \vec{\ell}_3\}$ of the Fano plane. Thus, by Proposition 14.3.18, G has an even subgraph cover $\{C_{\vec{\ell}_1}, C_{\vec{\ell}_2}, C_{\vec{\ell}_3}\}$, two of which are 2-factors (by Lemma 14.3.17).

14.3.8 Some proofs

Readers are referred to [259] for the proofs of most results in this chapter. In this section, we only present some typical proofs that were not included in [259].

⋆ Berge–Fulkerson conjecture and SCC_3

Theorem 14.3.5 (Fan and Raspaud [53]) *If the Berge–Fulkerson conjecture (Conjecture 7.3.6) is true, then every bridgeless graph G with m edges has a 3-even subgraph cover of total length at most $\frac{22m}{15}$.*

With the assumption of the Berge–Fulkerson conjecture, Fan and Raspaud [53] were able to find some unbalanced 3-even subgraph cover with total length $\geq 2|E(G)|$. This big even subgraph cover leads us to a better upper bound for the shortest cycle cover problem (Theorem 14.3.5). It was further conjectured by Fan and Raspaud [53] that such a 3-even subgraph cover does exist for every bridgeless graph (Conjecture 14.4.5).

For an even subgraph cover $\mathcal{F}$ of a graph G, let $S(\mathcal{F})$ be the set of edges of G which are contained in *precisely one member* of $\mathcal{F}$. The following lemma finds another 3-even subgraph cover whose total length may be shorter than a given 3-even subgraph cover.

Lemma 14.3.27 (Fan and Raspaud [53]) *Let $\mathcal{F}$ be a 3-even subgraph cover of a graph G. Then G has a 3-even subgraph cover of total length at most*

$$2|E(G)| + \frac{2|S(\mathcal{F})|}{3} - \frac{\mathcal{L}(\mathcal{F})}{3}.$$

Note that $\mathcal{L}(\mathcal{F})$ is the total length of the cover (see Notation 14.1.3).

Proof Let $\mathcal{F} = \{C_1, C_2, C_3\}$ and $S_i = S(\mathcal{F}) \cap C_i$, for each $i = 1, 2, 3$. We have that

$$\mathcal{L}(\mathcal{F}) = |C_1| + |C_2| + |C_3| \tag{14.5}$$

and

$$|S(\mathcal{F})| = |S_1| + |S_2| + |S_3|. \tag{14.6}$$

Let $\mathcal{F}' = \{C_1 \bigtriangleup C_2, C_1 \bigtriangleup C_3, C_1 \bigtriangleup C_2 \bigtriangleup C_3\}$. By Lemma 14.3.2, $\mathcal{F}'$ is also an even subgraph cover of G, and $\mathcal{F}'$ covers each edge of S_1 exactly three times, each edge of $C_1 - S_1$ exactly once, and each remaining edge exactly twice. Hence, the total length of $\mathcal{F}'$ is

$$\mathcal{L}(\mathcal{F}') = 3|S_1| + |C_1 - S_1| + 2|E(G) - C_1| = 2|E(G)| + 2|S_1| - |C_1|.$$

Similarly, there are 3-even subgraph covers $\mathcal{F}''$ and $\mathcal{F}'''$ such that

$$\mathcal{L}(\mathcal{F}'') = 2|E(G)| + 2|S_2| - |C_2|, \quad \mathcal{L}(\mathcal{F}''') = 2|E(G)| + 2|S_3| - |C_3|.$$

Substituting (14.5) and (14.6) into those equations, we have that

$$\mathcal{L}(\mathcal{F}') + \mathcal{L}(\mathcal{F}'') + \mathcal{L}(\mathcal{F}''') = 6|E(G)| + 2|S(\mathcal{F})| - \mathcal{L}(\mathcal{F}).$$

Hence, one of the three 3-even subgraph covers must be of length at most

$$2|E(G)| + \frac{2|S(\mathcal{F})|}{3} - \frac{\mathcal{L}(\mathcal{F})}{3}.$$

$\square$

Proof of Theorem 14.3.5. If the Berge–Fulkerson conjecture is true, then, by Exercise 7.3, there is a 6-even subgraph 4-cover $\mathcal{F}^6 = \{C_1, \ldots, C_6\}$ of G. For each subset $B \subset \{1, \ldots, 6\}$ with $1 \leq |B| \leq 4$, let

$$E_B = \cap_{\mu \in B} E(C_\mu).$$

Since $\mathcal{F}^6$ is a 6-even subgraph 4-cover of G, we have that

$$\mathcal{B}_3 = \{E_B : B \subset \{1, \ldots, 6\}, |B| = 3\}$$

is a cover of G such that every edge belongs to precisely $\binom{4}{3} = 4$ members of $\mathcal{B}_3$. So,

$$\sum_{E_B \in \mathcal{B}_3} |E_B| = 4|E(G)|.$$

Let E_{B_1} be a member of $\mathcal{B}_3$ with $|E_{B_1}|$ as small as possible. That is,

$$|E_{B_1}| \leq \frac{4|E(G)|}{|\mathcal{B}_3|} = \frac{4|E(G)|}{\binom{6}{3}} = \frac{|E(G)|}{5}. \tag{14.7}$$

Without loss of generality, let $B_1 = \{1, 2, 3\}$ and $B_2 = \{4, 5, 6\}$. Here, $\mathcal{F} = \{C_4, C_5, C_6\}$ is a 3-even subgraph cover of G.

Let E^1 be the set of edges of G covered by precisely one member of $\mathcal{F}$, E^2 be the set of edges of G covered by precisely two members of $\mathcal{F}$, and E^3 be the set of edges of G covered by all members of $\mathcal{F}$. Note that

$E^1 = E_{B_1}$ and $E^3 = E_{B_2}$. By the choice of B_1 and B_2 ($\subset \{1, \ldots, 6\}$), we have that

$$|E^1| = |E_{B_1}| \leq |E_{B_2}| = |E^3|. \tag{14.8}$$

Since

$$\mathcal{L}(\mathcal{F}) = |E^1| + 2|E^2| + 3|E^3| = 2|E(G)| + |E^3| - |E^1|,$$

by (14.8), we have that

$$\mathcal{L}(\mathcal{F}) \geq 2|E(G)|. \tag{14.9}$$

By Lemma 14.3.27, the graph G has another 3-even subgraph cover $\mathcal{F}'$ such that

$$\mathcal{L}(\mathcal{F}') \leq 2|E(G)| + \frac{2|E^1|}{3} - \frac{\mathcal{L}(\mathcal{F})}{3}.$$

By inequalities (14.8) and (14.9),

$$\mathcal{L}(\mathcal{F}') \leq 2|E(G)| + \frac{2|E_{B_1}|}{3} - \frac{2|E(G)|}{3}.$$

By (14.7),

$$\mathcal{L}(\mathcal{F}') \leq 2|E(G)| + \frac{2|E(G)|}{15} - \frac{2|E(G)|}{3} = \frac{22}{15}|E(G)|.$$

This completes the proof of the theorem. $\qquad\qquad\square$

⋆ Fano flows and SCC_3

Theorem 14.3.21 (Máčajová, Raspaud, Tarsi and Zhu [178]) *If a bridgeless cubic graph G admits an F_5-flow then $SCC_3(G) \leq \frac{24}{15}|E(G)|$.*

Proof Given an F_5-flow f on $G = (V, E)$, select (Proposition 14.3.10) the set of missing lines to be $L_{miss} = \{\ell\langle 1, 1, 1\rangle, \ell\langle 0, 1, 1\rangle\}$ (as labeled in Figure 14.4). Thus, the vertex set $V(G)$ is partitioned into *five* subsets, each of which is V_ℓ for $\ell \in L - L_{miss}$.

Notice that three consecutive lines in the sequence

$$(\ell\langle 1, 0, 0\rangle, \ell\langle 0, 0, 1\rangle, \ell\langle 0, 1, 0\rangle, \ell\langle 1, 0, 1\rangle, \ell\langle 1, 1, 0\rangle),$$

in cyclic order (so line $\ell\langle 1, 0, 0\rangle$ follows line $\ell\langle 1, 1, 0\rangle$), do not share a common point, and therefore each such set forms a basis (by Proposition 14.3.11). By Proposition 14.3.18 and Lemma 14.3.19, each of these five bases B_i ($i = 1, 2, 3, 4, 5$) corresponds to a 3-even subgraph cover of length

$$3|V(G)| - |\cup_{\ell \in B_i} V_\ell|. \tag{14.10}$$

For each line $\ell \in L - L_{miss}$, ℓ is contained in exactly three of these bases. Therefore, for their corresponding vertex subsets V_ℓ, we have that

$$\sum_{i=1}^{5} |\cup_{\ell \in B_i} V_\ell| = 3|\cup_{\ell \in L - L_{miss}} V_\ell| = 3|V(G)|.$$

Hence, the sum of lengths of the five covers is, by (14.10),

$$\sum_{i=1}^{5} [3|V(G)| - |\cup_{\ell \in B_i} V_\ell|] =$$

$$(5 \times 3)|V(G)| - 3|V(G)| = 12|V(G)| = 8|E(G)|.$$

The average length of these five covers is $\frac{8}{5}|E(G)|$. $\qquad \square$

Theorem 14.3.22 (Máčajová, Raspaud, Tarsi and Zhu [178]) *If a bridgeless cubic graph G admits an F_4-flow then $SCC_3(G) < \frac{14}{9}|E(G)|$.*

Proof Let f be an F_4-flow on G. Let L_{miss} be the set of three missing lines. If L_{miss} has a common point $\vec{p}$, then f is a point-incomplete Fano flow. In this case, by Proposition 14.3.26, $SCC_2(G) \leq \frac{4}{3}|E(G)|$. Hence, we will only consider the case that L_{miss} has no common point. By Propositions 14.3.10 and 14.3.11, select the set of missing lines to be $L_{miss} = \{\ell\langle 1,1,1\rangle, \ell\langle 1,1,0\rangle, \ell\langle 0,1,1\rangle\}$ (as labeled in Figure 14.4). Thus, the vertex set $V(G)$ is partitioned into *four* subsets, each of which is V_ℓ for $\ell \in L - L_{miss}$. We have three bases: $B_i = \{\ell_{\langle 0,1,0\rangle}, \ell_\alpha, \ell_\beta\}$ where

$$\{\ell_\alpha, \ell_\beta\} \subset \{\ell_{\langle 1,0,0\rangle}, \ell_{\langle 0,0,1\rangle}, \ell_{\langle 1,0,1\rangle}\} = L - L_{miss} - \{\ell_{\langle 0,1,0\rangle}\}.$$

Without loss of generality (by symmetry), assume that $V_{\ell_{\langle 1,0,1\rangle}}$ is the smallest among $\{V_{\ell_{\langle 1,0,1\rangle}}, V_{\ell_{\langle 0,0,1\rangle}}, V_{\ell_{\langle 1,0,0\rangle}}\}$. That is, $|V_{\ell_{\langle 1,0,1\rangle}}| < \frac{1}{3}|V(G)|$ (note that the strict inequality holds since $V_{\ell_{\langle 0,1,0\rangle}} \neq \emptyset$ for otherwise, f is point-incomplete).

By Lemma 14.3.19, $\{C_{\ell_{\langle 0,1,0\rangle}}, C_{\ell_{\langle 0,0,1\rangle}}, C_{\ell_{\langle 1,0,0\rangle}}\}$ is a 3-even subgraph cover with total length

$$2|V(G)| + |\cup_{\ell \notin B} V_\ell| = 2|V(G)| + |V_{\ell_{\langle 1,0,1\rangle}}| < \frac{7}{3}|V(G)| = \frac{14}{9}|E(G)|$$

where B is the corresponding basis $\{\ell_{\langle 0,1,0\rangle}, \ell_{\langle 0,0,1\rangle}, \ell_{\langle 1,0,0\rangle}\}$ and $\cup_{\ell \notin B} V_\ell = V_{\ell_{\langle 1,0,1\rangle}}$ since all others are missing lines. $\qquad \square$

14.4 Open problems

Conjecture 14.1.2 (Alon and Tarsi [1]; Jaeger, see [178]) *Every bridgeless graph with m edges has an even subgraph cover with total length at most $\frac{21}{15}m$.*

A recent computer aided search [21] showed that Conjecture 14.1.2 holds for all cubic graphs of order at most 36.

Conjecture 14.3.3 (Tarsi [222]) *Every bridgeless cubic graph with m edges has a 3-even subgraph cover with total length at most $\frac{22}{15}m$.*

Conjecture 14.4.1   [213] *Let G be a graph with odd-edge-connectivity λ_o. If $\lambda_o > 3$, then G has a 2-even subgraph cover of total length at most $\frac{(\lambda_o+1)m}{\lambda_o}$ where $|E(G)| = m$.*

Conjecture 14.4.1 strengthens Theorem 14.2.4 that G has a 3-even subgraph $(1,2)$-cover of total length at most $\frac{(\lambda_o+1)m}{\lambda_o}$. Conjecture 14.4.1 is a generalization of an open problem proposed by Fan (Conjecture 14.4.2).

Conjecture 14.4.2   (Fan [50]) *Every 4-edge-connected graph G with m edges has a 2-even subgraph cover $\mathcal{F}$ with total length at most $\frac{6m}{5}$.*

The following conjecture proposed by Jackson is a weak version of Conjecture 14.4.2 (see Remark after Exercise 14.4).

Conjecture 14.4.3   (Jackson [122]) *Let G be a 4-edge-connected and 5-regular graph. Then G contains a perfect matching M which contains no odd cut of G.*

It is not hard to see that Conjecture 14.4.1 is a corollary of the following conjecture by Seymour.

Conjecture 14.4.4   (Seymour [204]) *For each $r \geq 4$, every r-graph G has a perfect matching M such that $G - E(M)$ is an $(r-1)$-graph.*

For finding an unbalanced 3-even subgraph cover, Fan and Raspaud proposed the following conjectures.

Conjecture 14.4.5   (Fan and Raspaud [53]) *Every bridgeless graph G with m edges has a 3-even subgraph cover with total length at least 2m.*

Remark. The 3-even subgraph cover described in Conjecture 14.4.5 has the following unbalanced property:

$$|E_{\langle 1,1,1\rangle}| \geq |E_{\langle 1,0,0\rangle}| + |E_{\langle 0,1,0\rangle}| + |E_{\langle 0,0,1\rangle}|.$$

A special case of Conjecture 14.4.5 for cubic graphs is presented as follows.

Conjecture 14.3.24′ (Fan and Raspaud [53]) *Every bridgeless cubic graph G has a 3-even subgraph cover $\mathcal{F}$ such that each member of $\mathcal{F}$ is a 2-factor.*

Conjecture 14.3.24′ is the original version of Conjecture 14.3.24.

Remark. Conjecture 14.3.24 is a weak version of the Berge–Fulkerson conjecture (Conjecture 7.3.6): *there are three perfect matchings M_1, M_2, M_3 with $M_1 \cap M_2 \cap M_3 = \emptyset$.* So, their complements are 2-factors and form a 3-even subgraph cover (this is an F_4-flow, see Definition 14.3.20). This 3-even subgraph cover is also unbalanced:

$$|E(C_1)| + |E(C_2)| + |E(C_3)| = 2|E(G)|$$

since each 2-factor is of size $\frac{2}{3}|E(G)|$.

Conjecture 14.3.24 (Fan and Raspaud [53]) *Every bridgeless cubic graph admits an F_4-flow.*

Conjecture 14.3.23 (Máčajová and Škoviera [175]) *Every bridgeless cubic graph admits an F_5-flow.*

Problem 14.4.6 Let G be a cubic graph admitting an F_5-flow and let C_0 be a given 2-factor of G. Can we find a 3-even subgraph cover $\{C_0, C_1, C_2\}$ such that C_1 is also a 2-factor?

14.5 Exercises

Exercise 14.1 (Corollary 14.2.3-(2)) Let $G = (V, E)$ be a graph admitting a nowhere-zero 4-flow. Then G has a 2-even subgraph cover with total length at most $\frac{4m}{3}$ where $|E(G)| = m$.

Exercise 14.2 (Theorem 14.2.4-(2) [213]) Let G be a bridgeless graph. Then $SCC_3(G) \leq \frac{2h+2}{2h+1}|E(G)|$ if the odd-edge-connectivity of G is $2h+1$ for $h \geq 2$.

Exercise 14.3 Let G be a bridgeless graph and $\mathcal{F} = \{C_1, \ldots, C_k\}$ be a k-even subgraph cover of G with the total length as short as possible. Then the subgraph of G induced by $\{e \in E(G) : ced_{\mathcal{F}}(e) > \frac{k}{2}\}$ is acyclic.

Exercise 14.4 Show that the following statement is equivalent to Conjecture 14.4.2.

Let G be a 5-regular graph with odd-edge-connectivity 5, and let $w : E(G) \mapsto \mathbb{Z}^+$ be a positive weight of G. Then G has a parity subgraph decomposition $\{P_1, P_2, P_3\}$ such that the total weight of P_1 is at most $\frac{w(G)}{5}$.

Remark. Exercise 14.4 shows that Conjecture 14.4.3 is a relaxed version of Conjecture 14.4.2 without the weight w (which is considered as the lengths of induced paths after splitting operation).

Exercise 14.5 Let G be a bridgeless cubic graph and C_1 be a given 2-factor, C_2 be a given even subgraph. Then the graph G has a 3-even subgraph cover $\{C_1, C_2, C_3\}$ if and only if, for every odd edge-cut T of G/C_1, $T \cap C_2 \neq \emptyset$.

Exercise 14.6 Show that every bridgeless cubic graph admits an F_6-flow.

Exercise 14.7 Let G be a cubic graph with a Fano flow $f = \langle f_1, f_2, f_3 \rangle$. For each Fano line $\vec{\ell} = \langle \delta_1, \delta_2, \delta_3 \rangle$ consisting of three points $\{\vec{p}_1, \vec{p}_2, \vec{p}_3\}$, show that the even subgraph $C_{\vec{\ell}}$ is

$$\bigcup_{\vec{p} \in P - \vec{\ell}} E_{\vec{p}} = \sum_{i=1}^{3} \delta_i C_i$$

where $E(C_i)$ is the support of f_i (defined in Proposition 14.3.13) and the vector $\langle \delta_1, \delta_2, \delta_3 \rangle$ is orthogonal to $\{\vec{p}_1, \vec{p}_2, \vec{p}_3\}$.

15

Beyond integer (1, 2)-weight

The faithful covering problem is introduced as a natural approach for the CDC conjecture. Although the faithful covering problem does not hold for all eulerian $(1,2)$-weighted graphs $((P_{10}, w_{10})$ is a counterexample), there are some generalizations and relaxations. Theorem 12.1.2 can be considered as a general faithful covering problem for all positive weights, while Theorem 15.1.2, Propositions 15.2.1 and 15.2.2, etc. are all considered as relaxations of the faithful covering problem.

⋆ A uniform definition in linear/integer programming

Most of the related faithful covering problems have similar definitions in graph theory and in linear/integer programming.

Definition 15.0.1 Let $G = (V, E)$ be a bridgeless graph and Γ be an additive group or additive semi-group with "0" as the additive identity. A Γ-weight w of G is a mapping: $E(G) \mapsto \Gamma$. A Γ-weight w of G is eulerian if the total weight of each edge-cut is an element of 2Γ (where $2\Gamma = \{\alpha + \alpha : \alpha \in \Gamma\}$). A Γ-faithful cover of (G, w) is an ordered pair $(\mathcal{F}, \chi)$ such that $\mathcal{F}$ is a circuit cover of G and χ is a weight: $\mathcal{F} \mapsto \Gamma$ such that

$$\sum_{e \in E(C),\ C \in \mathcal{F}} \chi(C) = w(e).$$

Let $M = [a_{ij}]_{m \times s}$ be an $(m \times s)$-matrix where $m = |E(G)|$ and s is the number of circuits in G, and the entry $a_{ij} = 1$ if the ith edge e_i is contained jth circuit C_j, and $= 0$ otherwise. And let $w : E(G) \mapsto \Gamma$ where Γ is an additive group or semi-group. Most circuit covering problems may also be uniformly defined as a linear programming (or integer programming) problem:

$$M\vec{\chi} = \vec{w} \tag{15.1}$$

where the ith term of the vertical vector $\vec{w}$ is $w(e_i)$, and the vertical vector $\vec{\chi}$ is a solution of Equation (15.1), and the jth term of the vector $\vec{\chi}$ is $\chi(C_j)$ for the jth circuit C_j (defined in Definition 15.0.1).

For example, (1) for the circuit double cover problem, in Equation (15.1), the vector $\vec{w} \in \{2\}^m$ and $\vec{\chi} \in \{0,1,2\}^s$; (2) for the faithful covering problem with respect to $(1,2)$-weight, in Equation (15.1), the vector $\vec{w} \in \{1,2\}^m$ and $\vec{\chi} \in \{0,1,2\}^s$.

15.1 Rational weights

For most faithful covering problems we have discussed, the range of the weight w is either $\mathbb{Z}^+$ or $\mathbb{Z}^\star$, a semi-group. However, if the semi-group $\mathbb{Z}^\star$ or $\mathbb{Z}^+$ is replaced with another semi-group (such as $\mathbb{Q}^\star$) or group (such as $\mathbb{Z}$), then the problem is significantly changed. This type of relaxation was first studied in [205], and will be presented in this and the next sections.

In this subsection, we investigate the faithful covering problem with respect to rational weights.

Definition 15.1.1 Let G be a graph and $w : E(G) \mapsto \mathbb{Q}^\star$ be a weight of G (where $\mathbb{Q}^\star$ is the set of all non-negative rational numbers).

(1) The weight w is **admissible** if, for each edge-cut T and each edge $e \in T$, $w(e) \leq w(T - \{e\})$.

(2) An edge-cut T of G is **critical** if T has an edge e such that $w(e) = w(T - \{e\})$.

(3) A $\mathbb{Q}^\star$-faithful cover $(\mathcal{F}, \chi)$ of (G, w) is also called a *fractional faithful circuit cover.*

We know that (P_{10}, w_{10}) is a contra pair for the faithful cover problem. But, by Proposition B.2.31, (P_{10}, w_{10}) does have a fractional faithful circuit cover $(\mathcal{F}, \chi)$ where $\mathcal{F}$ is a set of five 8-circuits and $\chi(C_\mu) = \frac{1}{2}$ for every $C_\mu \in \mathcal{F}$.

Theorem 15.1.2 (Seymour [205]) *Let G be a graph and $w : E(G) \mapsto \mathbb{Q}^\star$ be an admissible weight of G. Then (G, w) has a fractional faithful circuit cover $(\mathcal{F}, \chi)$.*

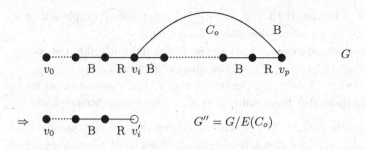

Figure 15.1 *Contraction of an "almost alternately colored circuit"* C_o

The following technical lemma will be used in the proof of Theorem 15.1.2.

Lemma 15.1.3 (Seymour [205]) *Let G be a bridgeless graph and ϕ be a mapping: $V(G) \mapsto E(G)$ such that $\phi(v) \in E(v)$ for each vertex $v \in V(G)$. Then G has a circuit C such that $\phi(v) \in E(C)$ if $v \in V(C)$.*

Proof Induction on $|V(G)| + |E(G)|$. If there is a vertex v such that $\phi(v) = vu$ and $\phi(v) \neq \phi(u)$, then contract the edge $\phi(v)$ and let the new vertex be u' with $\phi(u') = \phi(u)$. By induction, in $G' = G/\{\phi(v)\}$, there is a circuit C satisfying the lemma. If $u' \notin V(C)$, we are done. If $u' \in V(C)$ then $\phi(u') = \phi(u) \in E(C)$. Insert v and $\phi(v)$ into C if necessary. It is not hard to see that the (expanded) circuit C in G satisfies the lemma. So, we assume that $\phi(V(G))$ is a perfect matching of G.

Color the edges of $\phi(V(G))$ red and all others blue. We assume that G has no red-blue alternately bi-colored circuit. Let $P = v_0 \ldots v_p$ be a longest alternately colored (unclosed) path in G. Obviously, the last edge $v_{p-1}v_p$ of P must be red colored (for otherwise, either P can be extended with one more red edge or P can be extended to an alternately colored circuit). Hence every blue edge incident with v_p must be incident with some vertex v_i of P. Since G has no alternately red-blue bi-colored circuit, $C_o = v_i \ldots v_p v_i$ is a circuit of odd length and "almost alternately" colored (with the exception that two blue edges are incident with v_i). (See Figure 15.1.) Contract C_o to be a new vertex v_i' and let $\phi(v_i') = \phi(v_i)$. By induction, in $G'' = G/E(C_o)$, there is a circuit C'' satisfying the lemma. If $v_i' \notin V(C'')$, we are done. Let $v_i' \in V(C'')$ and therefore, $\phi(v_i') = \phi(v_i) \in E(C'')$. Insert a segment of C_o starting with a red-edge and ending at v_i into C'' if necessary; the (expanded) circuit in G satisfies the lemma. $\square$

Proof of Theorem 15.1.2. Let G be a bridgeless graph and w : $E(G) \mapsto \mathbb{Q}^*$ be an admissible weight such that

(1) the weighted graph (G, w) has no fractional faithful circuit cover,

(2) subject to (1), $|supp(w)|$ is as small as possible,

(3) subject to (1) and (2), the number of critical edge-cuts is as many as possible (note that the number of critical edge-cuts is bounded above).

We claim that (G, w) *has no non-trivial critical edge-cut*. Assume that $T = \{e_1, \ldots, e_t\}$ is a non-trivial critical edge-cut of G with $w(e_1) = \sum_{i=2}^{t} w(e_i)$. Let H_i be the graph obtained from G by contracting one component of $G - T$ ($i = 1, 2$). Since each H_i is smaller than G, let $(\mathcal{F}_i, \chi_i)$ be a fractional faithful circuit cover of (H_i, w), for each $i = 1, 2$. Denote $\mathcal{F}_i(e_\mu) = \{C \in \mathcal{F}_i : e_\mu \in E(C)\}$ for $\mu = 1, \ldots, t$ and $i = 1, 2$. Since T is critical, $\mathcal{F}_i(e_1) = \cup_{\mu=2}^{t} \mathcal{F}_i(e_\mu)$. Let N be an integer such that $N\chi_i(C)$ is an integer for every $C \in \mathcal{F}_i(e_1)$, $i = 1, 2$. In each $(\mathcal{F}_i, \chi_i)$ ($i = 1, 2$), replace each $C^* \in \mathcal{F}_i(e_1)$ with $N\chi_i(C^*)$ copies of C^*, and, for each new circuit C, let $\chi_i(C) = \frac{1}{N}$. Properly joining the circuits of $\mathcal{F}_1$ and $\mathcal{F}_2$ at the edges of T, we obtain a family $\mathcal{F}$ of circuits of G, and define a function $\chi : \mathcal{F} \mapsto \mathbb{Q}^*$ with

$$\chi(C) = \begin{cases} \chi_i(C) & \text{if } C \in \mathcal{F}_i \text{ and } E(C) \cap T = \emptyset, \\ \frac{1}{N} & \text{if } E(C) \cap T \neq \emptyset. \end{cases}$$

Obviously, $(\mathcal{F}, \chi)$ is a fractional faithful circuit cover of (G, w). Thus, we assume that (G, w) *has no non-trivial critical edge-cut.*

Let $\phi : V(G) \mapsto E(G)$ such that $\phi(v)$ is an edge of $E(v)$ with the heaviest weight. Let C be a circuit of G described in Lemma 15.1.3. Let t be the largest rational number such that (G, w') remains admissible where

$$w'(e) = \begin{cases} w(e) - t & \text{if } e \in E(C) \\ w(e) & \text{otherwise.} \end{cases}$$

Since every critical edge-cut of (G, w) is trivial and C contains the heaviest edge of $E(v)$ if it passes through a vertex v, every critical edge-cut of (G, w) remains critical in (G, w'). Obviously, one of the following must hold:

(1) there is some edge e such that $w'(e) = 0$ (in this case, $|supp(w')|$ is smaller than $|supp(w)|$),

(2) there is a new critical edge-cut (in this case, (G, w') has more critical edge-cuts than (G, w) does).

Then, by the choice of (G, w), (G, w') has a fractional circuit cover

$(\mathcal{F}', \chi')$. Adding C to $\mathcal{F}'$ and defining $\chi'(C) = t$, we obtain a fractional faithful circuit cover of (G, w). □

15.2 Group weights

In the previous section, the ranges of eulerian weights are semi-groups. If it is replaced with a group, *the existence of inverses of elements* makes the problem significantly different. This subject was also initially studied in [205] for the additive group $\mathbb{Z}$ of all integers.

Here we present a general result for all groups.

Proposition 15.2.1 *Let G be a 3-edge-connected graph and Γ be an abelian group. Then (G, w) has a Γ-faithful circuit cover for every eulerian Γ-weight w.*

Proof Denote

$$supp(w) = \{e \in E(G) : w(e) \neq 0\}.$$

Induction on $|supp(w)|$. Denote

$$E_{w=odd} = \{e \in E(G) : w(e) \in \Gamma - 2\Gamma\}.$$

Case 1. There is an edge $e_0 \in supp(w)$ with $w(e_0) \in 2\Gamma$.

Let $w(e_0) = 2\alpha$ for some $\alpha \in \Gamma - \{0\}$. Since G is 3-edge-connected, there are two edge-disjoint paths Q_1 and Q_2 connecting the endvertices of e_0 in $G - \{e_0\}$. Let $w^\star$ be a Γ-weight of G such that

$$w^\star(e) = \begin{cases} w(e) & \text{if } e \neq e_0 \\ 0 & \text{if } e = e_0. \end{cases}$$

It is clear that $w^\star$ is eulerian and with a smaller $supp(w^\star)$ ($supp(w^\star) = supp(w) - \{e_0\}$). By the inductive hypothesis, $(G, w^\star)$ has a faithful cover $(\mathcal{F}^\star, \chi^\star)$. Let $C_i = Q_i \cup \{e_0\}$ for $i = 1, 2$ and $C_3 = Q_1 \cup Q_2$, Then $(G, w^\star)$ has a faithful cover $(\mathcal{F}, \chi)$ where $\mathcal{F} = \mathcal{F}^\star \cup \{C_1, C_2, C_3\}$ and

$$\chi(C) = \begin{cases} \chi^\star(C) & \text{if } C \in \mathcal{F}^\star \\ \alpha & \text{if } C = C_1 \text{ or } C_2 \\ -\alpha & \text{if } C = C_3. \end{cases}$$

Case 2. For every $e \in supp(w)$, $w(e) \in \Gamma - 2\Gamma$.

It is evident that each vertex of G must be either incident with no edge of $E_{w=odd}$ or incident with at least two edges of $E_{w=odd}$. Thus the minimum degree of each non-trivial component of the subgraph induced by $E_{w=odd}$ is at least two and, therefore, let C_0 be a circuit contained in $G(E_{w=odd})$. Let $e_0 \in E(C_0) \subseteq E_{w=odd}$ with $w(e_0) = \beta$. Let $w^{\star\star}$: $E(G) \mapsto \Gamma$:

$$w^{\star\star}(e) = \begin{cases} w(e) & \text{if } e \notin E(C_0) \\ w(e) - \beta & \text{if } e \in E(C_0). \end{cases}$$

It is obvious that $w^{\star\star}$ is eulerian and with a smaller $supp(w^{\star\star})$ (since $supp(w^{\star\star}) \subseteq supp(w) - \{e_0\}$). By the inductive hypothesis, $(G, w^{\star\star})$ has a faithful cover $(\mathcal{F}^{\star\star}, \chi^{\star\star})$. Furthermore, (G, w) has a faithful cover $(\mathcal{F}, \chi)$ where $\mathcal{F} = \mathcal{F}^{\star\star} \cup \{C_0\}$ and

$$\chi(C) = \begin{cases} \chi^{\star\star}(C) & \text{if } C \in \mathcal{F}^{\star\star} \\ \beta & \text{if } C = C_0. \end{cases}$$

$\square$

The following is an immediate corollary of Proposition 15.2.1.

Proposition 15.2.2 *Let G be a graph and Γ be an abelian group. If w is an eulerian Γ-weight of G such that $w(e) = w(e')$ whenever $\{e, e'\}$ is a 2-edge-cut, then (G, w) has a faithful circuit cover.*

Proof Exercise 15.1. $\square$

15.3 Integer weights

15.3.1 Non-negative weights and Petersen minor

The following theorem is a generalization of Theorem 3.3.1 that the range of eulerian weights is extended to all non-negative integers. That is, in Equation (15.1), the vector $\vec{w} \in (\mathbb{Z}^{\star})^m$ and $\vec{\chi} \in (\mathbb{Z}^{\star})^s$ ($\mathbb{Z}^{\star}$ is the set of all non-negative integers).

Theorem 12.1.2 (Alspach, Goddyn and Zhang [4]) *A bridgeless graph G has a faithful circuit cover with respect to every admissible eulerian weight $w : E(G) \mapsto \mathbb{Z}^{\star}$ if and only if G does not contain a subdivision of the Petersen graph.*

15.3.2 Integer semi-group weights

For a positive integer k, the set $k\mathbb{Z}^{\star} = \{k\mu : \mu \in \mathbb{Z}^{\star}\}$ is a semi-group. We consider the following general problem.

Problem 15.3.1 Let G be a graph and k be a positive integer. Does G have a faithful circuit cover with respect to every admissible, eulerian weight $w : E(G) \mapsto k\mathbb{Z}^*$ of G?

For $k = 1$, Theorem 12.1.2 states that *Problem 15.3.1 is true if and only if G does not have a Petersen minor.*

For $k \geq 2$, it remains widely open. The following is a long standing open problem for $k = 2$.

Conjecture 2.4.1 (Seymour [205]) *Let G be a graph. Then G has a faithful circuit cover with respect to every admissible, eulerian weight $w : E(G) \mapsto 2\mathbb{Z}^*$ of G.*

Even with an arbitrary large integer k ($k \geq 2$), there is no solution yet for Problem 15.3.1. Problem 15.3.1 is reformulated as follows.

Problem 15.3.2 Find an integer k such that every graph G has a faithful circuit cover with respect to every admissible, eulerian weight $w : E(G) \mapsto k\mathbb{Z}^*$ of G.

Theorem 15.1.2 told us that, for any given *admissible eulerian weighted graph (G, w) with $w : E(G) \mapsto \mathbb{Z}^*$, there is an integer k such that (G, kw) has a faithful circuit cover.* (Note that the integer k is the least common multiple of the denominators of the rational function χ.) This integer k depends on both the graph G and the integer weight w. The goal of Problem 15.3.2 is to find *a universal upper bound* for all graphs and all integer weights.

15.3.3 Small range

Another variation of faithful cover and the double cover problem is the restriction of the integer weight w in a small range. The CDC conjecture restricts the eulerian weight in the single digit range $\{2\}$. For eulerian weights with ranges restricted in a single digit range, the faithful covering problem is completely solved for all integers other than 2. Let $w : E(G) \mapsto \{h\}$. If h is an odd integer, then it is trivial since G must be an even graph (Proposition 7.3.2). If h is even and ≥ 4, it is solved completely by the combination of Theorems 7.3.4 and 7.3.5.

Theorem 7.3.3 (Combination of Theorems 7.3.4 and 7.3.5) *Every bridgeless graph G has a circuit $2s$-cover if $2s \geq 4$.*

With a specified range of w, we may consider the following problem.

Problem 15.3.3   Let $w : E(G) \mapsto S$ be an admissible eulerian weight of G where $S \subseteq \mathbb{Z}^+$. For which S, (G, w) has a faithful circuit cover?

In this subsection, we present some solved cases for $|S| = 2$.

Theorem 15.3.4   (DeVos, Johnson and Seymour [40]) *Let G be a 3-edge-connected graph and $w : E(G) \mapsto \{2s, 2s + 2\}$. Then (G, w) has a faithful circuit cover if $2s = 84$ or $2s \geq 88$.*

Theorem 15.3.5   (DeVos, Johnson and Seymour [40]) *Let G be a 3-edge-connected graph and $w : E(G) \mapsto \{2s, 2s + 4\}$. Then (G, w) has a faithful circuit cover if $2s = 32$ or $2s \geq 36$.*

Theorem 15.3.6   (DeVos, Johnson and Seymour [40]) *Let G be a 3-edge-connected graph and $w : E(G) \mapsto \{2s, 2s + 6\}$. Then (G, w) has a faithful circuit cover if $2s = 48$ or $2s \geq 54$.*

These theorems will be proved after the presentation of the following lemmas. The following is one of the key lemmas in the proofs.

Lemma 15.3.7   (DeVos, Johnson and Seymour [40]) *Let G be a 3-edge-connected graph. Then $E(G)$ has a partition $\{E_1, \ldots, E_9\}$ such that $G - E_\mu$ is 2-edge-connected for every $\mu \in \{1, \ldots, 9\}$.*

By Lemma 15.3.7, we have the following lemma.

Lemma 15.3.8   (DeVos, Johnson and Seymour [40]) *Let G be a 3-edge-connected graph and $\{A, B\}$ be a partition of $E(G)$. Then G has families $\mathcal{F}_1$, $\mathcal{F}_2$ and $\mathcal{F}_3$ of circuits such that*

(1) $\mathcal{F}_1$ covers every edge of A 32 times and every edge of B 36 times,

(2) $\mathcal{F}_2$ covers every edge of A 48 times and every edge of B 54 times,

(3) $\mathcal{F}_3$ covers every edge of A 84 times and every edge of B 86 times.

Proof   (1) By Lemma 15.3.7, let $\{E_1, \ldots, E_9\}$ be a partition of $E(G)$. By Theorem 7.3.3 ($2s = 4$), each 2-edge-connected subgraph $G - (E_\mu \cap A)$ has a circuit 4-cover $\mathcal{C}_\mu$. Thus, $\bigcup_{\mu=1}^{9} \mathcal{C}_\mu$ is a circuit cover such that every edge of A is covered 32 (4×8) times, and every edge of B is covered 36 (4×9) times.

(2) Similar to (1). Applying Theorem 7.3.3 again with $2s = 6$, we have a family of circuits such that every edge of A is covered 48 (6×8) times, and every edge of B is covered 54 (6×9) times.

(3) This is a combination of (1) and (2). Let $\mathcal{F}_1^*$ be the circuit cover obtained from (1) such that every edge of A is covered 36 times, and

every edge of B is covered 32 times, and let $\mathcal{F}_2^*$ be the circuit cover obtained from (2) such that every edge of A is covered 48 times, and every edge of B is covered 54 times. Thus, the family $\mathcal{F}_1^* \cup \mathcal{F}_2^*$ of circuits is the one we needed. □

Theorems 15.3.4, 15.3.5 and 15.3.6 are immediate corollaries of the following observation and Lemma 15.3.8.

Lemma 15.3.9 *Let s, t, k be integers and G be a bridgeless graph. Let $\{A, B\}$ be a partition of $E(G)$. If G has a circuit cover such that every edge of A is covered s times and every edge of B is covered t times, then G has a circuit cover such that every edge of A is covered $(s + 2k)$ times and every edge of B is covered $(t + 2k)$ times, for every $k \geq 2$.*

Lemma 15.3.9 is a straightforward application of Theorem 7.3.3 (by adding a $2k$-cover).

Some other small range cases are also solved and are left as exercises at the end of this chapter (Exercises 15.5 and 15.6).

15.4 Open problems

Conjecture 2.4.1 (Seymour [205]) *Let G be a graph. Then G has a faithful circuit cover with respect to every admissible, eulerian weight $w : E(G) \mapsto 2\mathbb{Z}^*$ of G.*

Conjecture 2.4.2 (Goddyn [86]) *Let $w : E(G) \mapsto \mathbb{Z}^+$ be an admissible eulerian weight of a bridgeless graph G. If $w(e) \geq 2$ for every edge e of G, then (G, w) has a faithful circuit cover.*

Problem 15.3.2 *Find an integer k such that every graph G has a faithful circuit cover with respect to every admissible, eulerian weight $w : E(G) \mapsto k\mathbb{Z}^*$ of G.*

15.5 Exercises

⋆ Group weights

Exercise 15.1 (Proposition 15.2.2) Let G be a graph and Γ be an abelian group. If w is an eulerian Γ-weight of G such that $w(e) = w(e')$ whenever $\{e, e'\}$ is a 2-edge-cut, then G has a faithful circuit cover with respect to w.

⋆ Fractional weights

Exercise 15.2 Show that, for the problem of fractional faithful circuit cover (Theorem 15.1.2), it is necessary that the weight w is admissible.

Exercise 15.3 Show that, for the problem of fractional faithful circuit cover (Theorem 15.1.2), it is not necessary to assume that the weight w is eulerian.

⋆ Double covers

Exercise 15.4 Show the following *"proof"* of circuit double cover conjecture is wrong.

"Proof - an algorithm." *Input.* Let $w_0 : E(G_0) \mapsto \{1,2\}$ be an eulerian weight of a bridgeless cubic graph G_0.

Let $i \leftarrow 0$.

Step I. Define a mapping $\phi : V(G_i) \mapsto E(G_i)$ where $\phi(v) \in E(v) \cap E_{w_i=2}$ for every vertex $v \in V(G_i)$. Applying Lemma 15.1.3, we can find a circuit C_i which contains the edge $\phi(v)$ if $v \in V(C_i)$.

Step II. Define $w_{i+1} : E(G_i) \mapsto \{0,1,2\}$ where $w_{i+1}(e) = w_i(e)$ if $e \notin E(C_i)$, and, $= w_i(e) - 1$ if $e \in E(C_i)$.

Let $i \leftarrow i + 1$ and *repeat Step I* on the cubic graph $\overline{supp(w_i)}$ with respect to the new eulerian weight w_i until $G_i = \emptyset$.

Output. Then the collection of all those circuits $\{C_0, \ldots, C_i\}$ from Step I is a faithful circuit cover of (G_0, w_0).

⋆ Small ranges

Exercise 15.5 Let h_o be an odd integer and h_e be an even integer. Let G be a bridgeless graph and $w : E(G) \mapsto \{h_o, h_e\}$ be an admissible eulerian weight of G. Then (G, w) has a faithful circuit cover if $h_o > h_e \geq 4$.

Exercise 15.6 Let h_o be an odd integer and h_e be an even integer. Let G be a bridgeless graph and $w : E(G) \mapsto \{h_o, h_e\}$ be an admissible eulerian weight of G. Then (G, w) has a faithful circuit cover if $h_e - 1 = h_o \geq 3$.

16

Petersen chain and Hamilton weights

Let us start this chapter with a wrong "proof" of the CDC conjecture.

"Proof" Let G be a smallest counterexample to the CDC conjecture and $e_0 = x_0 y_0 \in E(G)$. Let $\mathcal{F}$ be a circuit double cover of $G - e_0$.

Let $\mathcal{C} = \{C_1, \ldots, C_t\}$ $(\subseteq \mathcal{F})$ be a chain of circuits joining x_0 and y_0. That is,

$$x_0 \in C_1, \quad y_0 \in C_t \text{ and } C_i \cap C_j \neq \emptyset \text{ if and only if } i = j \pm 1.$$

Suppose that

$$|E(C_i) \cap E(C_{i+1})| = 1$$

for every $i = 1, \ldots, t - 1$. (See Figure 16.1.) Then it is easy to see that the suppressed cubic graph $\overline{G[C_1 \cup \cdots \cup C_t \cup e_0]}$ is 3-edge-colorable. By Lemma 2.2.1, the induced subgraph $G[C_1 \cup \cdots \cup C_t \cup e_0]$ has a circuit cover $\mathcal{C}'$ such that each edge e is covered by μ members of $\mathcal{C}'$ if e is covered by μ members of $\mathcal{C}$ and the missing edge e_0 is covered twice. By replacing $\mathcal{C}$ with $\mathcal{C}'$ in $\mathcal{F}$, we obtain a circuit double cover of the entire graph G.

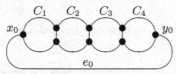

Figure 16.1 *Circuit chain joining x_0 and y_0*

Note that this *wrong proof* of the CDC conjecture uses the assumption that $|E(C_i) \cap E(C_{i+1})| = 1$ for every $i = 1, \ldots, t - 1$. It is very possible that $|E(C_i) \cap E(C_{i+1})| > 1$ for some i, and the intersection of C_i and C_j may "wind together" in a somehow complicated way (forming some subdivision of K_4). Furthermore, the complicated local structure of $C_i \cup C_j$ may cause non-3-edge-colorability of $\overline{G[C_1 \cup \cdots \cup C_t \cup e_0]}$. The Petersen graph P_{10} is an example: let $e_0 = x_0 y_0 \in E(P_{10})$, the subgraph $P_{10} - e_0$ has a circuit chain $\{C_1, C_2, C_3\}$ joining x_0 and y_0 and covering $E(P_{10}) - e_0$ (see Figure B.16). We notice that $|E(C_i) \cap E(C_{i+1})| = 2$ and the suppressed cubic graph $\overline{C_i \cup C_{i+1}} = K_4$ for each $i = 1, 2$.

From this wrong "proof," we may see that the structure of two incident circuits in a circuit cover plays a central role in some approaches to the CDC conjecture. This motivates the study of the "Hamilton weighted graph," which will provide some valuable structural information for circuit chains.

16.1 Local structures

Since a minimum counterexample to the circuit double cover conjecture is cubic and 3-connected (Theorem 1.2.4), we will discuss circuit covering problems for cubic graphs in most of this chapter.

Let G be a smallest counterexample to the circuit double cover conjecture and let $e_0 = x_0 y_0 \in E(G)$. Then $G - e_0$ has a circuit double cover $\mathcal{C}$. Let $\mathcal{P}$ ($\mathcal{P} \subseteq \mathcal{C}$) be a circuit chain joining the endvertices x_0, y_0 of the uncovered edge e_0 (see Figure 16.2).

Can we find a family $\mathcal{Q}$ of circuits of the induced subgraph $G[\{e_0\} \cup \bigcup_{C \in \mathcal{P}} E(C)]$ such that each edge $e \in \bigcup_{C \in \mathcal{P}} E(C)$ is covered by μ members of $\mathcal{Q}$ if e is covered by μ members of $\mathcal{P}$, and the edge e_0 is covered twice? (An example: Figure 16.3.)

If "yes," then $\mathcal{C} - \mathcal{P} + \mathcal{Q}$ is a circuit double cover of G. This contradicts that G is a counterexample to the CDC conjecture. So, the answer must be "NO."

Question 16.1.1 What is the structure of the induced subgraph $G[\{e_0\} \cup \bigcup_{C \in \mathcal{P}} E(C)]$?

The circuit chain is one of the most popular approaches to the CDC conjecture, and appeared originally in [205]. Motivated and promoted by this approach, some related structural studies, new concepts, and techniques have been introduced and studied ([205], [3], [4], etc.).

Our goal is to determine the structure of the graph described in Question 16.1.1. The main result (Theorem 16.5.7 in this chapter) further generalizes some earlier results in [3], [4], [261], and others.

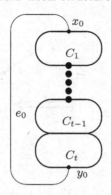

Figure 16.2 *Circuit chain $\mathcal{P}$ joining the endvertices of e_0*

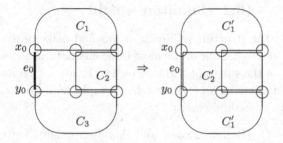

Figure 16.3 *Circuit cover adjustment for a circuit chain $\{C_1, C_2, C_3\}$ and a missing edge e_0*

The eulerian $(1,2)$-weighted graph (P_{10}, w_{10}) illustrated in Figure 2.2 (and Figure B.8) is the minimum contra pair. It was proved in Theorem 3.3.1 that every contra pair must contain a Petersen minor. However, that structural information is not sufficient for further study or final attack of the circuit double cover conjecture.

There are articles/results ([46], [198], [212]) providing powerful tools for finding a Petersen minor, but few applicable results yet for the determination of the precise structure of the Petersen graph, while many long standing open problems (Conjectures 3.5.1, 3.5.3, 3.5.4, etc.) demand the precise structure of the Petersen graph (instead of graphs with Petersen minor, the Petersen graph is expected to be the only exception to those conjectures about minimal or critical contra pairs). The determinations of the structures of the Petersen graph and the existence of the Petersen minor are significantly different in nature. One of the most important parts of the main theorems in this chapter is to determine the Petersen graph.

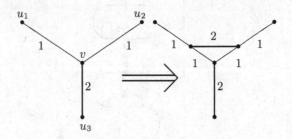

Figure 16.4 $(Y \to \triangle)$-*operation*

16.2 Hamilton weight

In order to study the structure of circuit chains and make possible adjustments of circuit covers, one of the most basic and natural steps is the characterization of the subgraph induced by two incident circuits. This motivates us to study eulerian $(1,2)$-weighted graphs with precisely two Hamilton circuits as faithful covers.

Definition 16.2.1 (Hamilton cover and Hamilton weight) Let G be a bridgeless cubic graph associated with an eulerian weight $w : E(G) \mapsto \{1,2\}$. A faithful circuit cover $\mathcal{F}$ of the eulerian weighted graph (G,w) is a Hamilton cover if $\mathcal{F}$ consists of pair of Hamilton circuits.

If every faithful circuit cover of (G,w) is a Hamilton cover, then w is a Hamilton weight of G, and (G,w) is a Hamilton weighted graph.

A special case of Hamilton weighted graphs, strong $3H$ CDC, was briefly mentioned in Section 5.4. Actually, cubic graphs with the property of strong $3H$ CDC have the most applications in the study of circuit chain structure since, in most cases, $E_{w=1}$ induces a Hamilton circuit.

Definition 16.2.2 $((Y \to \triangle)$-operation, see Figure 16.4) Let v be a degree 3 vertex of an eulerian weighted graph (G,w) incident with $E(v) = \{e_i = vu_i : i = 1,2,3\}$. A $(Y \to \triangle)$-operation of (G,w) at the vertex v is the construction of a new weighted graph (G',w') from (G,w) by splitting v into three degree 1 vertices $\{v_1,v_2,v_3\}$ where v_i is incident with e_i, and adding a triangle $v_1v_2v_3v_1$ and assigning $w'(e) = w(e)$ if $e \in E(G) - \{e_1,e_2,e_3\}$ and $w'(v_jv_i) = w'(v_hu_h) = w(e_h)$ to new edges for every $\{h,i,j\} = \{1,2,3\}$.

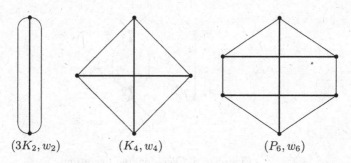

$$(3K_2, w_2) \qquad (K_4, w_4) \qquad (P_6, w_6)$$

Figure 16.5 *Three small* $\langle \mathcal{K}_4 \rangle^{HW}$*-graphs*

Observation 16.2.3 Let (G', w') be a weighted graph obtained from another weighted graph (G, w) via a $(Y \to \triangle)$-operation. Then w is a Hamilton weight of G if and only if w' is a Hamilton weight of G'.

Definition 16.2.4 The family of weighted graphs constructed from $(3K_2, w_2)$ via a series of $(Y \to \triangle)$-operations is denoted by $\langle \mathcal{K}_4 \rangle^{HW}$.

Three smallest weighted graphs in $\langle \mathcal{K}_4 \rangle^{HW}$ are illustrated in Figure 16.5. A Hamilton weight is also illustrated in the figure. Bold lines are weight 2 edges, and slim lines are weight 1 edges: $(3K_2, w_2)$ *is the graph* $3K_2$ *(three parallel edges on two vertices) with the eulerian weight* w_2 *that* $|E_{w_2=2}| = 1$; (K_4, w_4) *is the complete graph* K_4 *with the eulerian weight* w_4 *such that* $E_{w_4=2}$ *induces a perfect matching.*

Conjecture 16.2.5 [256] *Let* (G, w) *be a Hamilton weighted graph. If* G *is 3-connected, then* $(G, w) \in \langle \mathcal{K}_4 \rangle^{HW}$.

In [165], this conjecture was proved for the family of *Petersen minor free* graphs (Theorem 16.4.1). Its relation with the problem of *uniquely 3-edge-coloring* can be found in [256], [259].

Definition 16.2.6 (Edge-dividing, see Figure 16.6) Let (G, w) be an eulerian weighted graph and $e_0 \in E_{w=2}$ with end-vertices x_1 and x_2. Let G^* be the cubic graph obtained from G by deleting the edge e_0 and adding two new vertices $\{y_1, y_2\}$ and four new edges $\{e_1, e_2, f_1, f_2\}$ where, for each $i = 1, 2$, x_i and y_i are the endvertices of e_i, and y_1 and y_2 are the endvertices of the parallel edges f_1 and f_2. Let w^* be the weight of G^* obtained from w: $w^*(e) = w(e)$ if $e \notin \{e_1, e_2, f_1, f_2\}$, and $w^*(e_i) = 2$, $w^*(f_i) = 1$ for each $i = 1, 2$. Then (G^*, w^*) is the weighted graph obtained from (G, w) via an **edge-dividing operation** at e_0.

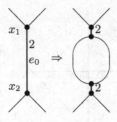

Figure 16.6 *Edge-dividing operation*

Observation 16.2.7 Let (G^*, w^*) be a weighted graph obtained from another weighted graph (G, w) via an edge-dividing operation. Then w is a Hamilton weight of G if and only if w^* is a Hamilton weight of G^*.

Definition 16.2.8 The family of eulerian $(1, 2)$-weighted graphs (G, w) constructed from $(3K_2, w_2)$ via a series of $(Y \to \triangle)$-operations and edge-dividing operations is denoted by $\langle \mathcal{K}_4 \rangle_2^{HW}$.

By Observations 16.2.3 and 16.2.7, the following lemma is straightforward.

Lemma 16.2.9 *Under the assumption of Conjecture 16.2.5, every Hamilton weighted graph is a member of $\langle \mathcal{K}_4 \rangle_2^{HW}$.*

16.3 $\langle \mathcal{K}_4 \rangle$-graphs

The following is a straightforward observation about small edge-cuts and small circuits of $\langle \mathcal{K}_4 \rangle_2^{HW}$-graphs.

Observation 16.3.1 For each $(G, w) \in \langle \mathcal{K}_4 \rangle_2^{HW}$ with $|V(G)| \geq 6$, we have the following properties.

(1) (G, w) has a non-trivial 2- or 3-edge-cut with total weight 4; and $(G, w) \in \langle \mathcal{K}_4 \rangle^{HW}$ if and only if G is 3-connected.

(2) For each non-trivial 2- or 3-edge-cut T with components Q_1 and Q_2, each Q_j contains a triangle with total weight 4 or a digon with total weight 2.

(3) All triangles of (G, w) are vertex disjoint if $(G, w) \in \langle \mathcal{K}_4 \rangle^{HW}$.

$\star$ $(Y \leftrightarrow \triangle)$-operations

The contraction of a triangle is called a $(\triangle \to Y)$-operation. And a $(Y \leftrightarrow \triangle)$-operation is either a $(Y \to \triangle)$- or a $(\triangle \to Y)$-operation.

Lemma 16.3.2 [261] *Let S be a triangle of a weighted graph (G, w). Then $(G, w) \in \langle \mathcal{K}_4 \rangle^{HW}$ if and only if the contracted weighted graph $(G/S, w) \in \langle \mathcal{K}_4 \rangle^{HW}$.*

Proof The "if" part is obvious by Definition 16.2.4. Here, we only prove the "only if" part.

Induction on $|V(G)| = 2n$. This is trivial if $G = 3K_2$ and $G = K_4$. Let G be obtained from $3K_2$ via the following sequence of $(Y \to \triangle)$-operations: $\pi_1, \ldots, \pi_{n-1}$. Let S^* be the triangle of G that is created by the last $(Y \to \triangle)$-operation π_{n-1}. By Definition 16.2.4, $(G/S^*, w) \in \langle \mathcal{K}_4 \rangle^{HW}$ since it is obtained from $3K_2$ via the sequence of $(Y \to \triangle)$-operations: $\pi_1, \ldots, \pi_{n-2}$.

If $S^* = S$, then $(G/S, w) = (G/S^*, w) \in \langle \mathcal{K}_4 \rangle^{HW}$. Hence, we assume that $S^* \neq S$. Since $(G/S^*, w) \in \langle \mathcal{K}_4 \rangle^{HW}$ and is smaller than G, by the inductive hypothesis, we have that $((G/S^*)/S, w) \in \langle \mathcal{K}_4 \rangle^{HW}$ (note that, S and S^* are disjoint by Observation 16.3.1-(3)). By Definition 16.2.4, $(G/S^*)/S$ is obtained from $3K_2$ via a sequence of $(Y \to \triangle)$-operations: $\pi_1^*, \ldots, \pi_{n-3}^*$. Note that $(G/S^*)/S = (G/S)/S^*$ since S and S^* are disjoint (by Observation 16.3.1). So, $(G/S, w)$ is obtained from $(3K_2, w_2)$ via the sequence of $(Y \to \triangle)$-operations: $\pi_1^*, \ldots, \pi_{n-3}^*, \pi_{n-1}$. $\qquad \square$

By Lemma 16.3.2, $(Y \leftrightarrow \triangle)$-operations expand or reduce $\langle \mathcal{K}_4 \rangle^{HW}$-graphs. Hence, they provide an inductive tool in the study of $\langle \mathcal{K}_4 \rangle^{HW}$-graphs.

Observation 16.3.3   (1) Let $(G_1, w_1), (G_2, w_2) \in \langle \mathcal{K}_4 \rangle_2^{HW}$, then

$$(G_1, w_1) \oplus_\mu (G_2, w_2) \in \langle \mathcal{K}_4 \rangle_2^{HW}$$

for each $\mu = 2, 3$, where an $\oplus_2$-sum is applied on a pair of weight 2 edges.
   (2) Let $(G_1, w_1), (G_2, w_2) \in \langle \mathcal{K}_4 \rangle^{HW}$, then

$$(G_1, w_1) \oplus_3 (G_2, w_2) \in \langle \mathcal{K}_4 \rangle^{HW}.$$

Note that $\oplus_\mu$-sums for weighted graphs are defined in Section 2.3.

By Observation 16.3.1 and Lemma 16.3.2, we further extend Observation 16.3.3 as follows.

Observation 16.3.4   Let $(G, w) \in \langle \mathcal{K}_4 \rangle_2^{HW}$ and $\mu = 2$ or 3. And let T

be a non-trivial edge-cut of G with components Q_1, Q_2. If $|T| = \mu$ and the total weight of T is 4, then both

$$(G/Q_1, w), \text{ and } (G/Q_2, w) \in \langle \mathcal{K}_4 \rangle_2^{HW}.$$

And,

$$(G, w) = (G/Q_1, w) \oplus_\mu (G/Q_2, w).$$

⋆ Unweighted $\langle \mathcal{K}_4 \rangle$-graphs

The $(Y \to \triangle)$-operation defined in Definition 16.2.2 can also be applied to unweighted cubic graphs. It was studied briefly in Chapter 5 (see Method 5.1 and Figure 5.8).

Definition 16.3.5 The family of cubic graphs constructed from $3K_2$ via a series of $(Y \to \triangle)$-operations is denoted by $\langle \mathcal{K}_4 \rangle$.

Note that the $\langle \mathcal{K}_4 \rangle$-graph is defined the same as Definition 16.2.4 but with no weight (also defined in Exercise 5.11).

Though there is no eulerian weight associated with graphs of $\langle \mathcal{K}_4 \rangle$, all observations and lemmas in this section remain valid for graphs of $\langle \mathcal{K}_4 \rangle$.

Observation 16.3.6 For an unweighted cubic graph G, the $(\triangle \to Y)$-operation and the $(Y \to \triangle)$-operation preserve each of the following properties of G:

(1) planarity,
(2) 3-edge-colorability,
(3) number of 1-factorizations,
(4) number of Hamilton circuits,
(5) number of Hamilton weights,
(6) number of Hamilton covers of an eulerian $(1, 2)$-weight.

⋆ Untouched edges

One may notice that, for a $\langle \mathcal{K}_4 \rangle$-graph G with a triangle S, the $(\triangle \to Y)$-operation on S is equivalent to the deletion of an edge of S and the suppression of the resulting graph. Thus, there are some edges e of G with the following property:

$$\overline{G - e} \in \langle \mathcal{K}_4 \rangle.$$

Which edge e of G has such a property? It is not necessary that e must be contained in a triangle. Definition 16.3.9 about *"untouched"* edges (and

Lemma 16.3.11, Lemma 16.3.13) classifies all such edges. Those structural lemmas characterize edges for the following two technical questions and will be applied in inductive processing.

Question 16.3.7 If $G \in \langle \mathcal{K}_4 \rangle$, characterize all edges e of G such that

$$\overline{G - e} \in \langle \mathcal{K}_4 \rangle.$$

Question 16.3.8 Let $G \in \langle \mathcal{K}_4 \rangle$. Let G^+ be the graph obtained from G by inserting two degree 2 vertices x, y into some pair of edges e_1, e_2 of G, and add a new edge e_0 joining x, y. Characterize all pairs of edges $\{e_1, e_2\}$ of G such that

$$G^+ \in \langle \mathcal{K}_4 \rangle.$$

The following definition and Lemma 16.3.11 give a response for Question 16.3.7.

Definition 16.3.9 For a $\langle \mathcal{K}_4 \rangle$-graph G of order at least 4, and for a series Ω of $(Y \to \triangle)$-operations that constructs the graph G from K_4, an edge $e \in E(G)$ is labeled as touched if one endvertex of e is expanded to a triangle by at least one $(Y \to \triangle)$-operation of Ω. Edges that are not "touched" are labeled as untouched.

Definition 16.3.10 Let G be a cubic graph and H_1, H_2 be subgraphs of G. An edge-attachment of H_2 in the suppressed graph $\overline{H_1}$ is an edge $e = uv$ of $\overline{H_1}$ such that the edge e corresponds to a maximal induced path $P = u \dots v$ (in H_1) and $V(H_2) \cap [V(P) - \{u, v\}] \neq \emptyset$.

Lemma 16.3.11 characterizes all edges e described in Question 16.3.7 such that *the suppressed cubic graph $\overline{G - e}$ remains in $\langle \mathcal{K}_4 \rangle$*.

Lemma 16.3.11 *Let $G \in \langle \mathcal{K}_4 \rangle$ of order at least 4, and $e = v_1 v_2 \in E(G)$. For each $i = 1, 2$, let e_i be the edge of $\overline{G - e}$ containing the vertex v_i (that is, e_i is an edge-attachment of e in the suppressed graph $\overline{G - e}$). Then the following statements are equivalent:*

(1) every 3-edge-cut of G containing e is trivial (that is, $E(v_1)$ and $E(v_2)$ are the only two 3-edge-cuts of G containing the edge e);

(2) the suppressed cubic graph $\overline{G - e} \in \langle \mathcal{K}_4 \rangle$;

(3) $\{e_1, e_2\}$ is contained in a 3-edge-cut of $\overline{G - e}$;

(4) e is untouched (see Definition 16.3.9).

Proof Induction on $|V(G)|$. It is trivial that the lemma is true for K_4. Hence, we assume that $|V(G)| \geq 6$.

(1) $\Leftrightarrow$ (4): By Definition 16.3.9. $\diamondsuit$

$(1) \Rightarrow (2)$ and (3): *(i)* We claim that *there is a triangle S in G such that e is not incident with any vertex of S.*

By Observation 16.3.1, G contains at least two disjoint triangles. Suppose that the edge e is incident with *every* triangle of G. Let $e = x_1 y_1$, and $S_1 = x_1 x_2 x_3 x_1$, $S_2 = y_1 y_2 y_3 y_1$ which are the only two triangles of G. Note that, the set T of all edges incident with S_1 but not in S_1 is a non-trivial 3-edge-cut of G. This contradicts the assumption that every 3-edge-cut of G containing e is trivial.

(ii) By (i), let S_3 be a triangle of G such that the edge e is not incident with any vertex of S_3. Note that, by Lemma 16.3.2, $G/S_3 \in \langle \mathcal{K}_4 \rangle$ and satisfies the description of the lemma since the contraction of S_3 does not create any new 3-edge-cut of G/S_3. By induction, Statements (2) and (3) hold for G/S_3. That is, *the suppressed graph $\overline{G/S_3 - e} \in \langle \mathcal{K}_4 \rangle$ and $\{e_1, e_2\}$ is contained in a 3-edge-cut T' of G/S_3.*

Note that $\overline{(G-e)/S_3} = \overline{G/S_3 - e}$. This implies that $\overline{G - e} \in \langle \mathcal{K}_4 \rangle$ since $\overline{G - e}$ is obtained from $\overline{(G-e)/S_3} = \overline{G/S_3 - e}$ via a $(Y \to \triangle)$-operation. This proves Statement (2).

(iii) Continue from (ii), the 3-edge-cut T' of $\overline{(G-e)}/S_3$ containing the edges e_1, e_2 remains as a 3-edge-cut in $\overline{G-e}$ (after a $(Y \to \triangle)$-operation) since e is not incident with any vertex of S_3. Statement (3) is therefore verified. $\Diamond$

$(3) \Rightarrow (1)$ and (2): Let $T' = \{e_1, e_2, f_0\}$ be a 3-edge-cut of $\overline{G - e}$ where e_1, e_2 are subdivided edges of $G - e$ containing endvertices of e. The cut T' corresponds to a 3-edge-cut T in G with components Q_1 and Q_2 (with $e \in Q_1$). By Observation 16.3.1, let S be a triangle of G/Q_1 not containing the contracted vertex. (1) and (2) are proved by applying the induction on $\overline{G/S}$. $\Diamond$

$(2) \Rightarrow (3)$: Since $\overline{G - e} \in \langle \mathcal{K}_4 \rangle$, the cubic graph $\overline{G - e}$ contains at least two disjoint triangles. If every triangle of $\overline{G - e}$ contains some of $\{e_1, e_2\}$, then $\overline{G - e}$ has only two triangles and, therefore, G is of girth ≥ 4. This contradicts that $G \in \langle \mathcal{K}_4 \rangle$. So, let S be a triangle of $\overline{G - e}$ not containing any of $\{e_1, e_2\}$.

In $\overline{G - e}/S$, by induction, (3) holds for the smaller $\overline{G - e}/S$. That is, $\{e_1, e_2\}$ is contained in a 3-edge-cut T' in $\overline{G - e}/S$. T' remains as a 3-edge-cut of $\overline{G - e}$ after a $(Y \to \triangle)$-operation. $\square$

Lemma 16.3.11 can be easily extended to Hamilton weighted graphs in $\langle \mathcal{K}_4 \rangle_2^{HW}$ as follows.

Lemma 16.3.12 *Let $(G, w) \in \langle\mathcal{K}_4\rangle_2^{HW}$ of order at least 4, and $e = v_1 v_2 \in E_{w=2}$. For each $i = 1, 2$, let e_i be the edge of $\overline{G - e}$ containing the vertex v_i (that is, e_i is an edge-attachment of e in the suppressed graph $\overline{G - e}$, and $e_i \in E_{w=1}$). Then the following statements are equivalent:*

(1) every 3-edge-cut of G containing e is trivial (that is, $E(v_1)$ and $E(v_2)$ are the only two 3-edge-cuts of G containing the edge e);

(2) the suppressed cubic graph $(\overline{G - e}, w) \in \langle\mathcal{K}_4\rangle_2^{HW}$;

(3) $\{e_1, e_2\}$ is contained in a 3-edge-cut of $(\overline{G - e}, w)$;

(4) e is untouched (see Definition 16.3.9).

The following Lemma 16.3.13 characterizes the addition of a new edge e_0 described in Question 16.3.8 (and more).

Lemma 16.3.13 *Let $(G, w) \in \langle\mathcal{K}_4\rangle_2^{HW}$. Let (G^+, w) be the weighted graph obtained from (G, w) by inserting two degree 2 vertices x_0, y_0 into some pair of edges e_1, e_2 of G, and add a new weight 2 edge e_0 joining x_0, y_0. Then*

(A) *G^+ is Petersen minor free and (G^+, w) has a faithful circuit cover; and*

(B) *the following statements are equivalent:*

(1) (G^+, w) is a Hamilton weighted graph;

(2) $(G^+, w) \in \langle\mathcal{K}_4\rangle_2^{HW}$;

(3) $e_1, e_2 \in E_{w=1}$ and $\{e_1, e_2\}$ is contained in a 3-edge-cut of G.

Proof of (A). By Proposition B.2.15, G^+ is Petersen minor free since G is planar. Hence, by Theorem 3.2.1, (G^+, w) has a faithful cover. □

Proof of (B). $(2) \Rightarrow (1)$: Trivial. ◇

$(1) \Rightarrow (2)$: This is true for if $|V(G)| \leq 4$. Suppose that (G, w) is a smallest counterexample. So, $|V(G)| \geq 6$.

(i) We claim that G^+ *has no non-trivial 2- or 3-edge-cut.* Suppose that T is a 2- or 3-edge-cut of G^+ with components Q_1, Q_2. By the inductive hypothesis, each $(\overline{G^+/Q_j}, w) \in \langle\mathcal{K}_4\rangle_2^{HW}$. By Observation 16.3.3, $(G^+, w) \in \langle\mathcal{K}_4\rangle_2^{HW}$.

Hence, the graph G does not have any 2-edge-cut and, therefore, by Observation 16.3.1-(1),

$$(G, w) \in \langle\mathcal{K}_4\rangle^{HW}.$$

Furthermore, G^+ is of girth ≥ 4.

(ii) We claim that (G^+, w) *has no non-Hamilton removable circuit* (a removable circuit that is not Hamilton). If C is a such removable circuit, then let

$$(G^+, w) = (C, 1) + (H, w_H)$$

(see Definition 3.1.1 for weight decomposition). By (A) and applying Theorem 3.2.1 to (H, w_H), (G^+, w) has a faithful over consisting of at least three circuits and one of them is C. This contradicts (1) that (G^+, w) is Hamilton weighted.

(iii) Since $(G, w) \in \langle K_4 \rangle^{HW}$, by Observation 16.3.1, (G, w) has at least two disjoint triangles. By (i), G^+ is of girth ≥ 4, and therefore, G has precisely two triangles S_1, S_2, each of which contains one of x_0, y_0 as an inserted vertex of some edge (e_1 or e_2).

(iv) Let $\{C_1, C_2\}$ be a faithful circuit cover of the Hamilton weighted graph (G, w). If both $e_1, e_2 \in E(C_1)$, then C_2 is a non-Hamiltonian removable circuit of (G^+, w). This contradicts (ii). Hence, let

$$e_1 \in E(C_1), \quad e_2 \in E(C_2).$$

(v) For a faithful circuit cover $\{C_1, C_2\}$ of the Hamilton weighted graph (G, w), let $C_1 = v_1 \ldots v_r v_1$ with triangles $S_1 = v_1 v_{r-1} v_r v_1$ containing $e_1 = v_r v_1$, and $S_2 = v_k v_{k+1} v_{k+2} v_k$ containing $e_2 = v_{k+2} v_k$ (by (iv)). (See Figure 16.7.) Here, $w(v_{r-1} v_r) = 2$ and there are two cases for the triangle S_2:

$$\text{either } w(v_{k+1} v_{k+2}) = 2, \text{ or } w(v_k v_{k+1}) = 2.$$

Let us consider the first case: $w(v_{k+1} v_{k+2}) = 2$ (the second case is similar).

Note that x_0 is inserted into the edge $e_1 = v_r v_1$ ($\in E(C_1)$), and y_0 is inserted into the edge $v_k v_{k+2}$ ($\in E(C_2)$), and the circuit C_2 contains the paths $v_{k+1} v_{k+2} v_k v_{k-1}$ and $v_2 v_1 v_{r-1} v_r$ (passing the triangles S_2 and S_1, respectively).

We are to find a removable circuit C in (G^+, w) as follows. Let $C = x_0 v_1 C_1 v_k y_0 x_0$ (where $x_0 v_1 C_1 v_k = x_0 v_1 \ldots v_{k-1} v_k$ is a segment of C_1). It is sufficient to show that (H, w_H) *is bridgeless* where

$$(G^+, w) = (H, w_H) + (C, 1)$$

(see Definition 3.1.1 for weight decomposition).

Let $P_j = C_j \cap H$ (for each $j = 1, 2$) where

$$P_1 = C_1 - \{v_1, \ldots, v_{k-1}\} = v_k v_{k+1} v_{k+2} C_1 v_{r-1} v_r x_0,$$

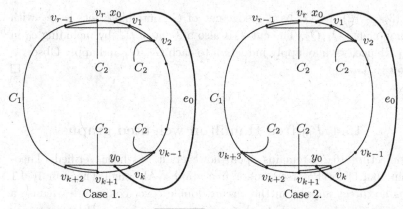

Figure 16.7 *The graph* $(\overline{(C_1 \cup C_2)}, w_{\{C_1,C_2\}})$ *plus one edge* $e_0 = x_0 y_0$

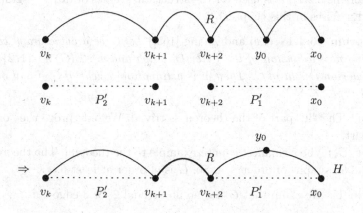

Figure 16.8 $(H, w_H) = (G, w) - (C, 1)$ *is the union of* P_1', P_2' *and* R

$$P_2 = C_2 - \{v_k y_0\} = y_0 v_{k+2} v_{k+1} C_2 v_k.$$

Here, $v_k P_1 x_0 y_0 P_2 v_k$ is a closed walk containing all edges of H.

Let $R = x_0 y_0 v_{k+2} v_{k+1} v_k$, and P_j' $(j = 1, 2)$ be the subpath of P_j by deleting the edges of R. (See Figure 16.8.) The graph H is bridgeless if we can prove that $V(P_1') \cap V(P_2') \neq \emptyset$. Certainly, it is obviously true since the edge $v_r v_{r-1}$ is contained in both P_1' and P_2'.

So, H is bridgeless and C is therefore a non-Hamilton removable circuit of (G^+, w). This contradicts (ii) and completes the proof. ◇

(2) ⇒ (3): This is a corollary of Lemma 16.3.12-(2) and (3). That is, e is a untouched edge of (G^+, w). ◇

(3) $\Rightarrow$ (2): Let T be a 3-edge-cut of G containing both e_1, e_2 with components Q_1, Q_2. The cut T is also a 3-cut of G^+ by including e_0 in Q_1. Hence, we may apply induction to each G^+/Q_j and apply Observation 16.3.4. $\qquad\qquad\square$

16.4 P_{10}-free Hamilton weighted graphs

Conjecture 16.2.5 remains open, however, it has been verified (Theorem 16.4.1) for Petersen minor free graphs. Although Theorem 16.4.1 has no direct impact on the circuit double cover conjecture because a possible counterexample to the conjecture must have a Petersen minor (Theorems 3.2.1), it is used for the structural studies of $\langle \mathcal{K}_4 \rangle^{HW}$-graphs (see Exercises in this chapter).

Theorem 16.4.1 (Lai and Zhang [165]) *Let G be a cubic graph containing no subdivision of the Petersen graph and $w : E(G) \mapsto \{1, 2\}$ be an eulerian weight of G. Then w is a Hamilton weight of G if and only if $(G, w) \in \langle \mathcal{K}_4 \rangle_2^{HW}$.*

Proof The "if" part of the theorem is trivial. We only prove the "only if" part.

Let (G, w) be a smallest counterexample to the theorem. The theorem is true for graphs of order ≤ 4. So, G is of order at least 6.

Claim 1 We claim that G has no non-trivial 2- or 3-edge-cut.

For otherwise, let T be a non-trivial edge-cut of G with $|T| \leq 3$ and components Q_1, Q_2. By the inductive hypothesis, $(\overline{G/Q_j}, w) \in \langle \mathcal{K}_4 \rangle_2^{HW}$ for each $j = 1, 2$. By Observation 16.3.3,

$$(G, w) = (G/Q_1, w) \oplus_\mu (G/Q_2, w) \in \langle \mathcal{K}_4 \rangle_2^{HW}$$

for $\mu = |T|$. $\diamond$

Furthermore, by Observation 16.3.1-(1), we are to show that $(G, w) \in \langle \mathcal{K}_4 \rangle^{HW}$ in the remaining part of the proof since G is 3-edge-connected.

Claim 2 We claim that G is of girth ≥ 4.

Corollary of Claim 1. $\diamond$

Claim 3 We claim that (G, w) has no non-Hamilton removable circuit (a non-Hamilton removable circuit is a removable circuit but not a Hamilton circuit).

For otherwise, applying Theorem 3.2.1 to (G^-, w_{G^-}) where $(G, w) = (C, 1) + (G^-, w_{G^-})$ and C is a removable circuit, we have a faithful circuit cover of (G, w) consisting of at least three circuits (C is one of them). This contradicts that (G, w) is Hamilton weighted and proves the claim. $\Diamond$

Claim 4 We claim that, for an arbitrary edge $e_0 = x_0 y_0 \in E_{w=2}$,
(1) $(G - e_0, w)$ has a faithful circuit cover $\mathcal{F}_{e_0}$,
(2) $\mathcal{F}_{e_0}$ is a circuit chain joining x_0, y_0.

Since $G - e_0$ is bridgeless (by Claim 1), (1) is proved by applying Theorem 3.2.1.

If a faithful circuit cover $\mathcal{F}_{e_0}$ is not a chain, then any member of $\mathcal{F}_{e_0}$ not contained in a chain is removable. This contradicts Claim 3. $\Diamond$

Claim 5 We claim that $E_{w=1}$ induces a Hamilton circuit of G.

Since w is a Hamilton weight of G, a faithful circuit cover of (G, w) consists of two circuits. By Exercise 3.1, $E_{w=1}$ induces an even 2-factor of G.

Suppose that $E_{w=1}$ has more than one components. Let $e \in E_{w=2}$ which links two components of $E_{w=1}$. By Claim 4, $(G - e, w)$ has a faithful circuit cover which is also a chain. Hence, by applying Exercise 3.1 to $\overline{G - e}$, $E_{w=1}$ induces an even 2-factor of the suppressed cubic graph $\overline{G - e}$. That is, in the original graph G, $E_{w=1}$ induces a 2-factor containing two odd length circuits, each of which contains an endvertex of e. This contradicts the conclusion of the last paragraph. $\Diamond$

Let $e_0 = x_0 y_0 \in E_{w=2}$, and, by Claim 4, let $\mathcal{F}_{e_0} = \{C_1, \ldots, C_t\}$ be a faithful circuit cover of $(G - e_0, w)$ with $|\mathcal{F}|$ as large as possible. Note that $\mathcal{F}_{e_0}$ is a circuit chain joining x_0 and y_0, let $x_0 \in V(C_1)$ and $y_0 \in V(C_t)$ (by Claim 4).

Let $e^* = y_0 v \in E_{w=1} \cap E(C_t)$ (with $v \in V(C_{t-1}) \cap V(C_t)$). Let H be the graph obtained from G by deleting all edges of $(E(C_t) - E(C_{t-1}) - e^*)$ and contracting the edge e^*. Define $w_H : E(H) \mapsto \{1, 2\}$ as follows: $w_H(e) = w_{\{C_1, \ldots, C_t\}}(e)$ if $e \neq e_0$, and $w_H(e_0) = 2$ (where $w_{\{C_1, \ldots, C_t\}}$ is the eulerian weight induced by the circuit set $\{C_1, \ldots, C_t\}$, see Definition 2.1.7).

By Theorem 3.2.1, $(\overline{H}, w_H)$ has a faithful circuit cover $\mathcal{F}^*$. Note that $|\mathcal{F}^*| = 2$, for otherwise, any member of $\mathcal{F}^*$ not containing e_0 is removable in (G, w). This contradicts Claim 3. Hence, by the inductive hypothesis, we have the following claim.

Claim 6

$$(\overline{H}, w_H) \in \langle \mathcal{K}_4 \rangle_2^{HW}.$$

For the faithful cover $\mathcal{F}_{e_0} = \{C_1, \ldots, C_t\}$ described above, we claim the following.

Claim 7

$$t = 3.$$

If $t = |\mathcal{F}_{e_0}| = 2$, $(\overline{C_1 \cup C_2}, w_{\{C_1, C_2\}})$ is a Hamilton weighted graph, hence, by induction,

$$(\overline{G - e_0}, w) = (\overline{C_1 \cup C_2}, w_{\{C_1, C_2\}}) \in \langle \mathcal{K}_4 \rangle_2^{HW}.$$

By Lemma 16.3.13, $(G, w) \in \langle \mathcal{K}_4 \rangle_2^{HW}$.

So, suppose that $t \geq 4$. By Claim 6, $(\overline{H}, w_H)) \in \langle \mathcal{K}_4 \rangle_2^{HW}$. (Note, the graph H was constructed in the paragraph before Claim 6.) By Lemma 16.3.12, we have two cases.

Case 1.

$$(\overline{C_1 \cup \cdots \cup C_{t-1}}, w_{\{C_1, \ldots, C_{t-1}\}}) = (\overline{H - e_0}, w_H) \in \langle \mathcal{K}_4 \rangle_2^{HW}$$

(Statement (2) of Lemma 16.3.12).

Case 2. $\overline{H}$ has a non-trivial 3-edge-cut T containing the edge e_0 (the negation of Statement (1) of Lemma 16.3.12).

For Case 1, $t - 1 = 2$ and this contradicts the assumption that $t \geq 4$.

For Case 2, by Exercise 16.6, $T - e_0 \subseteq E_{w=1}$. Therefore, $T - e_0 \subseteq E(C_j) \cap E_{w=1}$ for some $j \in \{1, \ldots, t-1\}$.

Suppose that $j = t - 1$. All vertices of $C_1 \cup \cdots \cup C_{t-2}$ are contained in one component of $H - T$ and, therefore, another component of $H - T$ must be trivial. This contradicts that T is a non-trivial cut.

So $j \leq t-2$. Then T is also a non-trivial 3-edge-cut of G (since all edge-attachments of C_t in $\overline{C_1 \cup \cdots \cup C_{t-1}}$ are contained $E(C_{t-1}) \cap E_{w_H=1}$). This contradicts Claim 1. $\diamondsuit$

Claim 8 We claim that G is bipartite.

By Claim 7, for *every* $e_0 = x_0 y_0 \in E_{w=2}$, $(G - e_0, w)$ has a faithful circuit cover (circuit chain) $\mathcal{F}_{e_0}$ consisting of precisely *three* circuits. Let $\mathcal{F}_{e_0} = \{C_1, C_2, C_3\}$. Color the edges of $(\overline{G - e_0}, w)$ as follows: *red* for $(C_1 \cup C_3) - C_2$, *blue* for $C_2 - (C_1 \cup C_3)$, and *purple* for $(C_1 \cup C_3) \cap C_2 = E_{w=2} - e_0$. Here, the Hamilton circuit $E_{w=1}$ is red-blue bi-colored. Note that both vertices x_0 and y_0 are contained in red-colored, subdivided edges of $\overline{G - e_0}$. Hence, *the endvertices x_0, y_0 of e_0 are joined by an odd length segment of the Hamilton circuit $E_{w=1}$.*

Note that the selection of the weight 2 edge e_0 is *arbitrary*. The above conclusion is valid for every weight 2 edge (chord of the Hamilton circuit $E_{w=1}$). Hence, $V(E_{w=1})$ has a proper 2-vertex coloring and every chord of $E_{w=1}$ joins a pair of differently colored vertices. That is, G is bipartite. $\diamondsuit$

In the remaining part of the proof, we are to find an odd circuit in $C_1 \cup C_2$ (for the purpose of finding a contradiction to Claim 8).

Claim 9 We claim that

$$(\overline{C_1 \cup C_2}, w_{\{C_1, C_2\}}) \in \langle \mathcal{K}_4 \rangle^{HW}.$$

Since $|\mathcal{F}_{e_0}|$ is maximized, $(\overline{C_1 \cup C_2}, w_{\{C_1, C_2\}})$ is a Hamilton weighted graph. By induction,

$$(\overline{C_1 \cup C_2}, w_{\{C_1, C_2\}}) \in \langle \mathcal{K}_4 \rangle_2^{HW}.$$

We will show that the suppressed cubic graph $(\overline{C_1 \cup C_2})$ is 3-connected since the difference between $\langle \mathcal{K}_4 \rangle^{HW}$-graphs and $\langle \mathcal{K}_4 \rangle_2^{HW}$-graphs is the existence of 2-cut (by Observation 16.3.1-(1)).

Let $(J, w_J) = (\overline{C_1 \cup C_2}, w_{\{C_1, C_2\}})$.

Suppose that $(J, w_J) \in \langle \mathcal{K}_4 \rangle_2^{HW} - \langle \mathcal{K}_4 \rangle^{HW}$. By Observation 16.3.1-(1), J has a 2-edge-cut T with $w(T) = 4$ and with components Q' and Q''. By Claim 1, let Q' contain the vertex x_0, and Q'' contain an edge-attachment e'' of C_3. Let D be the component of $E_{w_J=1}$ containing e''. Then, $\{C_1 \triangle D, C_2 \triangle D\}$ is another faithful cover of (J, w_J) consisting of precisely two circuits (since $|\mathcal{F}_{e_0}|$ is maximum). Hence, $\mathcal{F}_{e_0} - \{C_1, C_2\} + \{C_1 \triangle D, C_2 \triangle D\}$ is a faithful circuit cover of $(G - e_0, w)$, but not a circuit chain (since the circuit $C_1 \triangle D$ contains both x_0 and e'', and therefore, $C_2 \triangle D$ is removable). This contradicts Claim 3 and completes the proof of the claim. $\diamondsuit$

The final step. By Claim 9 and by Observation 16.3.1, the suppressed cubic graph $\overline{C_1 \cup C_2}$ has at least two disjoint triangles $S_1, S_2, \dots$. Since the girth of G is at least 4 (by Claim 2), each S_j contains some edge-attachment(s) of e_0 or C_t. Without loss of generality, let $x_0 \in V(S_1)$ (in G). The triangle S_2 of $\overline{C_1 \cup C_2}$ must contain some edge-attachment(s) of C_t (but not e_0). Hence, S_2 remains of odd length in the cubic graph

G since every intersection of C_t uses precisely two internal vertices of some edge of S_2. This contradicts Claim 8 that G is bipartite.   □

16.5 Circuit chain and Petersen chain

The main subject of this chapter is an attempt to characterize the weighted graph described in Question 16.1.1.

Definition 16.5.1   Let G be a bridgeless cubic graph with an eulerian weight $w : E(G) \mapsto \{1,2\}$, and let $e_0 = x_0 y_0$ be a weight 2 edge. The eulerian weighted graph (G, w) is called a circuit chain plus an edge e_0 (abbreviated as $\mathcal{CCPE}$-graph, see Figure 16.2 or 16.9-(a)), if $(G - e_0, w)$ has a faithful circuit cover $\{C_1, C_2, \ldots, C_t\}$ that forms a circuit chain connecting the vertices x_0 and y_0.

With the assumption of Conjecture 16.2.5, the following result is one of the main results of the chapter.

Theorem 16.5.2   *Let (G, w) be a $\mathcal{CCPE}$ graph. Then, under the assumption of Conjecture 16.2.5, one of the following conclusions must hold:*

(1) (G, w) has a removable circuit;

(2) (G, w) is a Petersen chain.

Petersen chains will be defined in the following definitions, and Theorem 16.5.2 is an immediate corollary of a structural result (Theorem 16.5.7) of this chapter.

Definition 16.5.3   Let G be a bridgeless cubic graph with an eulerian weight $w : E(G) \mapsto \{1,2\}$, and let $e_0 = x_0 y_0$ be a weight 2 edge. The eulerian weighted graph (G, w) is called a simple Petersen chain with a bowstring e_0 (see Figure 16.9-(a)) if (G, w) has a set of 3-edge-cuts $\{T_1, T_2, \ldots, T_c\}$ such that

(1) $e_0 \in T_\mu$ and $w(T_\mu) = 4$ for every $\mu = 1, \ldots, c$;

(2) let Q_μ, R_μ be components of $G - T_\mu$ with $x_0 \in Q_\mu$ and $y_0 \in R_\mu$,

$$
\begin{aligned}
\{x_0\} &= Q_1 \subset Q_2 \subset \cdots \subset Q_c = G - \{y_0\}, \\
G - \{x_0\} &= R_1 \supset R_2 \supset \cdots \supset R_c = \{y_0\};
\end{aligned}
$$

(3) for each $\mu = 1, \ldots, c - 1$, the contracted weighted graph $S_\mu = (G/[Q_\mu \cup R_{\mu+1}], w)$ is either (P_{10}, w_{10}) or (K_4, w_4) (the contracted graph S_μ is called a segment of the Petersen chain, see Figure 16.9-(b)).

Definition 16.5.4 A Petersen chain (G, w) with a bowstring e_0 is obtained from a simple Petersen chain with the bowstring $e_0 = x_0 y_0$ via a series of $(Y \rightarrow \triangle)$-operations at any vertex other than x_0 and y_0, and a series of edge-dividing operations at any weight 2 edge other than e_0.

The definition of a segment of a Petersen chain is similar to Definition 16.5.3 (see Figure 16.9-(b)). For each segment S_μ of a Petersen chain, (P_{10}, w_{10}) or (K_4, w_4) can be obtained from S_μ by recursively contracting triangles/digons (not containing x_0, y_0) and recursively suppressing degree 2 vertices.

Definition 16.5.5 A segment S_μ of a Petersen chain with bowstring $e_0 = x_0 y_0$ is called a K_4-segment (or P_{10}-segment, respectively) if (K_4, w_4) (or (P_{10}, w_{10}), respectively) can be obtained from S_μ by recursively contracting triangles/digons and recursively suppressing degree 2-vertices (see Figure 16.9-(b)).

Definition 16.5.6 A single segment Petersen chain is a Petersen chain with precisely one segment.

The following theorem characterizes the structure of weighted graphs induced by a circuit chain plus a missing edge, which will be proved later in this section after some preparations.

Theorem 16.5.7 *Let G be a bridgeless cubic graph with an eulerian weight $w : E(G) \mapsto \{1, 2\}$, and let $e_0 = x_0 y_0$ be a weight 2 edge. Assume that $(G - e_0, w)$ has a faithful circuit cover $\{C_1, C_2, \ldots, C_t\}$ and (G, w) has no removable circuit C with $e_0 \notin E(C)$, then, under the assumption of the Hamilton weight conjecture (Conjecture 16.2.5), (G, w) is a Petersen chain with e_0 as the bowstring edge.*

Note that the removable circuit is defined in Definition 3.1.3. Obviously, the faithful circuit cover $\{C_1, C_2, \ldots, C_t\}$ forms a circuit chain connecting the vertices x_0 and y_0 since any circuit of $\{C_1, C_2, \ldots, C_t\}$ not in a circuit chain is removable.

⋆ Observations about Petersen chain

The following is a straightforward observation from the definition of Petersen chain and Proposition B.2.28.

Lemma 16.5.8 *Let (G, w) be a Petersen chain with a bowstring $e_0 = x_0 y_0$. If $\mathcal{P} = \{C_1, \ldots, C_t\}$ is a circuit chain of (G, w) joining x_0, y_0 with*

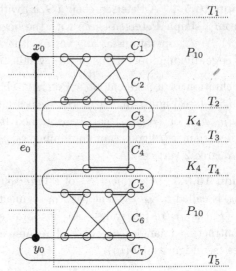

(a)   A circuit chain joined by e_0: Petersen chain

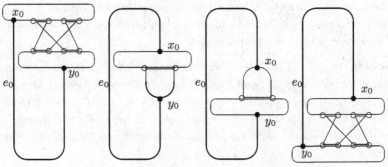

(b)   Segments of the Petersen chain in (a)

Figure 16.9 *A Petersen chain and its segments*

$|\mathcal{P}| = t$ *maximum, then each 3-edge-cut* T_i *($i = 1, \ldots, c$) containing* e_0 *must also contain two weight 1 edges of* $C_{\phi(i)}$ *where* $\phi : \{1, \ldots, c\} \mapsto \{1, \ldots, t\}$ *is a one-to-one mapping such that*

    (1) $1 = \phi(1) < \phi(2) < \cdots < \phi(c) = t$,

    (2) $\phi(\mu + 1) - \phi(\mu) = 1$ *or 2,*

    (3) if $\phi(\mu + 1) - \phi(\mu) = 1$ *then the segment* S_μ *is a* K_4*-segment,*

    (4) if $\phi(\mu + 1) - \phi(\mu) = 2$ *then the segment* S_μ *is a* P_{10}*-segment.*

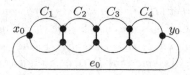

Figure 16.10 *Circuit chain joining x_0 and y_0*

Note that K_4- and P_{10}-segments S_μ are defined in Definition 16.5.5.

⋆ Lemmas for non-contra pair $\mathcal{CCPE}$

The following lemmas solve some special cases for Theorem 16.5.7 and are useful lemmas for further applications.

Lemma 16.5.9 *Let (H, w) be a $\mathcal{CCPE}$ graph consisting of a circuit chain $\mathcal{P} = \{C_1, \ldots, C_t\}$ plus a weight 2 edge $e_0 = x_0 y_0$ such that (H, w) has no removable circuit C with $e_0 \notin E(C)$ (the same description as in Theorem 16.5.7). If the eulerian weighted graph (H, w) itself has a faithful circuit cover, then, under the assumption of Conjecture 16.2.5,*
(1) $(H, w) \in \langle \mathcal{K}_4 \rangle_2^{HW}$,
(2) $|C_j \cap C_{j+1}| = 1$ if every triangle and digon of H contains either x_0 or y_0 (see Figure 16.10).
That is, (H, w) is a Petersen chain with e_0 as the bowstring and every segment of the Petersen chain is a K_4-segment.

Proof It is obvious that every faithful cover of (H, w) consists of precisely two circuits, for otherwise, the third one not containing e_0 is removable. So, w is a Hamilton weight of H and, therefore, $(H, w) \in \langle \mathcal{K}_4 \rangle_2^{HW}$ (by Lemma 16.2.9). This proves the conclusion (1).

Let (H, w) be a smallest counterexample to conclusion (2) of the lemma. It is easy to see that $|V(H)| \geq 6$. By Observation 16.3.1, (H, w) has a pair of vertex disjoint short circuits of length ≤ 3 (triangle(s) with total weight 4 or digon(s) with total weight 2). Since $\{C_1, \ldots, C_t\}$ is a circuit chain joining x_0, y_0, it is easy to see that, by Observation 16.3.1, (H, w) has precisely two triangles, each of which contains one of $\{x_0, y_0\}$ and, therefore, these two triangles are C_1 and C_t. Since $(H/C_t, w)$ is smaller than the smallest counterexample, conclusion (2) holds for $(H/C_t, w)$ (apply Lemma 16.3.2 here). That is, $|C_j \cap C_{j+1}| = 1$ for each $j = 1, \ldots, t - 2$. The proof of (2) is completed since C_t is a triangle that intersects C_{t-1} with precisely one edge. □

Lemma 16.5.10   *Let (H, w) be a CCPE graph consisting of a circuit chain $\mathcal{P} = \{C_1, \ldots, C_t\}$ plus a weight 2 edge $e_0 = x_0 y_0$ such that (H, w) has no removable circuit C with $e_0 \notin E(C)$ (the same description as in Theorem 16.5.7). Assume that $\mathcal{P}$ is a faithful cover of $(H - e_0, w)$ with $|\mathcal{P}|$ as large as possible. Let f_{x_0}, f_{y_0} be subdivided edges of $H - e_0$ containing x_0 or y_0, respectively. If*

$$|\mathcal{P}| = t = 2,$$

then, under the assumption of Conjecture 16.2.5,

(1) $(H, w) \in \langle \mathcal{K}_4 \rangle_2^{HW}$,

(2) $\{f_{x_0}, f_{y_0}\}$ is contained in a 3-edge-cut of $H - e_0$,

(3) every 3-edge-cut of H containing e_0 is trivial (that is, $E(x_0)$ and $E(y_0)$ are the only two 3-edge-cuts of H containing e_0).

Proof   By Lemma 16.2.9 and the choice of $\mathcal{P}$ that $|\mathcal{P}|$ is maximum, the weighted graph $(H - e_0, w) \in \langle \mathcal{K}_4 \rangle_2^{HW}$. Since every member of $\langle \mathcal{K}_4 \rangle_2^{HW}$ is planar, by Proposition B.2.15, H does not contain a subdivision of the Petersen graph. Hence, by Theorem 3.2.1, (H, w) has a faithful cover. Conclusion (1) of the lemma follows immediately from Lemma 16.5.9-(1).

Now, we are to prove the conclusions (2) and (3). By Lemma 16.5.9-(2), (K_4, w_4) can be obtained from (H, w) by recursively contracting all triangles/digons not containing x_0, y_0 and recursively suppressing all degree 2 vertices (along some subdivided weight 2 edges). Thus, the lemma holds for (K_4, w_4), so does for (H, w) (since none of those operations (and their inverses) affects the conclusions (2) and (3) of the lemma). $\square$

⋆ Proof of Theorem 16.5.7

Let (G, w) be a smallest counterexample to the theorem. And we choose $t = |\mathcal{P}|$ as large as possible.

A. First part of the proof: claim $|\mathcal{P}| = 3$

In this subsection, we are to show that $|\mathcal{P}| = 3$.

I. Since (G, w) has no removable circuit avoiding e_0, we have the following property for (G, w).

Claim 1   Every faithful circuit cover of $(G - e_0, w)$ is a circuit chain joining x_0 and y_0.

By Lemma 16.5.10, if $t = 2$, then $(G, w) \in \langle \mathcal{K}_4 \rangle_2^{HW}$ (a single segment Petersen chain). This contradicts that (G, w) is a counterexample. Hence, we have the following.

Claim 2 $t \geq 3$.

By Lemma 16.5.9, we have the following.

Claim 3 (G, w) is a contra pair.

Note that any circuit of length ≤ 3 can be contracted, and the resulting weighted graph is a smaller $\mathcal{CCPE}$ graph.

Claim 4 G is of girth at least 4.

Claim 5 G does not contain any non-trivial 3-edge-cut T consisting of e_0 and a pair of weight one edges.

For otherwise, let Q_1 and Q_2 be the components of $G - T$, one may apply the theorem to the smaller $\mathcal{CCPE}$ graphs $(G/Q_1, w)$ and $(G/Q_2, w)$. $\Diamond$

II.

Notation 16.5.11 For $1 \leq i < j \leq t$, let $(G_{i,j}, w_{i,j})$ be the induced subgraph $G[C_i \cup \cdots \cup C_j]$ associated with the eulerian weight $w_{i,j} = w_{\{C_i, \ldots, C_j\}}$ (weight induced by the circuit subchain $\{C_i, \ldots, C_j\}$). (See Figure 16.11. See Definition 2.1.7 for induced weight $w_{\{C_i, \ldots, C_j\}}$.)

Recall the definition of **edge-attachment** (Definition 16.3.10). Let G be a cubic graph and G^1, G^2 be subgraphs of G. An **edge-attachment** of G^2 in the suppressed graph $\overline{G^1}$ is an edge $e = uv$ of $\overline{G^1}$ such that the edge e corresponds to a maximal induced path $P = u \ldots v$ (in G^1) and $V(G^2) \cap [V(P) - \{u, v\}] \neq \emptyset$.

III.

Claim 6 For each $j < t$, the number of edge-attachments of C_{j+1} in $(\overline{G_{1,j}}, w_{1,j})$ is at least 2.

For otherwise, G has a 3-edge-cut consisting of e_0 and two weight 1 edges of C_j (part of an edge-attachment of C_{j+1} in $(\overline{G_{1,j}}, w_{1,j})$). This contradicts Claim 5. $\Diamond$

By Lemma 16.2.9 and the assumption that $|\mathcal{P}|$ is maximum, we have the following.

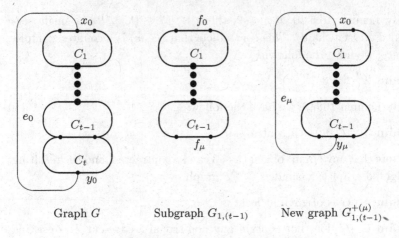

Graph G Subgraph $G_{1,(t-1)}$ New graph $G_{1,(t-1)}^{+(\mu)}$

Figure 16.11 *Circuit chain and subchain*

Claim 7 $(\overline{G_{j,(j+1)}}, w_{j,(j+1)}) \in \langle \mathcal{K}_4 \rangle_2^{HW}$, for each $j \in \{1, \ldots, t-1\}$.

Claim 8 G does not have any 2-edge-cut T separating e_0 from others.

Suppose that T is a 2-edge-cut of G with components Q' and Q'' and $e_0 \in Q'$. Let C_j, C_{j+1} be two members of $\mathcal{P}$ passing through T (we only deal with the case that $w(T) = 4$ since the case that $w(T) = 2$ is trivial). Since (G, w) has no removable circuit avoiding e_0, $\{C_j, C_{j+1}\}$ covers Q''. Hence, Q'' is a subgraph of $G_{j,(j+1)}$, for some j. By Claim 7 and Observation 16.3.1, Q'' contains a triangle or digon. This contradicts Claim 4. $\diamond$

Claim 9 For each $i = 1, \ldots, t-1$, the suppressed cubic graph $\overline{G_{i,(i+1)}}$ is 3-connected, and, therefore, $(\overline{G_{i,(i+1)}}, w_{i,(i+1)}) \in \langle \mathcal{K}_4 \rangle^{HW}$.

Let $(J, w_J) = (\overline{G_{i,(i+1)}}, w_{i,(i+1)}))$. By Claim 7, $(J, w_J) \in \langle \mathcal{K}_4 \rangle_2^{HW}$.

Suppose that $(J, w_J) \in \langle \mathcal{K}_4 \rangle_2^{HW} - \langle \mathcal{K}_4 \rangle^{HW}$. By Observation 16.3.1-(1), J has a 2-edge-cut T with $w(T) = 4$ and with components Q' and Q''. By Claim 8, let Q' contain an edge-attachment e' of C_{i-1} (or contain the vertex x_0 if $i = 1$), and Q'' contain an edge-attachment e'' of C_{i+2} (or contain the vertex y_0 if $i + 1 = t$). Let D be the component of $E_{w_J=1}$ containing e''. Then, $\{C_i \triangle D, C_{i+1} \triangle D\}$ is another faithful cover of (J, w_J) consisting of precisely two circuits (since $|\mathcal{P}|$ is maximum). Hence, $\mathcal{P} - \{C_i, C_{i+1}\} + \{C_i \triangle D, C_{i+1} \triangle D\}$ is a faithful cover of $(G - e_0, w)$, but not a circuit chain (since $C_i \triangle D$ contains both e' and e'', and therefore, $C_{i+1} \triangle D$ is removable). $\diamond$

IV. Let $F_t = \{f_1, \ldots, f_s\}$ be the set of all edge-attachments of C_t in $G_{1,(t-1)}$ (induced paths of $G_{1,(t-1)}$ containing vertices of C_t) and let f_0 be the edge-attachment of e_0 in $G_{1,(t-1)}$ (the induced path of length 2 containing the vertex x_0). Here, by Claim 6,

$$|F_t| = s \geq 2. \tag{16.1}$$

Notation 16.5.12 For each $f_\mu \in F_t$, construct $(G_{1,(t-1)}^{+(\mu)}, w_{1,(t-1)}^{+(\mu)})$ from $(G_{1,(t-1)}, w_{1,(t-1)})$ by adding a weight 2 edge e_μ joining x_0 and a degree 2 vertex y_μ in the edge-attachment f_μ and suppressing all degree 2 vertices (see Figure 16.11).

V. This is the final step of this subsection. We show that

$$t = 3. \tag{16.2}$$

Suppose that $t \geq 4$.

V-1. By applying the theorem to the smaller $\mathcal{CCPE}$ weighted graph $(G_{1,(t-1)}^{+(\mu)}, w_{1,(t-1)}^{+(\mu)})$ (for each $\mu = 1, \ldots, s$), it has the following properties.

(a) $(G_{1,(t-1)}^{+(\mu)}, w_{1,(t-1)}^{+(\mu)})$ is a Petersen chain with the bowstring e_μ (since any removable circuit of $(G_{1,(t-1)}^{+(\mu)}, w_{1,(t-1)}^{+(\mu)})$ avoiding e_μ is also removable in (G, w)).

(b) $(G_{1,(t-1)}^{+(\mu)}, w_{1,(t-1)}^{+(\mu)})$ does not have any 3-edge-cut T of (G, w) that consists of the bowstring e_μ and two weight 1 edges of C_i for some $i : 1 < i < t - 1$ (since $T - e_\mu + e_0$ would be a non-trivial 3-edge-cut of (G, w) and this contradicts Claim 5). Thus, $E(x_0)$ and $E(y_\mu)$ are the only 3-edge-cuts of $(G_{1,(t-1)}^{+(\mu)}, w_{1,(t-1)}^{+(\mu)})$ containing the bowstring e_μ.

(c) $(G_{1,(t-1)}^{+(\mu)}, w_{1,(t-1)}^{+(\mu)})$ must be a Petersen chain with single segment (by (b) and Lemma 16.5.8 since (G, w) is a smallest counterexample). Thus, $t - 1 = 3$.

(d) $G_{1,(t-1)}^{+(\mu)}$ is not 3-edge-colorable. (For otherwise, by Lemma 2.2.1, $(G_{1,(t-1)}^{+(\mu)}, w_{1,(t-1)}^{+(\mu)})$ has a faithful cover. Thus, by (a) and Lemma 16.5.8, the Petersen chain $(G_{1,(t-1)}^{+(\mu)}, w_{1,(t-1)}^{+(\mu)})$ with a circuit chain of length $t - 1 = 3$ must have a non-trivial 3-edge-cut containing the bowstring e_μ. This contradicts (b).)

Among these properties, Property (c) is the most important one: it indicates that, after a series of contractions of triangles/digons and a series of suppressing degree 2 vertices (with the bowstring e_μ untouched),

$(G_{1,(t-1)}^{+(\mu)}, w_{1,(t-1)}^{+(\mu)})$ becomes the weighted Petersen graph (P_{10}, w_{10}). Thus, we have the following corollary.

(e) In $(G_{1,(t-1)}^{+(\mu)}, w_{1,(t-1)}^{+(\mu)})$, the endvertex y_μ of the bowstring e_μ is not contained in any circuit of length ≤ 4 (by (c) and Definition 16.5.4).

V-2. By (c), $(G_{1,(t-1)}^{+(\mu)}, w_{1,(t-1)}^{+(\mu)})$ is a single segment Petersen chain. First we claim that

$$|V(G_{1,(t-1)}^{+(\mu)})| = 10 \tag{16.3}$$

for each $\mu = 1, \ldots, s$.

Suppose that $|V(G_{1,(t-1)}^{+(\mu)})| > 10$. By (c), $(G_{1,(t-1)}^{+(\mu)}, w_{1,(t-1)}^{+(\mu)})$ is a Petersen chain with single segment, but not simple (since the Petersen graph has 10 vertices). Hence, the weighted graph $(G_{1,(t-1)}^{+(\mu)}, w_{1,(t-1)}^{+(\mu)})$ has the following further properties.

(f) In $(G_{1,(t-1)}^{+(\mu)}, w_{1,(t-1)}^{+(\mu)})$, there must be some circuit(s) of length ≤ 3 (by Definition 16.5.4).

(g) Those triangle(s)/digon(s) described in (f) must contain some edge f_ν for $\nu \in \{1, \ldots, s\} - \{\mu\}$ (since, by Claim 4, G is of girth at least 4).

Hence, some triangle(s)/digon(s) described in (g) becomes circuit(s) of length ≤ 4 in $(G_{1,(t-1)}^{+(\nu)}, w_{1,(t-1)}^{+(\nu)})$ (for some $\nu \neq \mu$). This contradicts (e) in V-1 (by a symmetric argument for replacing μ with ν) and completes the proof of the claim.

V-3. Thus, by properties (c), (d) of V-1 and Equation (16.3), both $G_{1,(t-1)}^{+(\mu)}$ and $G_{1,(t-1)}^{+(\nu)}$ are snarks of order 10. Note that the Petersen graph is the only snark of order 10. This contradicts Proposition B.2.17, since at most one of $G_{1,(t-1)}^{+(\mu)}$ and $G_{1,(t-1)}^{+(\nu)}$ is the Petersen graph, the other one must be 3-edge-colorable. (See Figure 16.12.) This is a contradiction. ◇

B. Second part of the proof, *L*-graphs

Before the final step of the proof, we introduce a new concept, the *L*-graph, which is critical in the final determination of the Petersen graph structure.

Definition 16.5.13 A weighted *L*-graph is a cubic graph L of order $2n$ $(n \geq 2)$ associated with an eulerian weight $w : E(L) \mapsto \{1, 2\}$ and a weight 1 edge $e_0 = v_0 v_n$ (called a **diagonal crossing chord**) such that

(1) $(L, w) \in \langle \mathcal{K}_4 \rangle^{HW}$,

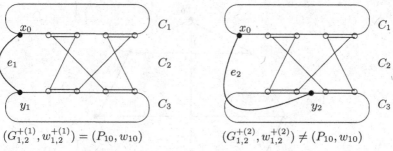

$$(G_{1,2}^{+(1)}, w_{1,2}^{+(1)}) = (P_{10}, w_{10})$$ $$(G_{1,2}^{+(2)}, w_{1,2}^{+(2)}) \neq (P_{10}, w_{10})$$

Figure 16.12 *One is P_{10}, while the other is not*

(2) every triangle of L must contain either v_0 or v_n.
(See Figure 16.13.)

The following lemma describes the structure of L-graphs.

Lemma 16.5.14 *The following two statements are equivalent:*
(1) (L, w) is an L-graph with a diagonal crossing chord $e^* = x^* y^*$;
(2) let e^* be a weight 1 edge of $(3K_2, w_2)$, (L, w) is constructed recursively from $(3K_2, w_2)$ by a series of $(Y \rightarrow \triangle)$-operations only at some endvertex of e^*. (Note that the edge e^* will remain as the diagonal crossing chord during the expansion of the L-graph.)

Proof (2) $\Rightarrow$ (1) is trivial. We prove (1) $\Rightarrow$ (2) by induction on $|V(L)|$. The lemma is true if $|V(L)| \leq 4$. So, by Observation 16.3.1-(3), L has precisely two triangles, each containing precisely one of $\{x^*, y^*\}$. Let S be a triangle of L containing x^* (but not y^*). By Lemma 16.3.2, $(L/S, w) \in \langle \mathcal{K}_4 \rangle^{HW}$ and, without causing any confusion, denote the new contracted vertex by x^* which remains as an endvertex of e^*. It is easy to see that $(L/S, w)$ is an L-graph (by Definition 16.5.13). Since any resulting triangle (after contraction of S) must contain the contracted vertex x^*, by induction, (1) $\Rightarrow$ (2) for $(L/S, w)$. The lemma is true for $(L/S, w)$, so it is true for (L, w) since (L, w) is obtained from $(L/S, w)$ via a $(Y \rightarrow \triangle)$-operation at x^*.   □

Drawing of L-graphs. Since an L-graph $(L, w) \in \langle \mathcal{K}_4 \rangle^{HW}$ and is a Hamilton weighted graph, let $\{C_1, C_2\}$ be the faithful circuit cover of (L, w). Each C_j is a Hamilton circuit. One may draw the L-graph (L, w) on the plane as follows (by Lemma 16.5.14, see Figure 16.13).

The Hamilton circuit $C_2 = v_0, \ldots, v_{2n-1} v_0$ is the boundary of the exterior region with a diagonal crossing chord $v_0 v_n$ and a set Z of parallel

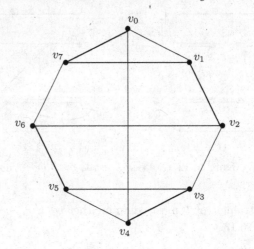

Figure 16.13 *An L-graph*

chords where $Z = \{v_{2n-\mu}v_\mu : \mu = 1, \ldots, n-1\}$. And another Hamilton circuit $C_1 = v_0v_{2n-1}v_1v_2v_{2n-2}v_{2n-3}v_3v_4 \ldots v_nv_0$, and w is a Hamilton weight with $E_{w=2} = \{v_{2i-1}v_{2i} : i = 1, \ldots, n, \pmod{2n}\}$ where C_1 and C_2 intersect.

Figure 16.13 is an illustration of a weighted L-graph with 8 vertices. Note that, in Figure 16.13, bold lines are edges in $E_{w=2}$ and slim lines are edges in $E_{w=1}$.

One can see that all parallel chords (also called Z-chords) do not cross each other, while the diagonal crossing chord v_0v_n crosses every parallel chord.

Lemma 16.5.15   Let $(H, w) \in \langle \mathcal{K}_4 \rangle^{HW}$ of order $2n$ (≥ 4). Let $\{C_1, C_2\}$ be a faithful circuit cover of (H, w). Let $e \in C_1 - C_2$ and $F \subseteq C_2 - C_1$. Assume that

(a) every triangle of H contains some edge of $F \cup \{e\}$ and,

(b) for every edge $f \in F$, H contains a 3-edge-cut T with both $f, e \in T$.
Then (H, w) must be a weighted L-graph described above with $e = v_0v_n$ as the diagonal crossing chord and

$$\{v_0v_1, v_\nu v_{\nu+1}\} \subseteq F \subseteq \{v_{2i}v_{2i+1} : i = 0, \ldots, n-1\} = E(C_2) - E(C_1)$$

where $\nu = n$ if n is even and $\nu = n - 1$ if n is odd.

Proof   By Definition 16.5.13, we only need to show that *every triangle of H must contain an endvertex of $e = v_0v_n$.*

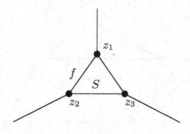

Figure 16.14 *Lemma 16.5.15*

Suppose that $S = z_1 z_2 z_3 z_1$ is a triangle of H that v_0 and $v_n \notin V(S)$. Hence, by (a), let $f = z_1 z_2 \in E(S) \cap F$. (See Figure 16.14.)

By (b), there is a 3-edge-cut T containing both e and f.

Note that $|S \cap T|$ must be even since one is a circuit, while another one is a cut. Since $f \in S \cap T$, $|S \cap T| = 2$. Note that S is a triangle, either $T = E(z_j)$ (for some $j \in \{1, 2\}$) or H has a 2-edge-cut $E(z_i) \triangle T$ (for some $i \in \{1, 2\}$). So, $T = E(z_j)$ since H is 3-connected. Hence, both $f, e \in E(z_j)$ for some $j \in \{1, 2\}$. This contradicts that v_0 and $v_n \notin V(S)$. □

Figure 16.15 is an illustration of a weighted L-graph with 8 vertices. Note that, in Figure 16.15, bold lines are edges in $E_{w=2}$ and slim lines are edges in $E_{w=1}$; the circuit $C_2 = v_0 \ldots v_7 v_0$, the circuit $C_1 = v_0 v_7 v_1 v_2 v_6 v_5 v_3 v_4 v_0$, edges labeled with f are possible locations of edges of F.

Continuation of the proof. We continue the proof of Theorem 16.5.7. By Lemma 16.5.10-(1), we have the following immediate corollary.

Claim 10 For each $f_\mu \in F_3$, $(G_{1,2}^{+(\mu)}, w_{1,2}^{+(\mu)}) \in \langle \mathcal{K}_4 \rangle_2^{HW}$

Note, $G_{1,2}$ is defined in Notation 16.5.11, and $(G_{1,2}^{+(\mu)}, w_{1,2}^{+(\mu)})$ and F_3 ($= F_t$) are defined in Notation 16.5.12.

Claim 11 $(\overline{G_{1,2}}, w_{1,2})$ is an L-graph in which the diagonal crossing chord is an edge-attachment of e_0 (the edge f_0 of $G_{1,2}$, defined in Notation 16.5.12).

Apply Lemma 16.5.10-(2) to $(G_{1,2}^{+(\mu)}, w_{1,2}^{+(\mu)})$, we have that, *for each $f_\mu \in F_3$, both f_0 and f_μ are contained in some 3-edge-cut of $\overline{G_{1,2}}$* (satisfying hypothesis (b) of Lemma 16.5.15). Since G is of girth at least 4

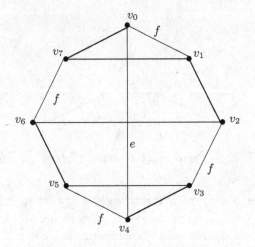

Figure 16.15 *An L-graph with the diagonal crossing chord e and F*

(by Claim 4), *every triangle/digon of* $\overline{G_{1,2}}$ *must contain some edge of* $F_3 \cup \{f_0\}$ (satisfying hypothesis (a) of Lemma 16.5.15).

By Lemma 16.5.15, $(\overline{G_{1,2}}, w_{1,2})$ must be a weighted L-graph with f_0 (an edge-attachment of e_0) as the diagonal crossing chord. $\diamondsuit$

Symmetrically, *the graph* $G_{2,3} = \overline{C_2 \cup C_3}$ *is also an L-graph with the another edge-attachment of* e_0 *as the diagonal crossing chord.*

C. Final part of the proof, Petersen graph or removable circuits

We continue the proof of the main theorem. This final step is the core of the proof: *determine that the graph G is the Petersen graph.*

C-(I) Preliminary. Recall Claim 11, *both* $(\overline{G_{1,2}}, w_{1,2})$ *and* $(\overline{G_{2,3}}, w_{2,3})$ *are L-graphs in which the diagonal crossing chords are the edge-attachments of* e_0.

A drawing of two L-graphs. Let $C_2 = v_0 \ldots v_{r-1} v_0$. Draw the graph $\overline{C_1 \cup C_2 \cup C_3} = \overline{G - e_0}$ on the plane such that C_2 is the boundary of the exterior region and all chords $((C_1 \cup C_3) - C_2)$ are in the interior region of C_2. (Note, this drawing is not a *planar embedding*: some crossing may occur inside the interior of C_2.)

See Figure 16.16 for an illustration of these circuits in G. Note that, in the first graph of Figure 16.16, bold lines are edges in $E_{w=2}$ and slim

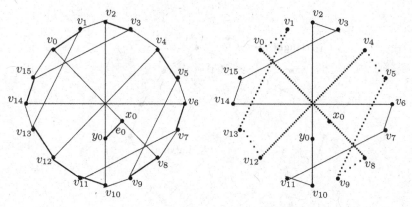

The circuit C_2: $v_0 \ldots v_{15}v_0$

Solid lines: the circuit C_3
Dotted lines: the circuit C_1

Figure 16.16 $(G - e_0, w)$ *is a pair of weighted L-graphs with C_2 as their overlapping part*

lines are edges in $E_{w=1}$. In the second graph of Figure 16.16, only C_1 and C_3 are illustrated.

Notation 16.5.16 (1) For each weighted L-graph $\overline{C_i \cup C_2}$ $(i = 1, 3)$, edges of $C_i - C_2$ are called C_i-chords.

(2) For each $i \in \{1, 3\}$, C_i-chords are classified into two types, *diagonal crossing chord and zigzag parallel chords ($Z(C_i)$-chords):* the subdivided edge containing the vertex x_0 or y_0 is the C_i-diagonal crossing chord, all other edges of $C_i - C_2$ are $Z(C_i)$-chords (zigzag parallel chords).

(3) For each $i \in \{1, 3\}$, each triangle of $\overline{C_i \cup C_2}$ not containing the C_i-diagonal crossing chord is called a C_i-triangle and the unique $Z(C_i)$-chord contained in a C_i-triangle is called a C_i-triangle chord (see Figure 16.17).

By Claim 4, G is of girth at least 4, we have the following property.

Claim 12 For each $\{i, j\} = \{1, 3\}$, let $S = v_\alpha v_\beta v_\gamma v_\alpha$ be a C_i-triangle with $v_\alpha v_\beta$ as the unique $Z(C_i)$-chord (the triangle chord), $v_\beta v_\gamma \in E(C_i) \cap E(C_2)$ (see Figure 16.17). Then S is a triangle in the suppressed graph $\overline{G[C_i \cup C_2]}$, but not a triangle in the original graph G since the edge $v_\alpha v_\gamma$ of S is subdivided at least once by some vertices of C_j in G.

Notation 16.5.17 Define a bijection

$$\lambda : \{0, 1, \ldots, (r-1)\} \mapsto \{0, 1, \ldots, (r-1)\}$$

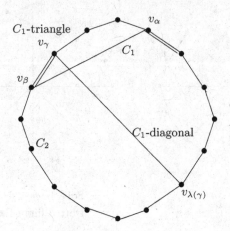

Figure 16.17 *A C_1-triangle $v_\alpha v_\beta v_\gamma \ldots v_\alpha$ is a circuit of length ≥ 5 in G*

such that, for each integer $\alpha \in \{0, \ldots, r-1\}$, $\lambda(\alpha) = \beta$ if there is a C_i-chord (for some $i \in \{1, 3\}$) joining v_α and v_β.

Notation 16.5.18   For the circuit $C_2 = v_0 \ldots v_{r-1} v_0$, and integers a, b: $0 \leq a < b \leq r-1$, the segment (subpath) $v_a v_{a+1} \ldots v_{b-1} v_b$ of C_2 between v_a and v_b is denoted by $v_a C_2 v_b$, while another segment $v_a v_{a-1} \ldots v_{b+1} v_b$ of C_2 between v_a and v_b is denoted by $v_a \overleftarrow{C_2} v_b$ (mod r).

Notation 16.5.19   For each $\{i, j\} = \{1, 3\}$ and each $Z(C_i)$-chord $e = v_\mu v_{\lambda(\mu)} \in E(C_i) - E(C_2)$ (a zigzag parallel chord), the **crossing degree** $d_X(e)$ of e is the number of C_j-chords crossing the edge e in the interior of C_2.

Since C_j is a circuit (for $\{i, j\} = \{1, 3\}$), it is easy to see that

$$d_X(e) \equiv 0 \quad (\text{mod } 2) \tag{16.4}$$

for every C_i-chord e. We further claim that,

$$d_X(e) > 0 \tag{16.5}$$

for every $Z(C_i)$-chord e.

Suppose that $d_X(e) = 0$, for some $Z(C_1)$-chord $e = v_\alpha v_{\lambda(\alpha)}$. Without loss of generality, let $V(C_3) \subseteq \{v_{\alpha+1}, v_{\alpha+2}, \ldots, v_{\lambda(\alpha)-2}, v_{\lambda(\alpha)-1}\}$. Here, $\{v_{\lambda(\alpha)}, v_{\lambda(\alpha)+1}, \ldots, v_{\alpha-1}, v_\alpha\} \subseteq V(C_1)$ (see Figure 16.18). Therefore,

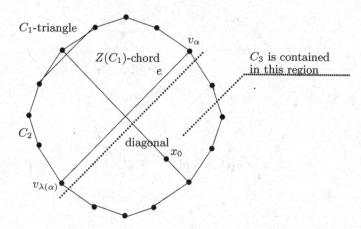

Figure 16.18 *If $d_X(e) = 0$ then a C_1-triangle is of length 3 in G*

the induced subgraph $G[\{v_{\lambda(\alpha)} \ldots v_\alpha\}]$ contains a C_1-triangle, which is not subdivided by C_3. This contradicts Claim 12.

C-(2) Quadruples and removable circuit

Notation 16.5.20 Let (a, b, c, d) be a quadruple (see Figure 16.19) such that

(1) v_a, v_b, v_c, v_d are around the circuit C_2 in this order,

(2) $v_a v_c$ is a $Z(C_1)$-chord and $v_b v_d$ is a $Z(C_3)$-chord,

(3) $v_a v_{a+1}, v_c v_{c+1} \in E(C_3) \cap E(C_2)$, and $v_b v_{b-1}, v_d v_{d-1} \in E(C_1) \cap E(C_2)$.

The proof will be completed after the proofs of the following two claims.

Claim 13 If $(G, w) \neq (P_{10}, w_{10})$, then a quadruple (described in Notation 16.5.20) exists.

Claim 14 If a quadruple (described in Notation 16.5.20) exists, then the circuit $D = v_a C_2 v_b v_d \overleftarrow{C_2} v_c v_a$ is a removable circuit of (G, w).

C-(3) Existence of a quadruple (Proof of Claim 13). In this subsection, we will prove that one of the following statements must be true:

(1) the existence of the quadruple described in Notation 16.5.20;

(2) $(G, w) = (P_{10}, w_{10})$.

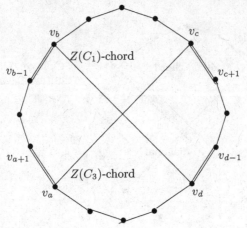

Figure 16.19 *A quadruple*

Suppose that $(G, w) \neq (P_{10}, w_{10})$ and *there is no such quadruple around the circuit* C_2.

I. *Let $v_0 v_p$ be a C_i-triangle chord ($i = 1$ or 3) such that the crossing degree d_X of $v_0 v_p$ is as large as possible (among all C_i-triangle chords for both $i = 1, 3$).* Note, C_i-triangle chords are defined in Notation 16.5.16-(3).

Without loss of generality, let $v_0 v_1 C_2 v_p v_0$ be a C_1-triangle (see Notation 16.5.16) with v_1 incident with the C_1-diagonal crossing chord (see Figure 16.20).

Since $v_0 v_p$ is a C_1-triangle chord, the path $v_2 C_2 v_{p-1}$ contains no vertex of C_1. By Claim 4, $v_1 C_2 v_p$ is not a single edge in G, which must contain some vertices of C_3. Therefore, by the definition of L-graph, edges in the path $v_2 C_2 v_{p-1}$ are alternately in $C_2 - C_3$ and $C_2 \cap C_3$. That is,

$$v_2 v_3, \quad \ldots, \quad v_{2i} v_{2i+1}, \quad \ldots, \quad v_{p-2} v_{p-1} \in C_3 \cap C_2 \qquad (16.6)$$

for $i = 1, \ldots, \frac{p-2}{2}$ and

$$p \geq 4 \text{ and } p \equiv 0 \pmod 2.$$

II. We claim that *for each odd integer $q \in \{2, \ldots, p-1\}$, the C_3-chord $v_q v_{\lambda(q)}$ is the C_3-diagonal crossing chord* (see Figure 16.20).

Suppose not, then $v_q v_{\lambda(q)}$ is a $Z(C_3)$-chord. If $\lambda(q) \in \{2, \ldots, p-1\}$, then the C_3-chord $v_q v_{\lambda(q)}$ is of zero crossing degree. This contradicts the inequality (16.5). So, $\lambda(q) \in \{p+1, \ldots, r-1\}$. Then $(0, q, p, \lambda(q))$

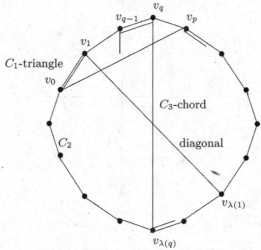

Figure 16.20 C_3-chord $v_q v_{\lambda(q)}$ with one end v_q inside a C_1-triangle $v_0 v_1 v_p v_0$

is the quadruple that we needed and contradicts our assumption (see Figure 16.20).

III. Since there is only one C_3-diagonal chord, by II, q is the only odd integer in $\{2, \ldots, p-1\}$. By Equation (16.6),

$$3 = q = p - 1.$$

IV. In summary, we have proved the following results:

IV-(1) $d_X(v_0 v_p) = 2$ *(by III)*;

IV-(2) $|\{v_2, \ldots, v_{p-1}\}| = 2$ *(by III)*;

IV-(3) both $v_2 v_{\lambda(2)}, v_3 v_{\lambda(3)}$ *are C_3-chords crossing the $Z(C_1)$-chord* $v_0 v_p$ *(by Claim 12)*;

IV-(4) $v_3 v_{\lambda(3)}$ *is the C_3-diagonal crossing chord, and,* $v_2 v_{\lambda(2)}$ *is a* $Z(C_3)$-chord *(by II)*;

IV-(5) $v_2 v_{\lambda(2)}$ *is a C_3-triangle chord (corollary of IV-(4))*.

V. By Equation (16.4) and inequality (16.5), the crossing degree of every C_i-triangle chord is positive and even $(i = 1, 3)$. By IV and the maximality of the crossing degree of the triangle chord $v_0 v_p$ (defined in I), the crossing degree of every C_i-triangle chord is precisely 2 (for each $i = 1, 3$). Hence, *all results we have had in IV for $v_0 v_p$ can be applied to each C_i-triangle chord $(i = 1, 3)$.*

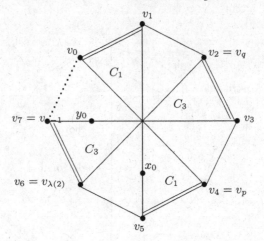

Figure 16.21 C_3-chords $v_q v_{\lambda(q)}$, $v_{q+1} v_{\lambda(q+1)}$ *crossing the* C_1-*triangle chord* $v_0 v_p$

First, we have some direct results from IV (see Figure 16.21):

$$p = 4 \text{ and } \lambda(3) > \lambda(2) > p + 1 = 5,$$

and $v_2 v_{\lambda(2)}$ *is a* C_3-*triangle chord (by IV-(5)), and* $v_3 v_{\lambda(3)}$ *is a* C_3-*diagonal crossing chord (by IV-(4)).*

Symmetrically (see Figure 16.21), applying the results of IV to the C_3-triangle $v_2 v_3 C_2 v_{\lambda(2)} v_2$, we have that $\lambda(2) = 6$ since $|\{v_4, \ldots, v_{\lambda(2)-1}\}| = 2$ (by IV-(2)). Furthermore, we have that $v_4 v_5 \in C_1 \cap C_2$, $v_0 v_4 v_5 C_2 v_0$ is a C_1-triangle (other than $v_0 v_1 C_2 v_4 v_0$) with $v_0 v_4$ as the C_1-triangle chord, $v_5 v_{\lambda(5)}$ is the C_1-diagonal crossing chord with $\lambda(5) = 1$ since there is only one C_1-diagonal chord $v_1 v_{\lambda(1)} = v_{\lambda(5)} v_5$.

Note that we have completely identified all edges of $C_1 = v_0 v_1 x_0 v_5 v_4 v_0$. That is, $\overline{C_1 \cup C_2} = K_4$.

Furthermore, applying the results of IV to the C_1-triangle $v_4 v_5 C_2 v_0 v_4$, we have that $|\{v_6, \ldots, v_{r-1}\}| = 2$ and $v_6 v_2$ is a C_3-triangle chord, $v_{\lambda(7)} v_7 = v_3 v_{\lambda(3)}$ is the C_3-diagonal crossing chord. Therefore, $\overline{C_3 \cup C_2} = K_4$, $r = 8$ and the graph G is the Petersen graph (see Figure 16.22 or Figure B.10). This contradicts the assumption that $(G, w) \neq (P_{10}, w_{10})$. $\Diamond$

C-(4) Existence of removable circuit (Claim 14). In this subsection, we will prove Claim 14 that

$$D = v_a v_{a+1} C_2 v_{b-1} v_b v_d v_{d-1} \overleftarrow{C_2} v_{c+1} v_c v_a$$

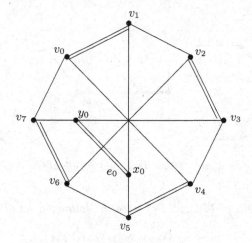

Figure 16.22 *The Petersen graph*

is a removable circuit. (See Figure 16.19.) We consider the following weight decomposition (Definition 3.1.1)

$$(G, w) = (G_X, w_1) + (D, w_D)$$

where $w_D(e) = 1$ for every $e \in E(D)$. Since D is a circuit, it is trivial that w_1 is an eulerian $(1, 2)$-weight of G_X. So, it is sufficient to show that G_X is bridgeless. *Assume that there is a bridge e^* of G_X with $w_1(e^*) = 2$* (since w_1 is eulerian).

I. Let $G_X^{(-)} = G_X - e_0$. It is obvious that $G_X^{(-)}$ is covered by four paths (see Figure 16.19):

$$\begin{cases} P_0 & = & & v_{b-1}v_bC_2v_cv_{c+1}, \\ P_2 & = & & v_{d-1}v_dC_2v_av_{a+1}, \\ P_1 & = & C_1 - \{v_b, v_d\} & = & v_{d-1}C_1v_{b-1}, \\ P_3 & = & C_3 - \{v_a, v_c\} & = & v_{a+1}C_3v_{c+1}. \end{cases}$$

Note that P_0 and P_2 are two segments (subpaths) of C_2 (by deleting some edges of D), and P_i is a segment of C_i for $i = 1$ and 3. It is easy to see that $v_{c+1}P_0v_{b-1}P_1v_{d-1}P_2v_{a+1}P_3v_{c+1}$ is a closed walk of $G_X^{(-)}$ covering every edge e once if $w_1(e) = 1$, twice if $w_1(e) = 2$.

By the discussion above and by the structure of weighted L-graphs $(\overline{C_1 \cup C_2}, w_{1,2})$ and $(\overline{C_2 \cup C_3}, w_{1,2})$, we have the following summary:

I-(1) $G_X^{(-)}$ is a connected graph, so is G_X;

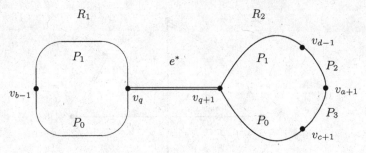

Figure 16.23 *Two components of* $G_X^{(-)} - e^*$

I-(2) *Paths of* $\{P_0, \ldots, P_3\}$ *have the following set of endvertices*

$$\{v_{a+1}, v_{b-1}, v_{c+1}, v_{d-1}\}$$

where

$$v_{b-1} \in P_0 \cap P_1, \ v_{d-1} \in P_1 \cap P_2, \ v_{a+1} \in P_2 \cap P_3, \ v_{c+1} \in P_3 \cap P_0;$$

I-(3) $P_i \cap P_j = \emptyset$ if $i \neq j \pm 1 \pmod 4$.

II. Let R_1, R_2 be components of $G_X - e^*$. (Recall: we assumed that e^* is a bridge of G_X.)

II-(1) *The edge* $e_0 = x_0 y_0$ *is not a bridge of* G_X (that is, $e^* \neq e_0$) since $G_X^{(-)} = G_X - e_0$ is spanning and connected (by I-(1)). Thus, $e^* \neq e_0$.

II-(2) Since $w_1(e^*) = 2$, by II-(1), let $P_\alpha, P_\beta \ (\in \{P_0, \ldots, P_3\})$ contain the edge e^*. By I-(3), we have that $\alpha = \beta \pm 1 \pmod 4$. Without loss of generality, let $e^* \in P_0 \cap P_1$. It is easy to see that P_2 and P_3 must be contained in the same component of $G_X - e^*$ since $v_{a+1} \in V(P_2) \cap V(P_3)$ (by I-(2)). So, without loss of generality, let $P_2 \cup P_3 \subseteq R_2$. (See Figure 16.23.) Therefore, by I-(2) again,

$$v_{c+1}, v_{d-1}, v_{a+1} \in R_2.$$

Since each of P_0 and P_1 passes through the bridge e^* precisely once,

$$v_{b-1} \in R_1.$$

II-(3) Let $e^* = v_q v_{q+1}$ where

$$v_q, v_{q+1} \in \{v_{b+1}, v_{b+2}, \ldots, v_{c-2}, v_{c-1}\} \subseteq P_0 \subset C_2.$$

For the path $P_0 = v_{b-1} C_2 v_{c-1}$ (by II-(2)), the segments $v_{b-1} C_2 v_q \subseteq R_1$ and $v_{q+1} C_2 v_{c-1} \subseteq R_2$. (See Figure 16.23.)

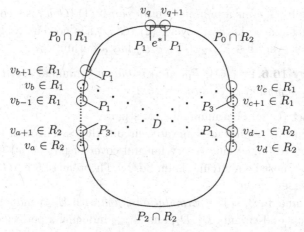

Figure 16.24 *The $Z(C_1)$-chord $v_{d-1}v_{\lambda(d-1)} = v_{d-1}v_{b+1}$*

Note that the segment $v_{b-1}C_2v_q$ is contained in R_1 while C_3 is contained in R_2. Thus, $v_{b-1}C_2v_q$ contains vertices of C_1, but not C_3. That is,

$$V(C_3) \subseteq V(R_2), \quad V(R_1) \subseteq V(C_1). \tag{16.7}$$

II-(4) We claim that $v_{d-1}v_{\lambda(d-1)}$ is not the C_1-diagonal (see Figure 16.24), for otherwise, $v_{d-1}v_dv_bC_2v_{d-1}$ is a C_1-triangle of $\overline{C_1 \cup C_2}$ and, therefore, the segment $v_bC_2v_{d-1}$ contains no edges of $C_1 \cap C_2$. This contradicts that v_qv_{q+1} ($\in P_1 \cap P_0 \subset C_1 \cap C_2$) lies in the segment of C_2 from v_b to v_{d-1}.

Since both v_dv_b and $v_{d-1}v_{\lambda(d-1)}$ are $Z(C_1)$-chords, the vertex $v_{\lambda(d-1)}$ must be in $\{v_{b+1}, v_{b+2}, \ldots, v_q\}$. That is, according to the structure of the L-graph and by Equation (16.7), $\lambda(d-1) = b + 1 \leq q$.

Furthermore, this edge $v_{d-1}v_{\lambda(d-1)} = v_{d-1}v_{b+1}$ joins the components R_1 and R_2 since $v_{d-1} \in R_2$ while $v_{\lambda(d-1)} = v_{b+1} \in R_1$. This contradicts that $e^* = v_qv_{q+1}$ is a bridge of $G - E(D)$. This completes the proof of Claim 14, and the proof of the theorem. □

16.6 Minimal contra pair

The first application of Theorem 16.5.7 is the verification of Conjecture 3.5.1 under the assumption of Conjecture 16.2.5.

Recall the definition of minimal contra pair. An eulerian $(1, 2)$-weighted cubic graph (G, w) is a *minimal contra pair* if (1) (G, w) is a contra pair (bridgeless and no faithful cover), (2) (G, w) has no removable circuit, and (3) for every $e \in E_{w=2}$, $(G - e, w)$ has a faithful cover.

Corollary 16.6.1 [261] (P_{10}, w_{10}) *is the only* minimal *contra pair under the assumption of Conjecture 16.2.5.*

Proof Let (G, w) be a minimal contra pair.

Note that (G, w) has no removable circuit. Hence, Theorem 16.5.7 can be applied to (G, w) since every faithful cover of $(G - e, w)$ (for every $e \in E_{w=2}$) must be a circuit chain. So, by Theorem 16.5.7, (G, w) must be *a Petersen chain*.

By Lemma 3.2.3, G is a *permutation graph* with $E_{w=1}$ inducing a pair of chordless odd-circuits D_1, D_2, and $E_{w=2}$ inducing a perfect matching joining D_1 and D_2.

As a permutation graph, G is of girth at least 4. Hence, (G, w) must be a *simple Petersen chain*.

By Definition 16.5.3, it is easy to see that (1) if the Petersen chain (G, w) has no P_{10}-segment then $(G, w) \in \langle \mathcal{K}_4 \rangle_2^{HW}$ and therefore is not a contra pair; (2) if (G, w) has more than one P_{10}-segment then $E_{w=1}$ induces more than two circuits. Hence, as a permutation graph, $E_{w=1}$ induces two circuits if and only if the Petersen chain (G, w) has precisely one P_{10}-segment. Furthermore, if (G, w) has a K_4-segment, then G must have a triangle. Hence, (G, w) must be a *single segment, simple Petersen chain* which is (P_{10}, w_{10}). □

16.7 Open problems

The following is an equivalent version of Conjecture 16.2.5.

Conjecture 16.2.5 [256] *Every 3-connected cubic graph admitting a Hamilton weight contains a triangle.*

Conjecture 16.7.1 [261] *Let G be a cubic graph with an eulerian $(1, 2)$-weight w and $e \in E_{w=2}$. If $(\overline{G - e}, w)$ is a Hamilton weighted graph with $E_{w=1}$ inducing a Hamilton circuit of G, and (G, w) has no removable circuit C such that $e \notin E(C)$, then (1) $(G, w) \in \langle \mathcal{K}_4 \rangle_2^{HW}$, (2) $(\overline{G - e}, w) \in \langle \mathcal{K}_4 \rangle_2^{HW}$.*

Remark. In Exercise 16.9, we will prove that conclusions (1) and (2) are equivalent. That is, it is sufficient to prove one of them.

Conjecture 16.7.1 is a corollary of Conjecture 16.2.5, and also implies Conjecture 3.5.1 [261].

Conjecture 3.5.1 (Goddyn [87]) (P_{10}, w_{10}) *is the only* minimal *contra pair.*

Recall the definition of critical contra pair. A contra pair (G, w) is critical if it has no removable circuit.

Conjecture 3.5.3 (Goddyn [87]) *Let G be a permutation graph such that M is a perfect matching of G and $G - M$ is the union of two chordless circuits C_1 and C_2. If (G, w) is a critical contra pair with $E_{w=2} = M$ and $E_{w=1} = E(C_1) \cup E(C_2)$, then $(G, w) = (P_{10}, w_{10})$.*

Conjecture 3.5.4 (Goddyn [85], or see [87]) *(P_{10}, w_{10}) is the only 3-connected, cyclically 4-edge-connected,* critical *contra pair.*

Recall Definition 5.4.2. A cubic graph G is uniquely 3-edge-colorable if it has precisely one 1-factorization.

Some early studies about the problem of unique 3-edge-coloring can be seen, for example, in [55], [56], [78], [89], [91], [93], [106], [223], [225], [226], [235], [244], etc.

Conjecture 16.7.2 *Every 3-connected cubic graph admitting a Hamilton weight is uniquely 3-edge-colorable.*

Remark. (1) Conjecture 16.7.2 has been verified for Petersen minor free graphs (Theorem 16.4.1); (2) the 3-connectivity in this conjecture cannot be relaxed since members of $\langle \mathcal{K}_4 \rangle_2^{HW}$ are also Hamilton weighted graphs (but not uniquely 3-edge-colorable).

In Exercise 16.2, some equivalent statements of Conjecture 16.7.2 are proved.

16.8 Exercises

⋆ Hamilton weight, unique 3-edge-coloring

Exercise 16.1 Let (G, w) be a Hamilton weighted cubic graph. For every 1-factorization $\{M_1, M_2, M_3\}$ of G, show that $E_{w=2} \in \{M_1, M_2, M_3\}$.

Exercise 16.2 [256] Let G be a cubic graph admitting a Hamilton weight w. Then the following statements are equivalent:

(i) G is uniquely 3-edge-colorable;

(ii) G has precisely three Hamilton circuits;

(iii) (G, w) has precisely one faithful even subgraph cover;

(iv) $E_{w=1}$ is a Hamilton circuit of G

Remark. Note that the proof of "(i)$\Rightarrow$(ii)" does not need the assumption that (G, w) is a Hamilton weighted graph. However, the assumption that (G, w) is a Hamilton weighted graph is necessary for "(ii)$\Rightarrow$(i)". (That is, (i) and (ii) are not always equivalent to each other without any extra assumption [226].) Without any extra assumption, Greenwell and Kronk ([93], also see [226]) conjectured that *if a cubic graph G has exactly three Hamilton circuits then G is uniquely 3-edge-colorable*. This conjecture was disproved by Thomason [226], who found a family of counterexamples: the generalized Petersen graphs $P(6k+3, 2)$ for $k \geq 2$, each of which has exactly three Hamilton circuits but is not uniquely 3-edge-colorable. By Exercise 16.2, the graphs constructed by Thomason do not admit Hamilton weights.

Exercise 16.3 [256] Let G be a cubic graph. If G admits at least two Hamilton weights, then G is uniquely 3-edge-colorable.

Exercise 16.4 Show that the generalized Petersen graph $P(9, 2)$ (see Figure 16.25) does not admit a Hamilton weight.

Remark. Tutte [235] discovered that the generalized Petersen graph $P(9, 2)$ is uniquely 3-edge-colorable but not a member of $\langle \mathcal{K}_4 \rangle$.

$$\star \ \langle \mathcal{K}_4 \rangle\text{-graphs}$$

Exercise 16.5 Let $G \in \langle \mathcal{K}_4 \rangle$. Let $w : E(G) \mapsto \{1, 2\}$ be an arbitrary eulerian weight of G. Then w is a Hamilton weight of G if and only if $E_{w=1}$ induces a Hamilton circuit of G.

Exercise 16.6 Let $(G, w) \in \langle \mathcal{K}_4 \rangle^{HW}$ and $e_0 \in E_{w=2}$. Then every 3-edge-cut of G containing e_0 must contain two edges of $E_{w=1}$.

Exercise 16.7 Let (G, w) be a Hamilton weighted graph and $e_0 = x_0 y_0 \in E_{w=2}$ that $G - e_0$ remains bridgeless. If G is Petersen minor free, then

(1) $(G - e_0, w)$ has a faithful circuit cover;

(2) (G, w) has a no removable circuit C that $e_0 \notin E(C)$;

(3) Every faithful cover $\mathcal{F}$ of $(G - e_0, w)$ is a circuit chain joining x_0, y_0.

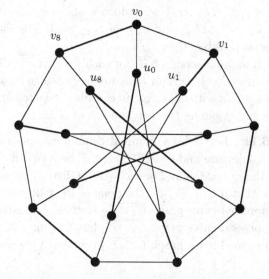

Figure 16.25 *A generalized Petersen graph* $P(9,2) = P(9,4)$

Note, prove Exercise 16.7 without applying Theorem 16.4.1.

Exercise 16.8   Let $(G, w) \in \langle \mathcal{K}_4 \rangle^{HW}$ and $e \in E_{w=2}$. Then
(1) $(G - e, w)$ has a unique faithful circuit cover $\mathcal{F}$;
(2) $|\mathcal{F}| = 1$ if and only if $G = 3K_2$;
(3) $|\mathcal{F}| = 2$ if and only if e is untouched;
(4) $|\mathcal{F}| \geq 3$ if and only if e is touched.

Exercise 16.9   Let G be a cubic graph with an eulerian $(1, 2)$-weight w and $e \in E_{w=2}$. If $(\overline{G - e}, w)$ is a Hamilton weighted graph and (G, w) has no removable circuit C such that $e \notin E(C)$, then the following statements are equivalent:
(1) $(G, w) \in \langle \mathcal{K}_4 \rangle_2^{HW}$;
(2) $(\overline{G - e}, w) \in \langle \mathcal{K}_4 \rangle_2^{HW}$.

⋆ Circuit cover

Exercise 16.10   Let (G, w) be a cubic graph and w be an eulerian $(1, 2)$-weight of G and $e_0 = xy \in E(G)$ with $w(e_0) = 2$. Suppose that (G, w) has no faithful circuit cover, while $(G - \{e_0\}, w)$ has a faithful circuit cover. Suppose further that every faithful circuit cover of $(G - \{e_0\}, w)$ is a circuit chain joining x and y. For a faithful circuit cover

$\mathcal{F} = \{C_1, \ldots, C_t\}$ of $(G - \{e_0\}, w)$ (which is also a circuit chain joining x and y), let $H_i = C_i \cup C_{i+1}$ and $w_i = w_{\{C_i, C_{i+1}\}}$. If $\mathcal{F}$ is chosen so that $|\mathcal{F}|$ is as large as possible, then

(1) w_i' is a Hamilton weight of $\overline{H_i}$ for each $i = 1, \ldots, t - 1$;

(2) the suppressed cubic graph $\overline{H_i}$ is uniquely 3-edge-colorable if G is a permutation graph with $E_{w=1}$ as its chordless 2-factor and $E_{w=2}$ (as a perfect matching) joining two components of $G[E_{w=1}]$.

Exercise 16.11   Let G be a minimal counterexample to the circuit double cover conjecture and $e \in E(G)$. Let $\mathcal{F}$ be a circuit double cover of $G - \{e\}$ with $|\mathcal{F}|$ maximum. For each pair of distinct members C_1 and $C_2 \in \mathcal{F}$, show that if $C_1 \cap C_2 \neq \emptyset$ then one of the following must hold:

(1) the suppressed cubic graph $\overline{C_1 \cup C_2}$ is uniquely 3-edge-colorable,

(2) the suppressed cubic graph $\overline{C_1 \cup C_2}$ has a 2-edge-cut,

(3) the suppressed cubic graph $\overline{C_1 \cup C_2}$ contains a Petersen minor.

Appendix A
Preliminary

A.1 Fundamental theorems

In this chapter, we present some fundamental theorems that are used in the book. Note that the proofs of some theorems are not presented here if they can be found in some popular graph theory textbooks (for instance, [15], [18], [19], [34], [41], [92], and [242], etc.).

⋆ Embedding

Theorem A.1.1 (Euler formula) *Let $G = (V, E)$ be a connected graph with a 2-cell embedding on a surface Σ. Then*

$$|V(G)| + |F(G)| = |E(G)| + \xi(\Sigma). \tag{A.1}$$

where $F(G)$ is the set of faces of G on Σ, $\xi(\Sigma)$ is the Euler characteristic *of the surface Σ such that*

(1) $\xi(\Sigma) \leq 2$ for any surface Σ,
(2) $\xi(\Sigma) = 2$ if Σ is a sphere,
(3) $\xi(\Sigma) = 1$ if Σ is a projective plane,
(4) $\xi(\Sigma) = 0$ if Σ is a torus or Klein bottle.

Proof See [19] p. 279, [13] p. 87, or [94] p. 27 and p. 121. ☐

Theorem A.1.2 (Kuratowski [160]) *A graph G is planar if and only if it contains no subdivision of K_5 or $K_{3,3}$.*

Proof See [18] p. 153, [34] p. 96 and p. 100, [92] p. 164, [242] p. 259, or [19] p. 268. ☐

⋆ Connectivity and spanning trees

Theorem A.1.3   (Menger [180]) *Let $D = (V, A)$ be a directed graph and $u, v \in V(D)$.*

(1) Let $uv \notin A(D)$. The graph D contains t internally disjoint directed paths from u to v if and only if, for each $X \subseteq V(D) - \{u, v\}$ with $|X| < t$, there is a directed path from u to v in $D - X$.

(2) The graph D contains t edge-disjoint directed paths from u to v if and only if, for each $F \subseteq A(D)$ with $|F| < t$, there is a directed path from u to v in $D - F$.

Proof   See [76] p. 55, [18] p. 203, [34] p. 158, [92] p. 51, [242] p. 149, or [19] p. 208.   □

Definition A.1.4   Let $\mathcal{P} = \{V_1, \ldots, V_t\}$ be a partition of $V(G)$. The graph $G/\mathcal{P}$ is obtained from G by identifying all vertices of each V_i to be a single new vertex for each $i = 1, \ldots, t$ and deleting all resulting loops. (This operation is called **vertex shrinking**.)

Theorem A.1.5   (Tutte [232] and Nash-Williams [184]) *Let G be a graph and k be a positive integer. The graph G contains k edge-disjoint spanning trees if and only if, for every partition $\mathcal{P}$ of $V(G)$, the shrunken graph $G/\mathcal{P}$ contains at least $k(|V(G/\mathcal{P})| - 1)$ edges.*

Proof   See [18] p. 31, [19] p. 570, [41] p. 48, or [259] p. 286.   □

The following is an immediate corollary of Theorem A.1.5.

Theorem A.1.6   (see [96], [159] and [191]) *Every $2k$-edge-connected graph contains at least k edge-disjoint spanning trees.*

⋆ Perfect matchings, r-graphs

Theorem A.1.7   (Petersen [189]) *Every 2-edge-connected cubic graph has a perfect matching.*

Proof   See [18] p. 79, [92] p. 213, or [242] p. 124.   □

Let G be a graph and $X \subset V(G)$. The set of edges of G with one end in X and another end in $V(G) - X$ is denoted by $(X, V(G) - X)_G$.

Definition A.1.8   An r-graph G is an r-regular graph of even order such that, for each vertex subset $X \subset V(G)$ with $|X| \equiv 1 \pmod 2$,

$$|(X, V(G) - X)_G| \geq r.$$

The following theorem is a corollary of Edmonds' matching polyhedron theorem [43].

Theorem A.1.9 (Edmonds [43], or see [204]) *Let G be an r-graph. Then there is an integer p and a family $\mathcal{M}$ of perfect matchings such that each edge of G is contained in precisely p members of $\mathcal{M}$. (Note that, it is not necessary that the members of $\mathcal{M}$ must be distinct.)*

Proof See Theorem A.3.2 in [259]. □

The following corollary of Theorem A.1.9 is useful in this book.

Corollary A.1.10 (see [204]) *Let G be an r-graph. Then G has a family $\mathcal{M}$ of perfect matchings such that there is an integer p, each edge of G is contained in precisely p members of $\mathcal{M}$, and for each edge-cut T of size r and for each $P \in \mathcal{M}$,*

$$|P \cap T| = 1.$$

Proof See Corollary A.3.3 in [259]. □

⋆ Petersen minor

Definition A.1.11 A cubic graph G is called a **permutation graph** if G has a 2-factor F such that F is the union of two chord-less circuits.

Theorem A.1.12 (Ellingham [46]) *If G is a permutation graph containing no subdivision of the Petersen graph, then*
(1) G contains a circuit of length 4,
(2) G contains a Hamilton circuit,
(3) G is 3-edge-colorable.

Proof See Theorem A.4.3 in [259]. Or see [90]. □

⋆ Vertex splitting

Definition A.1.13 Let G be a graph and v be a vertex of G and $F \subset E(v)$. The graph $G_{[v;F]}$ is obtained from G by splitting the edges of F away from v. That is, $G_{[v;F]}$ is constructed as follows: add a new vertex v' and change the endvertex v of the edges of F to be v'. (See Figure A.1.)

The following theorem is known as *the vertex splitting lemma*.

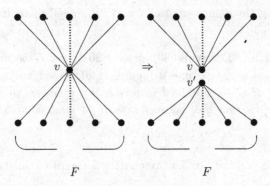

Figure A.1 G and $G_{[v;F]}$: *splitting F away from v*

Theorem A.1.14 (Fleischner [58], or see [71], [65]) *Let G be a 2-edge-connected graph, $v \in V(G)$ with $d(v) \geq 4$, and $e_0, e_1, e_2 \in E_G(v)$. Then one of the following statements must hold:*

(1) $G_{[v;\{e_0,e_1\}]}$ remains 2-edge-connected,

(2) $G_{[v;\{e_0,e_2\}]}$ remains 2-edge-connected,

(3) v is a cut-vertex and $\{e_0, e_1, e_2\}$ is an edge-cut of G.

Proof See Theorem A.5.2 in [259]. □

Definition A.1.15 The odd-edge-connectivity of a graph G, denoted by $\lambda_o(G)$, is the size of a smallest odd edge-cut of G.

The following is a generalization of Theorem A.1.14 about a vertex-splitting operation that maintains the odd-edge-connectivity and the embedding.

Theorem A.1.16 (Zhang [260]) *Let $G = (V, E)$ be a graph with odd-edge-connectivity λ_o. Assume that there is a vertex $v_a \in V(G)$ such that $d(v_a) \neq \lambda_o$ and $d(v_a) \neq 2$. Arbitrarily label the edges of G incident with v_a as $\{e_1, \ldots, e_b\}$ (where $b = d(v_a)$). Then there is an integer $i \in \{1, \ldots, b\}$ such that the new graph $G_{[v_a;\{e_i,e_{i+1}\}]}$ obtained from G by splitting e_i and e_{i+1} (mod b) away from v_a remains of odd-edge-connectivity λ_o.*

A.2 Even subgraphs and parity subgraphs

⋆ Even (sub)graphs

Definition A.2.1 A graph G is even if the degree of every vertex is even.

Lemma A.2.2 (Veblen [238]) *Every even graph has a circuit decomposition.*

Proof See [19] p. 56. □

⋆ Symmetric difference

Definition A.2.3 Let H_1 and H_2 be two subgraphs of a graph G. Then the symmetric difference of H_1 and H_2, denoted by $H_1 \triangle H_2$, is the subgraph of G induced by edges of $(E(H_1) \cup E(H_2)) - (E(H_1) \cap E(H_2))$.

Lemma A.2.4 *The symmetric difference of finitely many subgraphs $\{H_1, \ldots, H_t\}$ of G is the subgraph of G induced by the edges contained in an odd number of H_i.*

Lemma A.2.5 *Let $\{H_1, \ldots, H_t\}$ be a collection of subgraphs of a graph G and $H = H_1 \triangle \cdots \triangle H_t$. Then*

$$d_H(v) \equiv \sum_{i=1}^{t} d_{H_i}(v) \pmod 2.$$

Proof Corollary of Lemma A.2.4. □

Lemma A.2.6 *Let $\{C_1, \ldots, C_t\}$ be a collection of even subgraphs of a graph G. Then $C_1 \triangle \cdots \triangle C_t$ is also an even subgraph.*

Proof Corollary of Lemma A.2.5. Or, see [19] p. 65. □

⋆ Odd vertices

Lemma A.2.7 *Let G be a graph. The number of odd vertices in each component of G is even.*

Proof See [18] p. 10, [34] p. 7, [92] p. 4, or [242] p. 26. □

Lemma A.2.8 *Let G be a graph, $X \subset V(G)$. Then*

$$|(X, G - X)| \equiv |O(G) \cap X| \pmod 2$$

where $(X, G - X)$ is the edge-cut between X and $V(G) - X$, and $O(G)$ is the set of all odd degree vertices of G.

Proof Let G^* be the graph obtained from G by identifying all vertices of $G - X$ as a single vertex. Then apply Lemma A.2.7 to G^*. □

$\star$ Parity graphs, T-joins

Definition A.2.9 Let G be a graph. A spanning subgraph P is a parity subgraph of G if

$$d_P(v) \equiv d_G(v) \quad (\text{mod } 2)$$

for every $v \in V(G)$.

Lemma A.2.10 *Let G be a graph. A subgraph H is an even subgraph of G if and only if $G - E(H)$ is a parity subgraph of G.*

Lemma A.2.11 *The union of a pair of edge-disjoint parity subgraphs is an even subgraph.*

Lemma A.2.12 *Let $\{H_1, \ldots, H_t\}$ be a collection of parity subgraphs of a graph G. Then $\triangle_{j=1}^t H_j$ is a parity subgraph of G if t is odd; and $\triangle_{j=1}^t H_j$ is an even subgraph of G if t is even.*

Proof Corollary of Lemma A.2.5. □

Lemma A.2.13 *Let C be an even subgraph of a graph G. Then*

$$|T \cap E(C)| \equiv 0 \quad (\text{mod } 2)$$

for every edge-cut T of G.

Lemma A.2.14 *Let P be a parity subgraph of a graph G. Then*

$$|T \cap E(P)| \equiv |T| \quad (\text{mod } 2)$$

for every edge-cut T of G.

Proof By Lemma A.2.10, $C = G - E(P)$ is an even subgraph. By Lemma A.2.13,

$$|T \cap E(C)| \equiv 0 \quad (\text{mod } 2)$$

for every edge-cut T of G. Hence,

$$|T \cap E(P)| \equiv |T| \quad (\text{mod } 2).$$

□

Lemma A.2.15 (Itai and Rodeh [119]) *Every spanning tree of a graph G contains a parity subgraph of G.*

Proof Let T be a spanning tree of G. For each $e \in E(G) - E(T)$, there is a unique circuit C_e of G contained in $T \cup \{e\}$. The symmetric difference S of the circuits C_e for all $e \in E(G) - E(T)$ is an even subgraph (by Lemma A.2.6) of G. Since each edge $e \in E(G) - E(T)$ belongs to only one circuit of $\{C_e : e \in E(G) - E(T)\}$, we have that $E(G) - E(T) \subseteq E(S)$. Thus $G - E(S)$ is a parity subgraph contained in T. $\square$

For a graph G, the set of all odd degree vertices of G is denoted by $O(G)$.

Definition A.2.16 Let $T \subseteq V(G)$ with $|T| \equiv 0 \pmod 2$. A subgraph H of G is called a T-join of G if $O(H) = T$.

It is evident that the T-join is a generalized concept of parity subgraphs since a parity subgraph P of G is a T-join with $T = O(G)$. The following result generalizes Lemma A.2.15, which can be found in various articles.

Lemma A.2.17 *Every spanning tree of a graph G contains a T-join for every $T \subseteq V(G)$ of even order.*

⋆ Small parity graphs

Note that the odd-edge-connectivity is defined in Definition A.1.15. By applying the vertex splitting method in Theorem A.1.16 and the observation of Exercise A.8, we have the following lemma.

Lemma A.2.18 *Let G be a graph with odd-edge-connectivity λ_o. Then there is a series of vertex-splitting operations such that the graph G' which results from applying those operations on G has the following properties:*
(1) the odd-edge-connectivity of G' remains λ_o;
(2) $d_{G'}(v) = \lambda_o$ or 2 for every vertex of $v \in V(G')$;
(3) the suppressed graph $\overline{G'}$ is an r-graph with $r = \lambda_o$.

Note that r-graphs are defined in Definition A.1.8.

Lemma A.2.19 *Let G be a graph with odd-edge-connectivity λ_o, and $w : E(G) \mapsto \mathbb{Z}^+$ be a non-negative weight (length of edges). Then G contains a parity subgraph P with $w(P) \leq \frac{w(G)}{\lambda_o}$.*

Proof By Lemma A.2.18, there is a series of splitting operations of G such that the suppressed graph of the resulting graph G' is an r-graph with $r = \lambda_o$.

Define a mapping $h : E(G) \mapsto E(\overline{G'})$ such that, for each $e \in E(\overline{G'})$, $h^{-1}(e)$ is the set of all edges in the corresponding induced path of G' and define w' for $\overline{G'}$ as follows: $w'(e) = \sum_{e' \in h^{-1}(e)} w(e')$.

By Corollary A.1.10, let $\mathcal{M}$ be the set of 1-factors covering each edge of G' precisely p times. Since $|\mathcal{M}| = \lambda_o p$ and

$$\sum_{M \in \mathcal{M}} w(M) = p \times w'(\overline{G'}) = p \times w(G),$$

there is a 1-factor $M_0 \in \mathcal{M}$ with

$$w(M_0) \leq \frac{1}{\lambda_o} w(G).$$

Note that the subgraph of G induced by edges of $h^{-1}(M_0)$ is a parity subgraph of G of total weight at most $\frac{1}{\lambda_o} w(G)$. $\qquad\square$

A.3 Exercises

Exercise A.1 A minimal parity subgraph of a graph G is acyclic.

Exercise A.2 Let $\mathcal{F}$ be a family of even subgraphs of a graph G. Then the subgraph of G induced by the edges contained in odd numbers of members of $\mathcal{F}$ is an even subgraph.

Exercise A.3 Let $\mathcal{F}$ be a family of even subgraphs of a graph G. Then the subgraph of G induced by edges contained in even numbers of members of $\mathcal{F}$ is a parity subgraph of G.

Exercise A.4 If G is a 2-connected planar graph with minimum degree at least three, then G has a facial circuit of length at most five. Also the same result holds for graphs embedded on the projective plane.

Exercise A.5 If a *generalized Petersen graph* $P(n,t)$ is a permutation graph, show that
(1) $gcd(n,t) = 1$,
(2) $P(n,t)$ contains a subdivision of the Petersen graph unless $t \equiv \pm 1$ (mod n).

The generalized Petersen graph $P(n,t)$ is defined in Definition B.2.1.

Exercise A.6 Let the odd-edge-connectivity of a graph G be λ_o. If $\lambda_o \leq 5$, then G contains at least $\frac{\lambda_o - 1}{2}$ edge-disjoint parity subgraphs.

Remark. Exercise A.6 was further generalized in [214] for all graphs without the restriction $\lambda_o \leq 5$.

Exercise A.7 If an r-regular graph G is r-edge-colorable, then G is an r-graph.

Exercise A.8 Let r be an odd integer and G be an r-regular graph. Then G is an r-graph if and only if the odd-edge-connectivity of G is r.

Exercise A.9 Let G be a bridgeless cubic graph. Show that G has a 2-factor F such that the odd-edge-connectivity of the contracted graph G/F is at least 5.

Remark. Exercise A.9 is further generalized in [139]: *every 3-edge-connected cubic graph has a 2-factor F such that the contracted graph G/F is 5-edge-connected.*

Appendix B
Snarks, Petersen graph

Methods developed for 3-edge-colorings of cubic graphs play a central role in the study of circuit cover problems. In this chapter, we present some elementary and commonly used results, and methods in this subject.

B.1 3-edge-coloring of cubic graphs, snarks

The subject of 3-edge-colorings of cubic graphs has been extensively studied in graph theory because of its close relation with the map 4-coloring problem.

Theorem B.1.1 (Tait [220]) *Every bridgeless planar graph is 4-face colorable if and only if every bridgeless, cubic, planar graph is 3-edge-colorable.*

Proof See Theorem 9.12 of [18], or Theorem 11.4 of [19]. □

Theorem B.1.2 (The 4-Color Theorem, Appel and Haken [5], [7], [6] and [197], [224]) *Every bridgeless planar graph is 4-face colorable.*

For an obvious reason, the proof of the 4-color theorem (Theorem B.1.2) is not included in this book. From Theorems B.1.1 and B.1.2, we have that *every bridgeless, cubic, planar graph is 3-edge-colorable.*

Note that the determination of 3-edge-colorability of a cubic graph is an NP-complete problem [113].

B.1.1 Parity lemma

The following is a basic lemma that we will use frequently in this chapter.

Lemma B.1.3 (Blanuša [14] and Descartes [39], also see [118] or [146])
Let G be a cubic graph. If G has a 3-edge-coloring $c : E(G) \mapsto \{1, 2, 3\}$, then, for each edge-cut T,

$$|T \cap c^{-1}(i)| \equiv |T| \pmod 2$$

for each $i \in \{1, 2, 3\}$.

Proof A special case of Lemma A.2.14. $\qquad\qquad\square$

B.1.2 Snarks

Snarks were so named by Martin Gardner in 1976 [81], after the mysterious and elusive object of the poem *The Hunting of the Snark* by Lewis Carroll [23].

Traditionally, snarks are defined as *non-3-edge-colorable, cyclically 4-edge-connected, cubic graphs (and with girth ≥ 5, in some literature)*. It was remarked in [241] and [30] that "we should leave the definition of snarks as broad as possible" for the purpose of less confusion in studies of related areas. For example, any counterexample to any of the following conjectures must be a snark: *5-flow conjecture* (Conjecture C.1.10), *Berge–Fulkerson conjecture* (Conjecture 7.3.6), *circuit double cover conjecture, strong circuit double cover conjecture* (Conjecture 1.5.1), etc. However, the girth or cyclical edge-connectivity requirements for a minimum counterexample to each of those conjectures are achieved differently. The following table is a partial list of some known results.

Smallest counterexample to	girth		Cyclic edge-connectivity
5-flow conjecture	≥ 11	[150]	≥ 6 [149]
CDC conjecture	≥ 12	[115]	≥ 4
Strong CDC conjecture	≥ 5 (Ex. 6.1)		≥ 4
Berge-Fulkerson conjecture	≥ 4		≥ 4

Therefore, we simply define *snarks as non-3-edge-colorable, bridgeless, cubic graphs,* and all additional description/requirement will be specified whenever we need it.

Definition B.1.4 A 2-edge-connected, non 3-edge-colorable, cubic graph is called a **snark**.

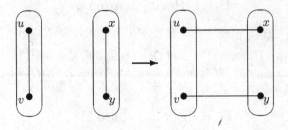

Figure B.1 *A new snark with a 2-edge-cut:* $\oplus_2$-*sum*

Structural characterization or construction of snarks has been one of the major approaches in the history of the map 4-color problem. Tutte [234] conjectured that *every snark must contain a subdivision of the Petersen graph*. It was announced by Robertson, Sanders, Seymour and Thomas that this conjecture is now a theorem. The proof consists of a series of papers: [197], [198], [195], [196], [199] and [200] (see [223]).

Because of the size limitation of this book, we are not able to list all snarks (see papers [37], [240], [241], [29], [30] for the most up-to-date and comprehensive surveys). In this section, we present a few basic and elementary methods to construct larger snarks from smaller snarks. These methods are also useful for constructing infinite families of counterexamples to some integer flow, circuit cover, circuit decomposition problems (for example, see [121], or construction methods in Sections 2.3 and 10.6, etc.).

B.1.3 Construction of snarks

In this section, we present three basic methods for construction of a large family of snarks. The *superposition method* introduced by Kochol ([145], [143] and [146]) is not presented in this book.

B.1.3.1 Method One. $\oplus_2$-sum

Let G_1, G_2 be two bridgeless cubic graphs and $e_1 = uv \in E(G_1), e_2 = xy \in E(G_2)$. Construct a new graph

$$G_1 \oplus_2 G_2 = [G_1 - \{e_1\}] \cup [G_2 - \{e_2\}] \cup \{ux, vy\}$$

(see Figure B.1).

Proposition B.1.5 *Let G_1 and G_2 be two bridgeless cubic graphs.*

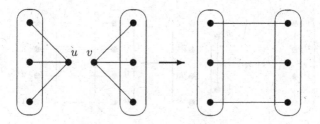

Figure B.2 *A new snark with a 3-edge-cut:* $\oplus_3$*-sum*

Then one of G_1, G_2 is a snark if and only if the new graph $G = G_1 \oplus_2 G_2$ constructed by Method One is a snark.

Proof If G is 3-edge-colorable, then, by Lemma B.1.3, both edges ux, vy are colored with the same color. By restricting the coloring of G to each G_1 and G_2, it is obvious that neither G_1 nor G_2 is a snark.

The proof of the other direction is trivial and similar. □

B.1.3.2 Method Two. $\oplus_3$-sum

Let G_1, G_2 be two bridgeless cubic graphs and $u \in V(G_1), v \in V(G_2)$ be two degree 3 vertices with $N(u) = \{u_1, u_2, u_3\}, N(v) = \{v_1, v_2, v_3\}$. Construct a new graph

$$G_1 \oplus_3 G_2 = [G_1 - \{u\}] \cup [G_2 - \{v\}] \cup \{u_1v_1, u_2v_2, u_3v_3\}$$

(see Figure B.2).

Proposition B.1.6 *Let G_1 and G_2 be two bridgeless cubic graphs. Then one of G_1, G_2 is a snark if and only if the new graph $G = G_1 \oplus_3 G_2$ constructed by Method Two is a snark.*

Proof If G is 3-edge-colorable, then, by Lemma B.1.3, all three edges u_1v_1, u_2v_2 and u_3v_3 are colored differently. By restricting the coloring of G to each G_1 and G_2, it is obvious that neither G_1 nor G_2 is a snark.

The proof of the other direction is trivial and similar. □

B.1.3.3 Method Three. Isaacs product

Let G_F, G_M be two cubic graphs and $e_1 = u_1u_2, e_2 = v_1v_2 \in E(G_F)$ and $e_3 = x_1x_2 \in E(G_M)$ with $N(x_1) = \{y_1, y_2, x_2\}$ and $N(x_2) = \{z_1, z_2, x_1\}$. Construct a new graph

$$G = [G_F - \{e_1, e_2\}] \cup [G_M - \{x_1, x_2\}] \cup \{u_1y_1, u_2y_2, v_1z_1, v_2z_2\}$$

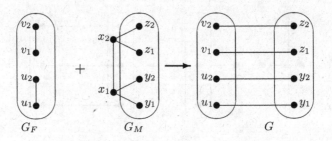

Figure B.3 *Isaacs product of snarks*

(see Figure B.3). This construction is called the Isaacs product [118].

Theorem B.1.7 (Isaacs [118]) *If both G_F and G_M are snarks, then the Isaacs product G of G_F and G_M is also a snark.*

Proof Let $T_1 = \{u_1y_1, u_2y_2\}$ and $T_2 = \{v_1z_1, v_2z_2\}$ and $T = T_1 \cup T_2$. Assume that G has a 3-edge-coloring $c : E(G) \mapsto \{1,2,3\}$. By Lemma B.1.3, $|c^{-1}(i) \cap T| = 0, 2$ or 4 for each $i \in \{1,2,3\}$. Thus, we assume that no edge of T is colored 3. That is, $c^{-1}(3) \cap T = \emptyset$. There are three cases here now:

(1) $|c^{-1}(1) \cap T| = 4$ and $c^{-1}(2) \cap T = \emptyset$;

(2) $|c^{-1}(i) \cap T| = 2$ for each $i \in \{1,2\}$ and $|c^{-1}(1) \cap T_1| = |c^{-1}(2) \cap T_2| = 2$;

(3) $|c^{-1}(i) \cap T| = 2$ for each $i \in \{1,2\}$ and $|c^{-1}(i) \cap T_j| = 1$ for each $i, j \in \{1,2\}$.

In the first two cases, $c(u_1y_1) = c(u_2y_2)$ and $c(v_1z_1) = c(v_2z_2)$. Then the coloring c induces a 3-edge-coloring of G_F. This contradicts that G_F is a snark. In the third case, the coloring c induces a 3-edge-coloring of G_M. This contradicts that G_M is a snark.   □

B.1.4 Girths and bonds of snarks

By Lemma B.1.3 and Propositions B.1.5 and B.1.6, one may immediately observe the following facts related to small edge-cut or small circuits.

Proposition B.1.8 *Let G be a snark, and let T be an edge-cut of G with $|T| \leq 3$. Let Q_1 and Q_2 be two components of $G - T$. Then either $\overline{G/Q_1}$ or $\overline{G/Q_2}$ is a snark.*

If G is a snark containing a 4-circuit C, then, by removing a pair of

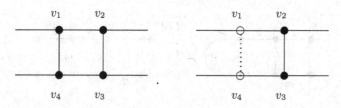

Figure B.4 *Delete edge $v_1 v_4$*

disjoint edges from C, the suppressed resulting graph is a smaller snark (see [241] p. 1134, or Lemma B.1.9). This is why most articles about snarks pay most attention to cyclically 4-edge-connected snarks with girth ≥ 5. The following lemma is a slightly stronger statement and is much more useful in various studies.

Lemma B.1.9 (Catlin [25]) *Let G be a 2-edge-connected cubic graph containing a circuit C of length ≤ 4. For a subset $F \subseteq E(C)$, if the suppressed cubic graph $\overline{G - F}$ is 3-edge-colorable, then G is also 3-edge-colorable.*

Proof Since the suppressed graph $\overline{G - F}$ is cubic, there are only two cases.

Case 1. $|E(C) - F| \leq 2$.

Case 2. $|E(C) - F| = 3$ (that is, $|C| = 4$ and $|F| = 1$).

The first case is rather trivial. Here, we only prove the lemma for the second case.

Assume that $C = v_1 v_2 v_3 v_4 v_1$ is a circuit of length 4 and $F = \{v_1 v_4\}$ (see Figure B.4). Let $H = \overline{G - \{v_1 v_4\}}$.

Let $c : E(H) \mapsto \{1, 2, 3\}$ be a 3-edge-coloring of H. Let $c(v_2 v_3) = 1$. Thus, $v_1 v_2, v_3 v_4 \in c^{-1}(2) \cup c^{-1}(3)$.

Subcase 2-1. If edges $v_1 v_2, v_3 v_4$ are contained in the same component Q of $c^{-1}(2) \cup c^{-1}(3)$. Then, by alternating the colors along a segment of the bi-colored circuit Q and coloring $v_1 v_4$ with 1, one is able to extend the coloring c to the entire graph G. This contradicts that G is a snark (see Figure B.5).

Subcase 2-2. If edges $v_1 v_2, v_3 v_4$ are contained in the different components Q_1 and Q_2 of $c^{-1}(2) \cup c^{-1}(3)$. First, if necessary, one may alternate the colors along a bi-colored circuit Q_1 so that both edges $v_1 v_2$ and $v_3 v_4$ are colored with the same color 2. Then one is able to extend

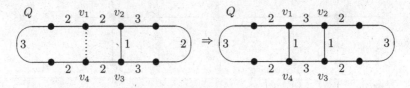

Figure B.5 *Subcase 2-1. Color adjustment along a segment $v_1v_2Qv_3v_4$ of Q*

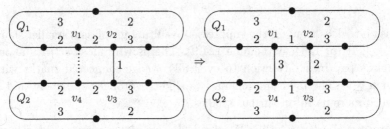

Figure B.6 *Subcase 2-2. Color adjustment along the path $v_1v_2v_3v_4$*

the coloring c to the entire graph G by alternating the colors along the $(1,2)$-bi-colored segment $v_1v_2v_3v_4$ and coloring the edge v_1v_4 with color 3 (see Figure B.6).   □

B.1.5 Small snarks

Snarks of small orders have been extensively searched, constructed, and studied [240], [241], [20], [29], [30]. These small snarks are, in addition to the Petersen graph, useful in the study of related subjects.

Proposition B.1.10   *The Petersen graph P_{10} is not 3-edge-colorable.*

Proof   See [19] p. 454, [183].   □

Proposition B.1.11   (See survey papers [240], [241], [20], [29], [30]) *The Petersen graph is the only snark with the least number of vertices.*

Proof   Let G be a smallest snark. By Proposition B.1.8 and Lemma B.1.9, G is of girth at least 5. It is also obvious that G has no Hamilton circuit.

By Theorem A.1.7, let F be a 2-factor of G. If G is of order < 10, then F must be a Hamilton circuit since G is of girth ≥ 5. Hence, as a smallest snark, G must be of order at least 10.

Furthermore, for the cubic graph G of order 10 without Hamilton circuit and of girth ≥ 5, the 2-factor F has precisely two components, each of which is a chordless circuit of length 5. Let $M = G - E(F)$ be the perfect matching joining the components Q_1 and Q_2 of F. Under some permutation, there is only one way to join Q_1 and Q_2 by edges of M (by avoiding any possible circuit of length 4), which is the Petersen graph. $\square$

Exercise B.8 is another proof of Proposition B.1.11.

Proposition B.1.12 (Fouquet [77]; also see [26], and survey papers [240], [241], [20], [29], [30]) *If G is a snark of order n: $12 \leq n \leq 16$, then*
(1) G must be of girth ≤ 4, and
(2) G must have a cyclic edge-cut of size ≤ 3.

Proposition B.1.13 (Preissmann [193]; also see survey papers [240], [241], [20], [29], [30]) *There are only two cyclically 4-edge-connected snarks of order 18, called* Blanuša *snarks, each of which is an Isaacs product of two copies of the Petersen graph.*

Proposition B.1.14 (See survey papers [240], [241], [20], [29], [30]) *Every snark of order ≤ 20 is of girth ≤ 5.*

For non-cubic graphs, the property of "3-edge-coloring" is equivalent to "nowhere-zero 4-flow." It is sometimes also useful to determine whether a small graph admits a nowhere-zero 4-flow (after applying certain reduction methods). The proof of the following result is omitted here.

Proposition B.1.15 (Lai [162]) *Let G be a 2-edge-connected graph of order at most 17. Then either G admits a nowhere-zero 4-flow or G can be contracted to the Petersen graph.*

Note that Blanuša's snarks are of order 18 ([14], or see [240], [241]), and can be constructed by Isaacs product (Method 3 in Section B.1.3) of two Petersen graphs, and are not contractible to the Petersen graph.

Readers are referred to [20] for a complete list of snarks of order ≤ 28 (cyclically 4-edge-connected and girth ≥ 5). This list was produced by a computer aided search, and was further verified by an independent algorithm in [29]. Also with a computer aided search [103], [21], some major conjectures in this book have been verified for small snarks of order ≤ 36 (see Proposition 6.3.3).

B.2 A mini encyclopedia of the Petersen graph

(The Petersen graph is) *"a remarkable configuration that serves as a counterexample to many optimistic predictions about what might be true for graphs in general."*

– Donald Knuth [142]

This section collects some properties of the Petersen graph [190]. Although no proposition in this section is very challenging, many of these properties are critically useful in the constructions of counterexamples to various circuit covering problems.

For other properties of the Petersen graph, readers are referred to some comprehensive reference books, such as *The Petersen Graph* by Holton and Sheehan [112] and *Graph Coloring Problems* by Jensen and Toft [136].

The most important and the most frequently used property of the Petersen graph is Proposition B.1.11 that *the Petersen graph is the only snark with the least number of vertices.*

There are several definitions for the Petersen graph. The following two definitions are most commonly used.

Definition B.2.1 (See Figure B.7 in Section B.4.) Let r and t be two positive integers. A generalized Petersen graph, denoted by $P(r,t)$, is defined as follows. Let $C = u_0 \ldots u_{r-1} u_0$ be a circuit of length r, and $M = \{u_i v_i : i = 0, \ldots, r-1\}$ be a perfect matching with one end u_i in $V(C)$, and add edges $\{v_j v_{j+t} : j = 0, \ldots, r-1, \pmod{r}\}$. The Petersen graph is $P(5,2)$. (The circuit $u_0 u_1 u_2 u_3 u_4 u_0$ (or $v_0 v_2 v_4 v_1 v_3 v_0$) is called the exterior pentagon (or the interior pentagon, respectively) of the Petersen graph.)

Definition B.2.2 (See [215], [170]) A Kneser graph, denoted by G_k^n, is defined as follows: each vertex represents a subset of cardinality k taken from $\mathbb{Z}_n = \{0, \ldots, n-1\}$ and two vertices are adjacent if and only if the intersection of their labels is empty. The Petersen graph $P_{10} = G_2^5$.

⋆ Automorphism

Note that Proposition B.2.3 will be used frequently in the proofs of most propositions in this section though we may not mention it.

Proposition B.2.3 *The Petersen graph is vertex transitive and edge transitive.*

Proof Use Definition B.2.2 for the Petersen graph. Let S_5 be the symmetric group on the set $\mathbb{Z}_5$. Each permutation $\pi \in S_5$ corresponds to an automorphism of P_{10}. Hence, S_5 corresponds to a subgroup of $Auto(P_{10})$, and therefore, P_{10} is vertex transitive.

For $i = 1, \ldots, 4$, let $h_i \subset \mathbb{Z}_5$ of order 2 (corresponds to a vertex of P_{10}), and h_i, h_{i+1} $(i = 1, 3)$ be a pair of disjoint subsets of $\mathbb{Z}_5$ (corresponds to an edge of P_{10}). Similarly, there is a permutation $\pi \in S_5$ such that $\{\pi(h_1), \pi(h_2)\} = \{h_3, h_4\}$. Hence, P_{10} is edge transitive. $\qquad\square$

⋆ Circuits

Proposition B.2.4 *No circuit of the Petersen graph is of length ≤ 4, 7 and 10.*

Proof Let M be a perfect matching of P_{10}. Every circuit of P_{10}, except for pentagons, must use two or four edges of M. (See Figure B.7.) $\qquad\square$

Proposition B.2.5 *Let C_1, C_2 be two circuits of the Petersen graph with the same length. Then there is an automorphism π of P_{10} such that $\pi(C_1) = C_2$.*

Proof By Proposition B.2.4, one needs to check only circuits of lengths $5, 6, 8$ and 9. (See Figure B.7.) $\qquad\square$

Proposition B.2.6 *For a given vertex $x \in V(P_{10})$, the graph $P_{10} - x$ contains precisely two Hamilton circuits, and all six edges of*

$$\bigcup_{v \in N(x)} [E(v) - \{vx\}]$$

are used in each of these Hamilton circuits.

Proof By Figure B.11 (or Figure B.12), $\overline{P_{10} - \{x\}} = K_{3,3}$. Since all vertices of $N(x)$ must be contained in every Hamilton circuit of $P_{10} - \{x\}$, the number of perfect matchings of the hexagon $P_{10} - \{x\} - N(x)$ equals to the number of Hamilton circuits of $P_{10} - \{x\}$. $\qquad\square$

Proposition B.2.7 *For each pair of hexagons C_1, C_2 of the Petersen graph, $|E(C_1) \cap E(C_2)| \geq 2$.*

Proof If $|E(C_1) \cap E(C_2)| \leq 1$, then either P_{10} has 12 vertices or has a Hamilton circuit. Both are contradictions. $\qquad\square$

Proposition B.2.8 [204] *For each circuit C of length 6 in the Petersen graph P_{10}, $P_{10} - V(C)$ is a star.*

Proof See Figure B.11 and apply Proposition B.2.5.   □

Proposition B.2.9 *For each edge e of the Petersen graph P_{10}, the graph $P_{10} - V(e)$ has a unique Hamilton circuit (corresponding to a 8-circuit of P_{10}).*

Proof See Figure B.10 and apply Proposition B.2.5.   □

⋆ Embeddings

Proposition B.2.10 *On the projective plane, the Petersen graph P_{10} has an embedding with six faces each of which is of degree 5 (see Figure B.13).*

Remark. On the projective plane, the embedding of P_{10} described in Proposition B.2.10 is unique (by the Euler formula (Theorem A.1.1)).

Proposition B.2.11 *On the torus, the Petersen graph has an embedding with five faces, each of which is of degree $9, 6, 5, 5, 5$ (see Figure B.14).*

Proposition B.2.12 *On the Klein bottle, the Petersen graph has an embedding with five faces, each of which is of degree $8, 6, 6, 5, 5$ (see Figure B.15).*

Remark. On either the torus or the Klein bottle, the Petersen graph P_{10} does not have an embedding such that every face is a hexagon (by Proposition B.2.25).

Proposition B.2.13 *For each vertex v of the Petersen graph P_{10}, the suppressed graph $\overline{P_{10} - \{v\}}$ is $K_{3,3}$.*

Proof See Figure B.11 and apply Proposition B.2.5.   □

Proposition B.2.14 *For each edge e of the Petersen graph P_{10}, the suppressed graph $\overline{P_{10} - \{e\}}$ is V_8.*

Proof See Figure B.10 and apply Proposition B.2.5.   □

Remark. Both $K_{3,3}$ and V_8 are Möbius ladders (M_3 and M_4).

Proposition B.2.15 *For any edge* $e \in E(P_{10})$ *or any vertex* $v \in V(P_{10})$, $P_{10} - e$ *and* $P_{10} - v$ *remain non-planar.*

Proof (See Figures B.11 and B.10.) By Propositions B.2.13 and B.2.14, $\overline{P_{10} - e} = V_8$ for any $e \in E(P_{10})$, and $\overline{P_{10} - v} = K_{3,3}$ for any $v \in V(P_{10})$. By Theorem A.1.2, neither of them is planar. $\square$

⋆ Edge switching

Proposition B.2.16 *Let* $v \in V(P_{10})$ *and* $e \in E(P_{10})$. P_{10} *contains a* 5-*circuit containing both* e *and* v.

Proof Without loss of generality, by Proposition B.2.3, let e be an edge on the exterior pentagon (see Figure B.7). Then, it is easy to see wherever the vertex v is, there is always a 5-circuit containing both of them. $\square$

Proposition B.2.17 *Let* G *be a graph with* 11 *vertices:* $d(v_i) = 2$ *for* $i = 0, 1, 2$ *and* $d(v_j) = 3$ *for* $j = 3, \ldots, 10$. *Construct a new graph* G_i *from* G *by adding a new edge* e_i *joining* v_0 *and* v_i *for each* $i = 1, 2$. *If* v_1 *and* v_2 *are not adjacent with each other, then at most one of* $\{\overline{G_1}, \overline{G_2}\}$ *is isomorphic to the Petersen graph.*

Proof Assume that $\overline{G_1} = P_{10}$. And let f_2 be the edge of $\overline{G_1}$ corresponding to the subdivided edge containing v_2. By Proposition B.2.16, $\overline{G_1}$ contains a 5-circuit C that contains both v_0 and f_2.

Let G^* be the graph obtained from G by adding both edges e_1, e_2. In G^*, this circuit C becomes a 6-circuit with a chord e_2. Thus, $C - e_1 + e_2$ contains a circuit of length at most 4 in $\overline{G_2}$ since $v_1 v_2 \notin E(G)$. Since the girth of P_{10} is 5, $\overline{G_2}$ cannot be P_{10}. $\square$

⋆ Perfect matchings

Proposition B.2.18 [204] *For each pair of distinct perfect matchings* M_1 *and* M_2 *of the Petersen graph* P_{10}, $|E(M_1) \cap E(M_2)| = 1$ *and* $M_1 \triangle M_2$ *is a circuit of length eight.*

Proof Since P_{10} is not 3-edge-colorable, $E(M_1) \cap E(M_2) \neq \emptyset$. Since $M_1 \triangle M_2$ is the union of disjoint even length circuits and every even circuit of P_{10} is of length either 6 or 8, $M_1 \triangle M_2$ is either a 6-circuit or an 8-circuit. Assume that $M_1 \triangle M_2$ is a 6-circuit. By Proposition B.2.8, $P_{10} - V(M_1 \triangle M_2)$ is a star. This contradicts that $E(M_1) \cap E(M_2)$ is a perfect matching of the graph $P_{10} - V(M_1 \triangle M_2)$ and therefore $M_1 \triangle M_2$ is an 8-circuit. $\square$

Proposition B.2.19 [204] *For each edge e of the Petersen graph P_{10}, there are precisely two distinct perfect matchings containing e.*

Proof Assume that M_1, M_2, M_3 are three distinct perfect matchings of P_{10} containing the edge $e = xy$. By Proposition B.2.18, for each pair of distinct $i, j \in \{1, 2, 3\}$, $M_i \triangle M_j$ is a Hamilton circuit of the graph $P_{10} - \{x, y\}$. It is not hard to see that $P_{10} - \{x, y\}$ has only one Hamilton circuit (by Proposition B.2.9, or see Figure B.10: $\overline{P_{10} - \{e\}}$ is a V_8-graph). So, $M_1 \triangle M_2 = M_1 \triangle M_3$. Thus, $M_2 = M_3$. This contradicts that M_1, M_2, M_3 are all distinct. □

Proposition B.2.20 [204] *The Petersen graph P_{10} does not have a family of perfect matchings that covers each edge of P_{10} an odd number of times.*

Proof Assume that $\mathcal{F}$ is a family of perfect matchings of P_{10} such that $\mathcal{F}$ covers each edge of P_{10} odd times and $|\mathcal{F}|$ is as small as possible. If M_i is a perfect matching that appears more than once in $\mathcal{F}$, then we can delete two copies of M_i from $\mathcal{F}$, and the resulting family of perfect matchings still covers each edge of P_{10} odd times. This would contradict the choice of $\mathcal{F}$ that $|\mathcal{F}|$ is minimum. So, all members of $\mathcal{F}$ are distinct. By Proposition B.2.19, each edge is covered by only one member of $\mathcal{F}$. This implies that $|\mathcal{F}| = 3$ and contradicts that P_{10} is not 3-edge-colorable. □

Proposition B.2.21 [204] *The Petersen graph P_{10} has precisely six distinct perfect matchings.*

Proof There is a one-to-one correspondence between $E(P_{10})$ and the set

$$\{\{M_i, M_j\} : M_i \text{ and } M_j \text{ are two distinct perfect matchings of } P_{10}\}$$

such that $\{M_i, M_j\}$ corresponds to an edge e if $E(M_i) \cap E(M_j) = \{e\}$ (by Propositions B.2.18 and B.2.19). Note that $|E(P_{10})| = 15$. Hence, it is easy to see that P_{10} has precisely 6 perfect matchings since $15 = \binom{6}{2}$. □

Proposition B.2.22 [204] *The Petersen graph does not have a family of 9 perfect matchings covering every edge precisely 3 times.*

Proof Immediate corollary of Proposition B.2.20. □

⋆ Circuit covering

Proposition B.2.23 *The Petersen graph has a circuit double cover consisting of six pentagons.*

Proof Look at the embedding of P_{10} on the projective plane (see Figure B.13).

An alternative proof: In Figure B.7, let $C_\mu = u_\mu v_\mu v_{\mu+3} v_{\mu+1} u_{\mu+1} u_\mu$ for each $\mu \in \mathbb{Z}_5$. Then

$$\{C_\mu : \mu \in \mathbb{Z}_5\} \cup \{u_1 u_2 u_3 u_4 u_0 u_1\}$$

is a 6-pentagon double cover. □

Proposition B.2.24 *The Petersen graph has a circuit double cover $\mathcal{F}_1$ consisting of five circuits with lengths: 9, 6, 5, 5, 5, and another circuit double cover $\mathcal{F}_2$ consisting of five circuits with lengths: 8, 6, 6, 5, 5.*

Proof See Figures B.14 and B.15. □

Proposition B.2.25 *The Petersen graph does not have a circuit double cover consisting of five hexagons.*

Proof Suppose that $\{C_1, \ldots, C_5\}$ is a circuit double cover of the Petersen graph consisting of five hexagons. By Proposition B.2.7, for each $\alpha, \beta \in \{1, \ldots, 5\}$, $|E(C_\alpha) \cap E(C_\beta)| \geq 2$. Hence, for each $i \in \{1, \ldots, 5\}$, edges of C_i must have a partition into four disjoint parts, each of which is covered by one of other four circuits, and is of size at least 2. This implies that the length of C_i must be at least 8. This is a contradiction. □

Let $w_{10} : E(P_{10}) \mapsto \{1, 2\}$ such that $E_{w_{10}=1}$ induces a pair of edge-disjoint chordless pentagons (see Figure B.8). (See Section 2.1 for definitions of *eulerian weights, faithful covers,* etc.)

Proposition B.2.26 (P_{10}, w_{10}) *does not have a faithful circuit cover.*

Proof Let $E_{w_{10}=2} = M$ which is a perfect patching of P_{10}. Assume $\mathcal{C}$ is a faithful cover. Then, for each $C \in \mathcal{C}$, edges around C are alternately in $E_{w_{10}=1}$ and M.

Since M is an edge-cut separating two pentagons, by Lemma A.2.13, we have that $|C \cap M| \equiv 0 \pmod 2$. Therefore, $|C| = 4k$. Furthermore, $|C| = 8$ since the girth of P_{10} is 5. But $\sum_{e \in E(P_{10})} w_{10}(e) = 20$ is not a multiple of 8. □

How about any eulerian $(1, 2)$-weight w other than w_{10}? The following proposition shows that (P_{10}, w) always have a faithful cover if $w \neq w_{10}$.

Proposition B.2.27 *Let w be an eulerian $(1,2)$-weight of the Petersen graph P_{10}. Then (P_{10}, w) has no faithful circuit cover if and only if $E_{w=1}$ induces a 2-factor of P_{10}. Furthermore, if $E_{w=1}$ does not induce a 2-factor of P_{10} and is not empty, then (P_{10}, w) has a faithful circuit cover consisting of four circuits.*

Proof The first part of the proposition is also proved in Exercise 6.6 (using circuit extension technique). Here is an alternative proof (by applying Proposition B.2.24).

Let F be the even subgraph of P_{10} induced by $E_{w=1}$. By Proposition B.2.4, it is easy to see that (1) either F is a single circuit, (2) or F is a 2-factor consisting of a pair of pentagons.

For Case (2), it is the weight w_{10} and is proved in Proposition B.2.26. For Case (1), it is a strong circuit double covering problem (a CDC containing a given circuit F). By Proposition B.2.4, the circuit F must be of length $5, 6, 8$ or 9. Then it is a corollary of Proposition B.2.24. □

Proposition B.2.28 *For each $e_0 \in E_{w_{10}=2}$, the weighted graph $(P_{10} - e_0, w_{10})$ has two faithful circuit covers $\mathcal{F}_1, \mathcal{F}_2$ that $|\mathcal{F}_1| = 3$ and $|\mathcal{F}_2| = 2$ (see Figure B.16).*

Proposition B.2.29 *Let F be a 2-factor of P_{10}. Let $w^+ : E(P_{10}) \mapsto \{1, 2, 3\}$ such that $w^+(e) = 2$ for every $e \notin F$, and $w^+(e) = 1$ or 3 for every $e \in F$ with precisely one edge e_0 of F that $w^+(e_0) = 3$. Then (P_{10}, w^+) has a faithful cover.*

Proposition B.2.30 ([119], [133]) *For every circuit cover $\mathcal{F}$ of P_{10} with the shortest total length, we have the following properties:*
(1) $\mathcal{F}$ is of total length 21,
(2) $\mathcal{F}$ consists of four circuits,
(3) $\mathcal{F}$ covers every edge at most twice.
(4) $E(C') \cap E(C'') \neq \emptyset$ for every pair $C', C'' \in \mathcal{F}$.

Proof By Proposition B.2.26, a shortest circuit cover of P_{10} must be of length greater than 20. By Proposition B.2.24, P_{10} has a 5-circuit double cover $\mathcal{F}_1$. A shortest circuit cover is obtained from $\mathcal{F}_1$ by removing the longest one (of length 9). This proves conclusion (1). ◇

Proof of conclusion (3). Let P be the subset of edges covered by an even number of members of $\mathcal{F}$ and C be the subset of edges covered by an odd number of members of $\mathcal{F}$. It is trivial that P is a parity subgraph and C is an even subgraph. And it is also trivial that

$$|P| + |C| = |E(P_{10})| = 15, \quad |P| \geq 5 \text{ and } |C| \leq 10. \qquad \text{(B.1)}$$

Suppose that some edge of C is covered 3 times by $\mathcal{C}$. Then, by inequality (B.1), the total length of $\mathcal{F}$ would be at least

$$2|P| + (|C| - 1) + 3 = (|P| + |C|) + (|P| + 2) = 15 + |P| + 2 \geq 22.$$

This contradicts (1). It is similar if some edge of P is covered at least 4 times by $\mathcal{F}$. $\Diamond$

Proof of conclusion (2). Since the girth of P_{10} is 5, the total length of $\mathcal{F}$ is at least $5|\mathcal{F}|$. By (1), C contains at most four circuits. Note that if $|\mathcal{F}| = 3$, then it is 3-edge-colorable (by Lemma 2.2.1 and conclusion (3)). This proves (2). $\Diamond$

Proof of conclusion (4). Since P_{10} is not 3-edge-colorable, conclusion (4) is proved by applying Lemma 2.2.1 and conclusion (3). $\qquad \Box$

Proposition B.2.31 $(P_{10}, 2w_{10})$ *has a faithful circuit cover $\mathcal{F}$ consisting of five 8-circuits.*

Note, w_{10} is the eulerian weight of P_{10} defined in Proposition B.2.26 or Section 2.1 (see Figure B.8). And $2w_{10} : E(P_{10}) \mapsto \{2, 4\}$ that $E_{2w_{10}=4}$ is a perfect matching and $E_{2w_{10}=2}$ is a pair of chordless pentagons.

Proof Let $M = E_{2w_{10}=4} = E_{w_{10}=2} = \{e_1, \ldots, e_5\}$ be the perfect matching. By Proposition B.2.9, let C_i be the unique 8-circuit of $P_{10} - V(e_i)$ (where $V(e_i)$ is the set of two endvertices of e_i). Note that the circuit C_j contains all edges of $M - e_j$. (Or see Figure B.16-(a): the 8-circuit C_2.) Hence, the set of all those five 8-circuits C_j forms a faithful cover of $(P_{10}, 2w_{10})$. $\qquad \Box$

Proposition B.2.32 *Let C_0 and C_1 be two distinct 2-factors of P_{10}. Then $P_{10} - (E(C_0) - E(C_1))$ has a bridge.*

Proof Let $M_1 = P_{10} - E(C_1)$ be a perfect matching. Since $|E(C_0) \cap M_1)| = 4$, the edge $M_1 - E(C_0)$ is a bridge of $P_{10} - (E(C_0) \cap M_1) = P_{10} - (E(C_0) - E(C_1))$. $\qquad \Box$

B.3 Exercises

Exercise B.1   Show that the Petersen graph P_{10} is not 3-edge-colorable.

Definition B.3.1   Let G be a graph. The crossing number $c(G)$ of G is the least integer k such that G can be drawn on the plane with at most k pairs of edges crossing each other.

Exercise B.2   (Jaeger [128], Celmins, Fouquet and Swart [32] or see [131], Loupekine and Watkins [169] or see [240]) If G is a bridgeless cubic graph with the crossing number $c(G) \leq 1$, then G is 3-edge-colorable (prove by applying the 4-color theorem).

Exercise B.3   Let G be a bridgeless cubic graph and e be an edge of G such that the contracted graph G/e is planar. Then G is 3-edge-colorable (prove by applying the 4-color theorem).

Exercise B.4   Let G be a cubic graph with a 3-edge-coloring c. Assume that G has a perfect matching M such that the edges of M are colored with at most two colors. Then G has a 3-edge-coloring c' such that all edges of M are colored with the same color.

Exercise B.5   Let G be a snark. If G is not cyclically 4-edge-connected, then G has a proper subgraph which is a subdivision of a smaller snark.

Exercise B.6   Let G ($G \neq K_2$) be a connected graph such that the degree of each vertex is either 3 or 1. Let E_1 be the set of all edges incident with a degree 1 vertex. If G has a 3-edge-coloring $c : E(G) \mapsto \mathbb{Z}_2^2 - \{(0,0)\}$, then

$$\sum_{e \in E_1} c(e) = (0,0).$$

Exercise B.7   (Chetwynd and Wilson [37], revised) Let G be a snark. If G is not cyclically 5-edge-connected, then G has a proper subgraph which is a subdivision of a smaller snark.

Exercise B.8   Let G be a 3-connected cubic graph of order ≤ 10. Then either G has a Hamilton circuit or is the Petersen graph.

B.4 Various drawings of the Petersen graph

Figure B.7 is the most standard drawing of the Petersen graph. However, in order to understand better the structure of the Petersen graph, different drawings of P_{10} are illustrated in this section for reference.

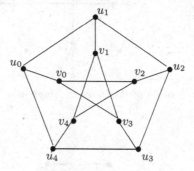

Figure B.7 *A standard drawing of the Petersen graph*

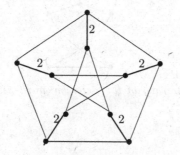

Figure B.8 *Contra pair* (P_{10}, w_{10})

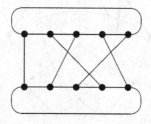

Figure B.9 *A drawing of the Petersen graph as a permutation graph*

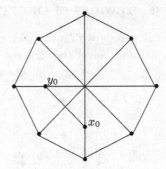

Figure B.10 *A drawing of the Petersen graph as a V$_8$ "plus" an edge* $x_0 y_0$

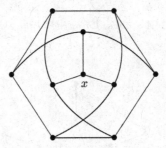

Figure B.11 *A drawing of the Petersen graph as a K$_{3,3}$ "plus" a vertex x*

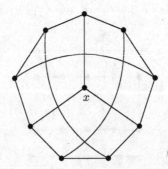

Figure B.12 *A drawing of the Petersen graph as a K$_{3,3}$ "plus" a vertex x*

Figure B.13 *An embedding of the Petersen graph on the projective plane*

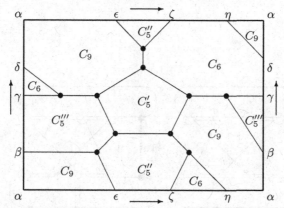

Figure B.14 *The Petersen graph embedded on the torus: five faces of lengths 9, 6, 5, 5, 5*

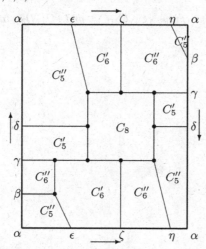

Figure B.15 *The Petersen graph embedded on the Klein bottle: five faces of lengths 8, 6, 6, 5, 5*

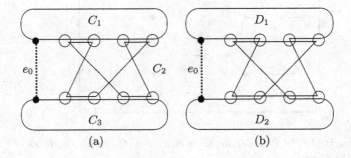

Figure B.16 *Circuit chains $\{C_1, C_2, C_3\}$ and $\{D_1, D_2\}$ in $P_{10} - e_0$*

Appendix C
Integer flow theory

The concept of integer flow was introduced by Tutte [229], [230] as a generalization of the map coloring problems (see Theorem C.1.9). This chapter is a brief survey of the integer flow theory. For further study in this area (and proofs of most theorems in this chapter), readers are referred to [259].

C.1 Tutte's integer flows

Let $G = (V, E)$ be a graph. For an orientation D of $E(G)$ (the graph G under the orientation D is sometimes denoted by $D(G)$),

(1) an oriented edge of G is called an arc,

(2) for a vertex $v \in V(G)$, let $E^+(v)$ (or $E^-(v)$) be the set all arcs of $D(G)$ with their tails (or heads, respectively) at the vertex v.

Definition C.1.1 Let G be a graph, D be an orientation of G, Γ be an abelian group (an additive group with "0" as the identity), and $f : E(G) \mapsto \Gamma$ be a mapping. The ordered pair (D, f) is a flow (or Γ-flow) of G if

$$\sum_{e \in E^+(v)} f(e) = \sum_{e \in E^-(v)} f(e) \qquad (C.1)$$

for every vertex $v \in V(G)$.

For the sake of simplification, we denote

$$f^+(v) = \sum_{e \in E^+(v)} f(e)$$

and

$$f^-(v) = \sum_{e \in E^-(v)} f(e).$$

In this book, we are only interested in *finite abelian groups, infinite groups* $\mathbb{Z}$ *(the set of integers),* $\mathbb{Q}$ *(the set of rational numbers),* $\mathbb{R}$ *(the set of real numbers).*

Definition C.1.2 Let (D, f) be a Γ-flow of a graph G and k be an integer.

(1) In general, (D, f) is called a **group** Γ**-flow** where Γ is an abelian group.

(2) (D, f) is called an **integer flow** if $\Gamma = \mathbb{Z}$ (that is, $f : E(G) \mapsto \mathbb{Z}$); and an integer flow (D, f) is a k**-flow** if $|f(e)| < k$ for every edge $e \in E(G)$.

(3) (D, f) is called a **mod-k-flow** if $f : E(G) \mapsto \mathbb{Z}$ such that

$$f^-(v) \equiv f^+(v) \pmod{k}$$

for every $v \in V(G)$ (that is, (D, f) is a $\mathbb{Z}_k$-flow where $\mathbb{Z}_k$ is the cyclic group of order k).

Definition C.1.3 Let G be a graph, and Γ be an abelian group with "0" as the identity.

(1) The **support** of a Γ-flow (D, f) is the set of all edges of G with $f(e) \neq 0$ and is denoted by $supp(f)$.

(2) A flow (D, f) is **nowhere-zero** if $supp(f) = E(G)$.

The following fundamental theorem is due to Tutte [229], [230].

Theorem C.1.4 (Tutte [229], [230]) *Let G be a graph, k be a positive integer, and Γ be an abelian group of order k. Then the following statements are equivalent:*

(1) G admits a nowhere-zero integer k-flow;

(2) G admits a nowhere-zero mod-k-flow;

(3) G admits a nowhere-zero group Γ-flow.

Readers are also referred to Theorem 1.3.3 and Theorem 2.2.3 in [259] for the proof of the above fundamental theorem.

The equivalency of (1) and (2) is also proved in Theorem C.2.1.

The equivalency of (2) and (3) was originally proved by Tutte by using the flow polynomial technique in the following lemma. Note that a mod-k-flow is also a group $\mathbb{Z}_k$-flow where $\mathbb{Z}_k$ ($\mathbb{Z}_k = \mathbb{Z}/k\mathbb{Z}$) is the cyclic group of order k.

Lemma C.1.5 (Tutte [230]) *Let G be a graph, k be a positive integer, and Γ_i be an abelian group of order k ($i = 1, 2$). Then G admits a nowhere-zero group Γ_1-flow if and only if G admits a nowhere-zero group Γ_2-flow.*

Proof Let G be a graph, k be a positive integer, Γ be an arbitrary abelian group of order k. Let $\phi(G, k)$ be the number of nowhere-zero Γ-flows (D, f) of G.

We claim that $\phi(G, k)$ *is independent from the structure of* Γ.

Induction on $|E(G)|$. For the single edge graph K_2, $\phi(K_2, k) = 0$; and for the single loop graph L, $\phi(L, k) = k - 1$. And the values of $\phi(K_2, k)$, $\phi(L, k)$ are independent from the structure of Γ.

For a graph G and an edge e, it is easy to see that

$$\phi(G, k) = \phi(G/e, k) - \phi(G - e, k) \text{ if } e \text{ is not a loop,}$$

$$\phi(G, k) = (k - 1)\phi(G - e, k) \text{ if } e \text{ is a loop.}$$

$\square$

By Theorem C.1.4, all nowhere-zero flows defined in Definition C.1.2 are equivalent. Hence, whenever we say *"a graph G admits a nowhere-zero k-flow,"* it always means that the graph G admits a nowhere-zero integer k-flow, or a nowhere-zero group Γ-flow with $|\Gamma| = k$, or a nowhere-zero mod-k-flow.

Furthermore, beyond the equivalency of different definitions of flows, each definition of nowhere-zero flows has its own special advantages depending on the topic being studied.

By the definition of integer flow, the following observation is straightforward.

Lemma C.1.6 (Tutte [230]) *If a graph G admits a nowhere-zero integer k-flow, then G admits a nowhere-zero integer h-flow for each $h \geq k$.*

Proof Exercise C.10. $\square$

By the definition of integer flows (Definition C.1.1), we have the following very useful lemma.

Lemma C.1.7 *Let (D, f) be an integer flow of a graph G. Then the subgraph of G induced by the edges with odd weights is an even subgraph of G.*

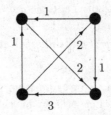

Figure C.1 *A nowhere-zero 4-flow in the complete graph* K_4

Proof Exercise C.8.   □

An immediate corollary of Lemma C.1.7 is as follows,

Theorem C.1.8 (Tutte [231]) *A graph G admits a nowhere-zero 2-flow if and only if the graph G is even.*

The following theorem by Tutte indicates the important relation between map coloring and integer flow, and therefore motivates and promotes the study of the theory of integer flow.

Theorem C.1.9 (Tutte [230]) *For a planar bridgeless graph G, G is k-face-colorable if and only if G admits a nowhere-zero k-flow.*

Proof Readers are referred to Theorem 1.4.5 in [259].   □

The following are the most famous conjectures in the theory of integer flows proposed by Tutte.

Conjecture C.1.10 (5-flow conjecture, Tutte [230]) *Every bridgeless graph admits a nowhere-zero 5-flow.*

Conjecture C.1.11 (4-flow conjecture, Tutte [233]) *Every bridgeless graph containing no subdivision of the Petersen graph admits a nowhere-zero 4-flow.*

Conjecture C.1.12 (3-flow conjecture, Tutte; see [216] and unsolved problem 48 in [18], unsolved problem 98 in [19]) *Every bridgeless graph containing no 3-edge-cut admits a nowhere-zero 3-flow.*

Though four decades have passed and some very significant and important approaches have been made toward these conjectures, they remain essentially open. In next section, some basic properties about integer flow will be introduced.

C.2 Fundamental lemmas

⋆ Mod-k-flows

The proof of "(1) $\Rightarrow$ (2)" of Theorem C.1.4 is trivial since a nowhere-zero integer k-flow is also a nowhere-zero mod-k-flow. The proof of "(2) $\Rightarrow$ (1)" of Theorem C.1.4 is a corollary of the following stronger (and very useful) result.

Theorem C.2.1 (Tutte [229]) *If a graph G admits a mod-k-flow (D, f_a), then the graph G admits an integer k-flow (D, f_b) such that $f_a(e) \equiv f_b(e)$ (mod k) for every edge $e \in E(G)$.*

Note that both flows in Theorem C.2.1 are under the same orientation D.

Proof Readers are referred to Theorem 1.3.4 in [259]. $\qquad\square$

⋆ Product of flows

The following lemma has been used in the proofs of some landmark theorems (see [127], [207]).

Lemma C.2.2 *Let G be a graph and k_1, k_2 be two integers. If G admits a k_1-flow (D, f_1) and a k_2-flow (D, f_2) such that $supp(f_1) \cup supp(f_2) = E(G)$, then both $(D, k_2 f_1 + f_2)$ and $(D, f_1 + k_1 f_2)$ are nowhere-zero $(k_1 k_2)$-flows of G.*

Proof Exercise C.18. $\qquad\square$

Lemma C.2.2 can be generalized to be an "if and only if" result.

Theorem C.2.3 *Let G be a graph and k_1, k_2 be two integers. Then G admits a nowhere-zero $(k_1 k_2)$-flow if and only if G admits a k_1-flow (D, f_1) and a k_2-flow (D, f_2) such that $supp(f_1) \cup supp(f_2) = E(G)$.*

Proof Readers are referred to Theorem 2.1.2 in [259]. $\qquad\square$

⋆ Sum of flows

Theorem C.2.4 (Little, Tutte and Younger [168]) *For each non-negative integer k-flow (D, f) of a graph G, G admits a set of $(k-1)$ non-negative 2-flows (D, f_μ) $(\mu = 1, \ldots, k-1)$ such that $f = \sum_{\mu=1}^{k-1} f_\mu$.*

Proof Readers are referred to Theorem 2.6.2 in [259]. $\qquad\square$

Since the support of a non-negative 2-flow is a directed even subgraph under the orientation D, the following theorem about directed even subgraph covering is an equivalent statement of Theorem C.2.4.

Corollary C.2.5 (Little, Tutte and Younger [168]) *Let G be a graph and D be an orientation of G. The graph G admits a positive k-flow (D, f) if and only if the directed graph $D(G)$ contains $(k-1)$ directed even subgraphs such that every arc of $D(G)$ is contained in at least one of them.*

⋆ Bounded orientations

Let $\{A, B\}$ be a partition of the vertex set $V(G)$, and D be an orientation of $E(G)$. The set of the arcs of G (under the orientation D) with tails in A and heads in B is denoted by $[A, B]_D$ (or simply $[A, B]$ if no confusion occurs).

Theorem C.2.6 (Hoffman [107] (revised), or see [11] p. 88, [124], [108], [88]) *Let G be a bridgeless graph, D be an orientation of G and a, b be two positive integers $(a \leq b)$. The following statements are equivalent:*

(1)

$$\frac{a}{b} \leq \frac{|[A, B]_D|}{|[B, A]_D|} \leq \frac{b}{a} \tag{C.2}$$

for every edge-cut (A, B) of G;

(2) G admits a nowhere-zero integer flow (D, f_1) such that $a \leq f_1(e) \leq b$ for each $e \in E(G)$;

(3) G admits a nowhere-zero real-valued flow (D, f_2) such that $a \leq f_2(e) \leq b$ for each $e \in E(G)$.

Proof Readers are referred to Theorem 2.3.1 in [259]. □

Corollary C.2.7 *A graph G admits a nowhere-zero k-flow if and only if G has an orientation D such that*

$$\frac{1}{k-1} \leq \frac{|[A, B]_D|}{|[B, A]_D|} \leq k - 1 \tag{C.3}$$

for every edge-cut (A, B) of G.

Corollary C.2.8 *A graph G admits a nowhere-zero k-flow (D, f_1) if and only if G admits a nowhere-zero real-valued flow (D, f_2) such that*

$$\frac{\max\{|f_2(e)| : e \in E(G)\}}{\min\{|f_2(e)| : e \in E(G)\}} \leq k - 1.$$

⋆ Circular flows

Definition C.2.9 Let k, d be two integers such that $0 < d \leq \frac{k}{2}$. An integer flow (D, f) of a graph G is called a circular $\frac{k}{d}$-flow if $f : E(G) \mapsto \{\pm d, \pm(d+1), \ldots, \pm(k-d)\} \cup \{0\}$.

The concept of circular flow, introduced in [88], is a generalization of integer flows, and a dual version of the circular coloring problem ([239], [17])[1]. Readers are referred to [263], [264] for comprehensive surveys in this area.

Theorem C.2.10 (Goddyn, Tarsi and Zhang [88]) *Let G be a bridgeless graph, D be an orientation of G and $k, d \in \mathbb{Z}^\star$, $q \in \mathbb{Q}^\star$ such that $q = \frac{k}{d} \geq 2$. Then the following statements are equivalent:*

(1) G admits a positive circular $\frac{k}{d}$-flow (D, f_1);

(2) G admits a rational-valued flow (D, f_2) such that $f_2 : E(G) \mapsto [1, q-1]$;

(3)

$$q - 1 = \frac{k-d}{d} \geq \frac{|[U, \, V(G) - U]_D|}{|[V(G) - U, \, U]_D|} \geq \frac{d}{k-d} = \frac{1}{q-1}$$

for each $U : \emptyset \subset U \subset V(G)$.

Proof Corollary of Theorem C.2.6. □

An immediate corollary of Theorem C.2.10 is the following analogy of Lemma C.1.6.

Lemma C.2.11 (Goddyn, Tarsi and Zhang [88]) *Let G be a graph and $p \in \mathbb{Q}^\star$. If G admits a nowhere-zero circular p-flow, then G admits a nowhere-zero circular q-flow for every $q \in \mathbb{Q}^\star$ with $q \geq p$.*

C.3 Exercises

Exercise C.1 For a given integer k ($k \geq 2$), if G is a smallest bridgeless graph not admitting a nowhere-zero k-flow, then G must be 3-edge-connected.

Exercise C.2 Let G be a graph, Γ be an abelian group and π be an automorphism of Γ. If (D, f) is a Γ-flow of G, then $(D, \pi f)$ is also a Γ-flow of G with $supp(f) = supp(\pi f)$.

---

[1] Circular flow was called *fractional flow* in [88] and [259]

⋆ Weights and flow parity

Exercise C.3 Let $G = (V, E)$ be a graph and D be an orientation of $E(G)$ and $f : E(G) \mapsto \Gamma$ where Γ is an abelian group (an additive group with "0" as the identity). For each $U \subseteq V(G)$ and each $v \in U$, define

$$f^+(v) = \sum_{e \in E^+(v)} f(e), \quad f^-(v) = \sum_{e \in E^-(v)} f(e)$$

and

$$f^+(U) = \sum_{v \in U} f^+(v), \quad f^-(U) = \sum_{v \in U} f^-(v).$$

Prove the following equations:

$$f^+(V(G)) = f^-(V(G)),$$

$$f^+(U) - f^-(U) = f^-(V(G) - U) - f^+(V(G) - U) \qquad \text{(C.4)}$$

and

$$f^+(U) - f^-(U) = \sum_{e \in [U, V(G) - U]} f(e) - \sum_{e \in [V(G) - U, U]} f(e). \qquad \text{(C.5)}$$

Exercise C.4 Note that the weight f described in Exercise C.3 is not necessarily a flow (that is, Equation (C.1) in Definition C.1.1 may not be satisfied). If (D, f) is a Γ-flow, show that both Equations (C.4) and (C.5) are zero in Γ.

Exercise C.5 If a graph G has a bridge, then G does not admit nowhere-zero k-flow for any k.

Exercise C.6 (Theorem C.1.8) A graph G admits a nowhere-zero 2-flow if and only if the graph G is an even graph.

Exercise C.7 Let (D, f) be a flow of a graph G. If the subgraph of G induced by $supp(f)$ is acyclic, then $f(e) = 0$ for every edge $e \in E(G)$.

Exercise C.8 (Lemma C.1.7) Let (D, f) be an integer flow of a graph G. Then the subgraph of G induced by the edges with odd weights is an even subgraph of G.

Exercise C.9 Let k be an even integer and (D, f) be a mod-k-flow of a graph G. Then the subgraph of G induced by the edges with odd weights is an even subgraph of G.

⋆ Basic properties

Exercise C.10 (Lemma C.1.6) If a graph G admits a nowhere-zero k-flow, then G admits a nowhere-zero h-flow for each $h \geq k$.

Exercise C.11 Let G be a graph and D be an orientation of $E(G)$. Let $\mathcal{F}$ be the set of all real-valued functions of $E(G)$ such that, for each $f \in \mathcal{F}$, (D, f) is a flow of G. Then $\mathcal{F}$ is a linear space over the field $\mathbb{R}$ where $\mathbb{R}$ is the set of all real numbers.

Exercise C.12 Let (D, f) be a Γ-flow of G and let E_0 be a subset of $E(G)$. Let D' be the orientation of G obtained from D by reversing the direction of every arc in E_0, and let f' be a weight of $E(G)$:

$$f'(e) = \begin{cases} f(e) & \text{if } e \notin E_0 \\ -f(e) & \text{if } e \in E_0. \end{cases}$$

Then (D', f') is also a Γ-flow of G with $supp(f) = supp(f')$.

Exercise C.13 Let (D, f) be a positive k-flow of G. Let C be a directed circuit of $D(G)$. Show that G admits a positive k-flow (D', f') where the orientation D' is obtained from D be reversing the direction of each arc of C. And determine f'.

Exercise C.14 Let (D, f) be a mod-k-flow of G and E_0 be a subset of $E(G)$. Let D' be an orientation of G obtained from D by reversing the direction of every edge in E_0, and let f' be the weight obtained from f as follows,

$$f'(e) = \begin{cases} f(e) & \text{if } e \notin E_0 \\ k - f(e) & \text{if } e \in E_0. \end{cases}$$

Show that (D', f') is also a mod-k-flow of G such that $supp_k(f) = supp_k(f')$.

Exercise C.15 Let G be a graph admitting a mod-k-flow (D_1, f_1). Then G admits a mod-k-flow (D_2, f_2) such that
 (1) $0 \leq f_2(e) \leq \frac{k}{2}$ for each $e \in E(G)$,
 (2) $f_2(e) \equiv f_1(e) \pmod{k}$ if the directions of e are the same under D_1 and D_2, and
 (3) $f_2(e) \equiv k - f_1(e) \pmod{k}$ if the directions of e are opposite under D_1 and D_2.

Exercise C.16 (Tarsi) Let (D, f) be an integer flow of a graph G.

Then G admits a pair of integer flows (D, f_1), (D, f_2) such that, for each $i = 1, 2$,

$$f_i(e) = \lfloor \frac{f(e)}{2} \rfloor \text{ or } \lceil \frac{f(e)}{2} \rceil,$$

and

$$f_1(e) + f_2(e) = f(e).$$

Exercise C.17   Find a graph G such that G admits two 2-flows (D, f_1) and (D, f_2) such that $supp(f_1) \cup supp(f_2) = E(G)$ but neither $(D, f_1 + f_2)$ nor $(D, f_1 - f_2)$ is a nowhere-zero 3-flow.

Exercise C.18   (Lemma C.2.2) Let G be a graph, D be an orientation of $E(G)$ and k_α, k_β be two integers. If G admits a k_α-flow (D, f_α) and a k_β-flow (D, f_β) such that $supp(f_\alpha) \cup supp(f_\beta) = E(G)$, then both $(D, k_\beta f_\alpha + f_\beta)$ and $(D, f_\alpha + k_\alpha f_\beta)$ are nowhere-zero $(k_\alpha k_\beta)$-flows of G.

Exercise C.19   Let (D, f) be a positive k-flow of G with $k \geq 2$. Then the directed graph $D(G)$ is strongly connected and therefore, every arc of $D(G)$ is contained in a directed circuit of $D(G)$.

Exercise C.20   Let G be a bridgeless graph and $e \in E(G)$. If $G - \{e\}$ admits a nowhere-zero k-flow, then G admits a nowhere-zero $(k+1)$-flow.

Exercise C.21   (Jaeger [131]) Let G be a bridgeless graph, T be a spanning tree of G, D be an orientation of G, and Γ be an abelian group. Then, for each mapping $c : E(G) - E(T) \mapsto \Gamma$, G has a Γ-flow (D, f) such that $f(e) = c(e)$ for each edge $e \in E(G) - E(T)$.

Exercise C.22   Let G admit a nowhere-zero k-flow ($k \geq 2$) and $T = \{e_1, \ldots, e_t\}$ be a minimal edge-cut of G with t edges ($t \leq 3$). Let Q_1, Q_2 be components of $G - T$. For each orientation D_T of T such that neither $[Q_1, Q_2]_{D_T}$ nor $[Q_2, Q_1]_{D_T}$ is empty, show that we can always find a positive k-flow (D, f) such that the orientation D agrees with D_T at every edge of T.

⋆ 4-flows and 4-colorings

Exercise C.23   A cubic graph G is 3-edge-colorable if and only if G admits a nowhere-zero 4-flow. (Note: do not use any result in Chapter 8.)

Exercise C.24   Let G be a graph containing a Hamilton circuit. Then G admits a nowhere-zero 4-flow. (Note: do not use any result in Chapter 8.)

Exercise C.25   Every planar graph containing a Hamilton circuit is 4-face-colorable on the plane (without using Exercise C.24 and Theorem C.1.9).

Exercise C.26   The statement of Exercise C.25 is not true if G is not embedded on the plane. Find a graph embedded on the torus that has a Hamilton circuit, but is not 4-face colorable.

Exercise C.27   Let G be a planar graph. Show that G has a 2-even subgraph cover if and only if G is 4-face-colorable (give a direct proof without using the 4-color theorem or any other integer flow theorem of the book).

Exercise C.28   Let G be a graph admitting a nowhere-zero 4-flow, $e_1, e_2 \in E(G)$. Then there is a nowhere-zero 4-flow (D, f) of G such that $f(e_1) = f(e_2) = 1$.

Exercise C.29   Every bridgeless graph with at most eight odd vertices admits a nowhere-zero 4-flow. (Give a direct proof without using any results in Section 7.4.)

⋆ 3-flows and 3-colorings

An orientation D of a graph G is called a *mod-3-orientation* if $d_D^+(v) \equiv d_D^-(v) \pmod 3$.

Exercise C.30   (Tutte [229], and see [217]) Let G be a graph. Show that the following statements are equivalent:
   (1) G admits a nowhere-zero 3-flow;
   (2) G has a mod-3-orientation.

Exercise C.31   (Tutte [229]) Let G be a cubic graph. Show that the following statements are equivalent:
   (1) G admits a nowhere-zero 3-flow;
   (2) G is bipartite.
(Exercise C.31 is proved in Theorem 8.5.1. Here, prove it as a corollary of Exercise C.30.)

Exercise C.32   (Franklin [79]) A plane triangulation G is 3-vertex-colorable if and only if G is an even graph.

Exercise C.33   (Ore [188]) Let G be a planar graph such that the number of edges in the boundary of each face is a multiple of three. If G is an even graph, then G is 3-vertex-colorable.

Exercise C.34   (Heawood [104]) A planar graph is 3-vertex-colorable if and only if it is a subgraph of a planar triangulated even graph.

Exercise C.35   A graph G admits a nowhere-zero 3-flow if and only if G is a graph obtained from a subdivision of a cubic bipartite graph via a series of vertex-identification operations. (Note, let u and v be two vertices of G, the operation that merges u and v as a single vertex is called a *vertex-identification operation*.)

Exercise C.36   Let G be a counterexample to the 3-flow conjecture (Conjecture C.1.12) with the least number of edges. Then G is a 5-regular, 4-edge-connected, simple graph.

$\star$ k-flows ($k \geq 5$)

Exercise C.37   The Petersen graph admits a nowhere-zero 5-flow but not a nowhere-zero 4-flow.

Exercise C.38   If Tutte's 3-flow conjecture (Conjecture C.1.12) is true, then every 2-connected cubic graph admits a nowhere-zero 6-flow. (Note: do not use any result in Section 8.3.)

Exercise C.39   (Jaeger [127]) Let G be a graph admitting a nowhere-zero 4-flow and $e_1 \in E(G)$ and $h \in \{1, 2, 3\}$. Then there is a nowhere-zero 4-flow (D, f) of G such that $f(e_1) = h$.

Exercise C.40   (Jaeger [125]) If a bridgeless graph G has a Hamilton path, then G admits nowhere-zero 5-flows.

Exercise C.41   (Jaeger [126]) Let G be a graph and $e \in E(G)$. If G admits a nowhere-zero 4-flow and $G - \{e\}$ is bridgeless, then $G - \{e\}$ admits a nowhere-zero 5-flow.

Exercise C.42   (Celmins [31]) Let G be a graph and $e \in E(G)$. If $G - \{e\}$ admits a nowhere-zero 4-flow, then G admits a nowhere-zero 5-flow.

Appendix D
Hints for exercises

Chapter 1

Hint for Exercise 1.1. See Section 1.4 and Conjecture 1.4.4.

Hint for Exercise 1.2. It is possible that the circuit decomposition $\mathcal{F}$ of $2G$ may contain a digon, which corresponds to a single edge of the original graph G (not a circuit).

Hint for Exercise 1.3. The union of $\mathcal{F}^*$ and the symmetric difference of all members of $\mathcal{F}^*$ forms an even subgraph double cover of the graph. (See Definition A.2.3 for the definition of symmetric difference.)

Hint for Exercise 1.5. Let $\mathcal{F}_i$ be a k-even subgraph double cover of H_i ($i = 1, 2$). Properly joining each pair of even subgraphs of $\mathcal{F}_1, \mathcal{F}_2$ containing the same pair of edges of T, we obtain a k-even subgraph double cover of G (see Figure 1.1).

Hint for Exercise 1.6. A block of a graph is called a circuit-block if it is a circuit. The number of circuit-blocks in a graph G is denoted by $cb(G)$. For a circuit double cover $\mathcal{F}$ of a graph G, the number of repeated circuits in $\mathcal{F}$ is denoted by $rc(\mathcal{F})$. It is obvious that $rc(\mathcal{F}) \geq cb(G)$ for any circuit double cover of G. Here, we will prove a more general statement as follows.

$(\star)$ *Let G be a bridgeless graph. If G and all its bridgeless subgraphs have circuit double covers, then G has a circuit double cover $\mathcal{F}$ such that*

$$rc(\mathcal{F}) = cb(G).$$

Let G be a smallest counterexample to $(\star)$. Obviously, G is 2-connected and with minimum degree at least 3. Hence, $cb(G) = 0$. Let $\mathcal{F}$ be a circuit double cover of G with $rc(\mathcal{F})$ minimum ($rc(\mathcal{F}) \geq 1$). Let C_0 be a repeated member of $\mathcal{F}$.

Case 1. $G - E(C_0)$ has at least two components. Let H_1 be the union of C_0 and a component of $G - E(C_0)$. Here, H_1 is 2-connected since G is 2-connected. We have that $rc(\mathcal{F}|H_1) \geq 1 > cb(H_1) = 0$. Since H_1 is a proper subgraph of G, let $\mathcal{F}_1$ be a circuit double cover of H_1 such that $rc(\mathcal{F}_1) = cb(H_1) = 0$. Thus, $\mathcal{F}' = [\mathcal{F} - [\mathcal{F}|H_1]] \cup \mathcal{F}_1$ is a circuit double cover of G with a smaller $rc(\mathcal{F}')$. This contradicts the choice of $\mathcal{F}$.

Case 2. $G - E(C_0)$ is connected. Let $e \in E(C_0)$ and $H_2 = [G - E(C_0)] \cup \{e\}$. Since $\mathcal{F} - \{C_0\}$ is a circuit double cover of $G - E(C_0)$, $rc(\mathcal{F} - \{C_0\}) \geq cb(G - E(C_0))$. Let $\mathcal{F}_2$ be a circuit double cover of H_2 such that $cb(H_2) = rc(\mathcal{F}_2)$. Note that $cb(G - E(C_0)) \geq cb(H_2)$ since $G - E(C_0)$ is 2-edge-connected. We have that

$$rc(\mathcal{F}_2) = cb(H_2) \leq cb(G - E(C_0)) \leq rc(\mathcal{F} - \{C_0\}) = rc(\mathcal{F}) - 1.$$

Let C'' be a circuit of $\mathcal{F}_2$ containing e. Let $\mathcal{Q}$ be a circuit decomposition of $C'' \triangle C_0$. Then $\mathcal{F}^* = [\mathcal{F}_2 - \{C''\}] \cup \{C_0\} \cup \mathcal{Q}$ is a circuit double cover of G. Since both C_0 and C'' are circuits and $|E(C_0) \cap E(C'')| = 1$, each member of $\mathcal{Q}$ must contain some edge of $C_0 - \{e\}$. Thus, no member of $\mathcal{Q}$ is contained in $\mathcal{F}_2$. Therefore,

$$rc(\mathcal{F}^*) \leq rc(\mathcal{F}_2) < rc(\mathcal{F}).$$

This contradicts the choice of $\mathcal{F}$.

Hint for Exercise 1.7. For an element $\gamma \in S$, we say that an edge e is **generated** by γ if the endvertices of e are x and $x\gamma$ (or simply, e is a γ-edge). If $\alpha \in S$ with $|\alpha| \geq 3$ then the subgraph induced by α-edges is a 2-factor. If $\beta', \beta'' \in S$ with $\beta' \neq \beta''$ and $|\beta'| = |\beta''| = 2$ then the subgraph induced by β'- and β''-edges is also a 2-factor. So a minimum counterexample must be cubic with $S = \{\alpha, \alpha^{-1}, \beta\}$ where $|\alpha| \geq 3, |\beta| = 2$.

First, we claim that Γ is not an abelian group. If Γ is abelian, then $\alpha\beta = \beta\alpha$. It is easy to see that the subgraph of $G(\Gamma, S)$ induced by all α-edges is a 2-factor with at most two components. Thus $G(\Gamma, S)$ is a hamiltonian graph. (This claim also can be proved simply by applying a theorem of Chen and Quinpo [35].) And a circuit double cover of G is not hard to find (Theorem 1.3.2).

Assume that Γ is not abelian. Then Γ is not a cyclic group, and therefore, $\Gamma_1 = \langle \alpha\beta \rangle$ is a proper subgroup of Γ. Obviously $\alpha \notin \Gamma_1$ for otherwise, $\beta \in \Gamma_1$ and we have that $\Gamma = \langle \alpha, \beta \rangle \subseteq \langle \alpha\beta \rangle$. So, Γ_1 and $\Gamma_1\alpha$ are two distinct cosets of Γ_1. Let $h = |\alpha\beta|$. We obtain a circuit

$$1, \alpha, \alpha\beta, \alpha\beta\alpha, \ldots, (\alpha\beta)^i, (\alpha\beta)^i\alpha, \ldots, (\alpha\beta)^h = 1$$

consisting of all vertices of $\Gamma_1 \cup \Gamma_1\alpha$ (where 1 is the identity of Γ). Therefore, each vertex sequence $x_0, x_1, x_2 \ldots x_{h-1}, x_h$ with $x_{2i} = x_{2i-1}\alpha$, $x_{2i+1} = x_{2i}\beta$ $(i \geq 1)$ is a circuit of $G(\Gamma, S)$. Then it is obvious that the collection of all such circuits is a circuit cover of $G(\Gamma, S)$ which covers each α-edge once and each β-edge twice.

Chapter 2

Hint for Exercise 2.5. Let $\mathcal{F}_i$ be a faithful circuit cover of (H_i, w) for each $i = 1, 2$. Let $T = \{e_1, e_2, e_3\}$ and let x_{iab} be the number of circuits of $\mathcal{F}_i$ containing the edges e_a and e_b. The following system of linear equations for the variables $\{x_{i12}, x_{i13}, x_{i23}\}$ has a unique solution for each $i = 1, 2$,

$$\begin{cases} x_{i12} + x_{i13} & = w(e_1) \\ x_{i12} + x_{i23} & = w(e_2) \\ x_{i13} + x_{i23} & = w(e_3). \end{cases}$$

Thus $x_{1ab} = x_{2ab}$ for $1 \leq a < b \leq 3$. Properly joining the x_{1ab} circuits of $\mathcal{F}_1$ and the x_{2ab} circuits of $\mathcal{F}_2$ pair by pair, we obtain a faithful circuit cover of (G, w).

Hint for Exercise 2.7. Induction on $|V(G)|$. By Lemma 2.2.1, we may assume that G is not 3-edge-colorable. By Propositions B.2.27 and B.1.11, we further assume that $|V(G)| \geq 12$. By Proposition B.1.12, let T be a cyclical 2- or 3-edge-cut with components Q_1 and Q_2. By induction, let $\mathcal{F}_\mu$ be a faithful circuit cover of $(G/Q_\mu, w)$. Then we are able to construct a faithful cover for (G, w) from $\mathcal{F}_1, \mathcal{F}_2$ by merging some members of them along the cut T.

Hint for Exercise 2.8. See Figure D.1 (Hägglund [102]) in which $E_{w=2}$ induces a perfect matching (bold lines in the figure, some of them are labeled). The graph in Figure D.1 is of order 28 and is constructed as follows:

$$((((P_{10}, w_{10}) \oplus_3 (K_4, w_4)) \oplus_3 (P_{10}, w_{10})) \otimes_{IFJ} (P_{10}, w_{10})$$

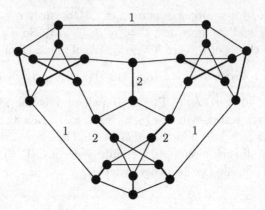

Figure D.1 *A strong contra pair of order 28 ($E_{w=2}$ induces a 1-factor: bold lines)*

where the last (P_{10}, w_{10}) is the one on the top-left (or top-right, symmetrically) and the first (P_{10}, w_{10}) is the lower-center one.

Hint for Exercise 2.9. Let $T = \{e_1, \ldots, e_4\}$ be a cyclic edge-cut of (G, w) with components Q_1, Q_2. And let $e_i = u_i v_i$ $(i = 1, \ldots, 4)$ with $u_i \in V(Q_1), v_i \in V(Q_2)$. For each $\{i, j\} = \{1, 2\}$, let $F\ell_i$ be a subset of $\{2, 3, 4\}$ such that $\mu \in F\ell_i$ if $(G/Q_j, w)$ has a faithful circuit cover $\mathcal{F}_i$ and one member of $\mathcal{F}_i$ contains the edges e_1 and e_μ. It is obvious that

$$F\ell_1 \cap F\ell_2 = \emptyset$$

since (G, w) is a contra pair. Without loss of generality, let $|F\ell_1| \leq |F\ell_2|$, and, say,

$$F\ell_1 \subseteq \{2\}.$$

Then (see Figure D.2) $(G, w) = (G_F, w_F) \otimes_{IFJ} (G_M, w_M)$ where
(1) (G_F, w_F) is constructed from (Q_1, w) adding a weight one edge joining u_1, u_2 and a weight one edge joining u_3, u_4,
(2) (G_M, w_M) is constructed from $(G - E(Q_1), w)$ by identifying u_1 and u_2 as a single vertex x and identifying u_3 and u_4 as a single vertex y and adding a weight 2 edge e_0 between x and y.
Since $F\ell_1 \subseteq \{2\}$, we can see that (G_M, w_M) is a contra pair.

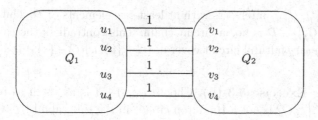

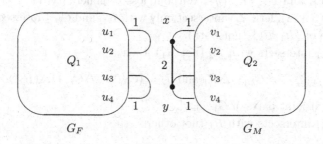

$$G_F \qquad\qquad\qquad\qquad\qquad G_M$$

Figure D.2 *Reduction along a weight 4 edge-cut*

Chapter 3

Hint for Exercise 3.1. Let $\mathcal{F} = \{C_1, \ldots, C_t\}$ be a chain. Color $c :$ $E(G) \mapsto \mathbb{Z}_3$ as follows: $c(e) = 1$ if $e \in \cup_k E(C_{2k-1}) - \cup_k E(C_{2k})$, $c(e) = 2$ if $e \in \cup_k E(C_{2k}) - \cup_k E(C_{2k-1})$, $c(e) = 0$ if $e \in [\cup_k E(C_{2k-1})] \cap [\cup_k E(C_{2k})]$. Here, $E_{w=1} = [\cup_k E(C_{2k-1})] \bigtriangleup [\cup_k E(C_{2k})]$ is an even 2-factor.

Hint for Exercise 3.2. An example: for each weight two edge e of (P_{10}, w_{10}), $(P_{10} - \{e\}, w_{10})$ has a faithful circuit cover consisting of two circuits (see Proposition B.2.28).

Hint for Exercise 3.3-(a). Let

$$D_1 = \bigcup_{\nu=0}^{\lfloor \frac{j-i}{2} \rfloor} C_{i+2\nu}, \quad D_2 = \bigcup_{\nu=1}^{\lceil \frac{j-i}{2} \rceil} C_{i+2\nu-1}.$$

Assume that R_1 and R_2 are two components of $\bigtriangleup_{\mu=i}^j C_\mu$ intersecting with C_{j+1}. Then $\{D_1' = D_1 \bigtriangleup R_1, D_2' = D_2 \bigtriangleup R_1\}$ is a faithful even subgraph cover of $(\bigcup_{\mu=i}^j C_\mu, w_{\{C_i, \ldots, C_j\}})$. Let $\mathcal{D}$ be the union of circuit decompositions of D_1', D_2'. Since C_{j+1} intersects with two components of $\triangle^j_{\mu=i} C_\mu$, C_{j+1} intersects with at least *two* elements of $\mathcal{D}$. Thus, $[\mathcal{F} - \{C_i, \ldots, C_j\}] \cup \mathcal{D}$ is not a circuit chain. This contradicts the condition (2) that *every* faithful circuit cover of $(G - \{e\}, w|(G - \{e\}))$ is a circuit chain.

Hint for Exercise 3.3-(b). Choose i, j: $j - i$ is as small as possible so that $\triangle^j_{\mu=i} C_\mu$ is not a Hamilton circuit in the subgraph $\bigcup^j_{\mu=i} C_\mu$. Let $R_1, \ldots, R_t$ be the components of $\triangle^j_{\mu=i} C_\mu$. For a technical reason, let $C_0 = \{x\}$, and $C_{k+1} = \{y\}$. Without loss of generality, let $j < k$. By Exercise 3.3-(a), let C_{j+1} intersect only with R_1, and C_{i-1} intersect with only one of $\{R_1, R_2\}$ (not both).

If C_{i-1} intersects with R_2 (not R_1), then

$$J_1 = (C_{j+1} \triangle \cdots \triangle C_{k+1}) \triangle R_1 \text{ and } J_2 = (C_0 \triangle \cdots \triangle C_{i-1}) \triangle (R_2 \cup \cdots \cup R_t)$$

are two disjoint parts of $G[E_{w=1}]$.

If C_{i-1} intersects with R_1 (not others), then

$$J_1 = (C_0 \triangle \cdots \triangle C_{i-1}) \triangle (C_{j+1} \triangle \cdots \triangle C_{k+1}) \triangle R_1 \text{ and } J_2 = (R_2 \cup \cdots \cup R_t)$$

are two disjoint parts of $G[E_{w=1}]$.

Both cases contradict that G is a permutation graph since the edges of $E(C_j) \cap E(C_{j+1}) = E(C_{j+1}) \cap E(R_1)$ are incident only with J_1, but not both components of $G[E_{w=1}]$.

Hint for Exercise 3.4. (See Figures 3.4 and D.3.) Let (G, w) be a counterexample such that the total weight $w(G)$ is as small as possible.

Let $\{z_i\} = N(v_i) - \{v_{i-1}, v_{i+1}\}$ for every $i \in \mathbb{Z}_r$.

Let $\mathcal{F}$ be a faithful circuit cover of $(G - e_0, w)$ with $|\mathcal{F}|$ as large as possible.

I. We claim that, *for every $C \in \mathcal{F}$, either $C \cap [C_0 \cap E_{w=1}] \neq \emptyset$ or $\{v_0, v_1\} \cap V(C) \neq \emptyset$.*

Suppose that there is a member $C' \in \mathcal{F}$ with $C \cap [C_0 \cap E_{w=1}] = \emptyset$ and $\{v_0, v_1\} \cap V(C) = \emptyset$.

Let G' be the graph obtained from G by deleting edges of $E(C') \cap E_{w=1}$ and let $w' : E(G') \mapsto \{1, 2\}$ such that

$$w'(e) = \begin{cases} w(e) & \text{if } e \notin E(C') \\ w(e) - 1 & \text{if } e \in E(C'). \end{cases}$$

(That is, C' is a removable circuit.) Then $(\overline{G'}, w')$ is an eulerian weighted graph with total weight smaller than (G, w) and with $C_0 \subseteq supp(w')$ and $P_0 \subseteq E_{w'=2}$. Since (G, w) is a smallest counterexample, (G', w') has a faithful cover $\mathcal{F}'$. Thus, $\mathcal{F}' + \{C'\}$ is a faithful cover of (G, w). This contradicts that (G, w) is a counterexample.

II. By Lemma 3.4.1-(4), there is no circuit chain of length ≤ 2 joining v_0 and v_1.

And, by Lemma 3.4.1-(5), we will only consider the case of $r = 5$ or 6.

III. Let $C_1, C_2, D_1, D_2 \in \mathcal{F}$ containing a vertex of $\{v_0, v_1\}$ such that (see Figure D.3)

C_1 contains the path $z_1 v_1 v_2 z_2$,

C_2 contains the path $z_1 v_1 v_2 v_3$,

D_1 contains the path $z_0 v_0 v_{r-1} z_{r-1}$, and

D_2 contains the path $z_0 v_0 v_{r-1} v_{r-2}$.

Note, C_1 exists if and only if $w(v_1 v_2) = 2$, while D_1 always exists since $r \geq 5$ and $E_{w=2}$ contains the path $v_{r-1} v_0 v_1$.

Obviously, $V(C_i) \cap V(D_j) = \emptyset$ for any $i, j \in \{1, 2\}$ (by II). Thus, $r - 2 > 3$ since $V(C_2) \cap V(D_2) = \emptyset$. That is,

$$r = 6$$

and

$$z_1 v_1 v_2 v_3 z_3 \subseteq C_2, \quad z_0 v_0 v_5 v_4 z_4 \subseteq D_2,$$

and both C_1, D_1 exist.

IV. Let $C_3 \in \mathcal{F}$ containing the edge $v_3 v_4$. By II, C_3 is distinct from any of $\{C_1, C_2, D_1, D_2\}$.

If $z_3 v_3 v_4 z_4 \subseteq C_3$ then the circuit chain $\mathcal{P} = \{C_2, C_3, D_2\}$ contradicts Lemma 3.4.1-(2).

If $z_2 v_2 v_3 v_4 v_5 z_5 \subseteq C_3$ then the circuit chain $\mathcal{P} = \{C_1, C_3, D_1\}$ contradicts Lemma 3.4.1-(2).

If $z_2 v_2 v_3 v_4 z_4 \subseteq C_3$ then the circuit chain $\mathcal{P} = \{C_1, C_3, D_2\}$ contradicts Lemma 3.4.1-(2).

If $z_3 v_3 v_4 v_5 z_5 \subseteq C_3$ then the circuit chain $\mathcal{P} = \{C_2, C_3, D_1\}$ contradicts Lemma 3.4.1-(2). $\qquad\square$

Hint for Exercise 3.5. (See Figure 3.5.) We may use some part of the proof of Exercise 3.4: continue the proof from parts I and II.

III. We claim that $w(v_1 v_2) = 1$, for otherwise, it is solved already in Exercise 3.4. So, define C_2, D_1, D_2 as in Exercise 3.4 (note that C_1 does not exist since $w(v_1 v_2) = 1$) and we also have that $z_1 v_1 v_2 v_3 z_3 \subseteq C_2$ and $z_0 v_0 v_5 v_4 z_4 \subseteq D_2$ (by II).

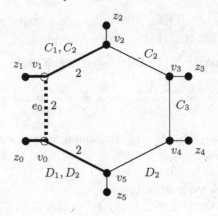

Figure D.3 *A 6-circuit C_0 containing the edge $e_0 = v_0v_1$*

IV. Note that $w(v_2v_3) = w(v_3v_4) = w(v_3z_3) = 2$. There are two members, say C_3, C_4, of $\mathcal{F}$ containing the edge v_3v_4, each of which is distinct from $\{C_2, D_1, D_2\}$ (by II).

Since C_2 contains $z_1v_1v_2v_3z_3$, let C_3 contain $z_3v_3v_4$.

If $z_3v_3v_4v_5 \subseteq C_3$ then the circuit chain $\mathcal{P} = \{C_2, C_3, D_1\}$ contradicts Lemma 3.4.1-(2).

If $z_3v_3v_4z_4 \subseteq C_3$ then $z_0v_0v_5v_4z_4 \subseteq D_2$, and furthermore, $z_2v_2v_3v_4v_5z_5 \subseteq C_4$. Thus, the circuit chain $\mathcal{P} = \{C_2, C_4, D_1\}$ contradicts Lemma 3.4.1-(2). □

Hint for Exercise 3.6. Suppose that $E_{w=2} \cap \{v_2v_3, v_3v_4\} = \emptyset$ and $\mathcal{F}$ is a circuit chain. Then it contradicts Lemma 3.4.1-(2).

Hint for Exercise 3.7. Let H' be the subgraph of G induced by edges of circuits of $\mathcal{F}'$. Since $|\mathcal{F}'| \leq 3$, by Lemma 2.2.1, $\overline{H'}$ is 3-edge-colorable. Note that $H = H' \cup F$. By Lemma B.1.9, $\overline{H}$ is 3-edge-colorable, as well.

Let $w_H : E(H) \mapsto \{1, 2\}$ such that, for each edge e contained in some member of $\mathcal{F}'$, $w_H(e)$ is the number of members of $\mathcal{F}'$ containing e; for each $e \in F$, $w_H(e) = 2$. By Lemma 2.2.1 again, $(\overline{H}, w_H)$ has a faithful circuit cover $\mathcal{F}''$. Thus, $\mathcal{F} - \mathcal{F}' + \mathcal{F}''$ is a faithful circuit cover of (G, w).

Hint for Exercise 3.8. Let $w_{H_j} : E(H_j) \mapsto \{1, 2\}$ such that, for each edge e contained in some member of $\mathcal{F}_j$, $w_{H_j}(e)$ is the number of members of $\mathcal{F}_j$ containing e, and, for each $e_j \in F$, $w_{H_j}(e_j) = 2$. Note that the faithful cover $\mathcal{F}$ was selected with $|\mathcal{F}|$ as large as possible. By

Lemma 3.4.1-(4), each (H_j, w_{H_j}) has a faithful circuit cover $\mathcal{F}'_j$. Thus, $\mathcal{F} - \cup_{i=1}^{t} \mathcal{F}_i + \cup_{j=1}^{t} \mathcal{F}'_j$ is a faithful circuit cover of (G, w).

Hint for Exercise 3.9. Let $|V(G)| = n$. Let M be a perfect matching ($|M| = n/2$) of G and let w be an eulerian $(1, 2)$-weight of G with $\{e \in E(G) : w(e) = 2\} = M$. Let $\mathcal{F}$ be a faithful even subgraph cover of (G, w) (by Theorem 3.2.1) with the least number of even subgraphs (say, m even subgraphs). Then each pair of distinct even subgraphs in $\mathcal{F}$ must intersect with each other. So

$$\frac{m(m - 1)}{2} \leq |M| = \frac{n}{2}$$

and hence,

$$(m - 1)^2 \leq n.$$

Thus, with an additional even subgraph $G - M$, we obtain an even subgraph double cover consisting of at most $\sqrt{n} + 2$ members.

Chapter 4

Hint for Exercise 4.1. Similar to the proof of Lemma 4.1.1 and apply Lemma 2.2.2 to G_1.

Hint for Exercise 4.2. Let $c : E(G_1) \mapsto \{1, 2, 3\}$ be a 3-edge-coloring of G_1 where $c(e) = 1$. Let $e = x_1 x_2$. And let Q_j be the component of $c^{-1}(2) \cup c^{-1}(3)$ containing the vertex x_j for each j. Obviously, either G_2 is 3-edge-colorable if $Q_1 = Q_2$ or is of oddness ≤ 2 if $Q_1 \neq Q_2$.

Hint for Exercise 4.3. (See Chapter 5 for further study.) Let P_0, P_1, P_2 be three induced paths of H. Let $M_{(i,j)}$ be the set of all edges $e = xy \in E(G) - E(H)$ such that $x \in V(P_i)$ and $y \in V(P_j)$. For each $i \in \mathbb{Z}_3$, apply Lemma 2.2.2 to $G_{(i,i+1)} = \overline{G[P_i \cup P_{i+1} \cup M_{(i,i)} \cup M_{(i,i+1)}}$ (in which, $P_i \cup P_{i+1}$ is a Hamilton circuit) where the sum of subscripts is subject to (mod 3).

Hint for Exercise 4.4. "$\Rightarrow$": Let $\{C_1, \ldots, C_5\}$ be a 5-even subgraph double cover of G, and $M = E(C_1) \cap E(C_2)$. Here, $\overline{G - M}$ is double covered by $\{C_1 \triangle C_2, C_3, C_4, C_5\}$, and, therefore, is 3-edge-colorable (by Theorem 1.3.2).

"$\Leftarrow$": By Lemma 2.2.1, $\overline{G - M}$ has a 4-even subgraph double cover $\mathcal{F}_1$ containing $C_1 \triangle C_2$ as a member. Then $\mathcal{F}_1 \cup \{C_1, C_2\}$ is a 5-even subgraph double cover.

Hint for Exercise 4.5. Similar to Exercise 4.4.

Hint for Exercise 4.6. Induction on $\sum_{v \in V(\overline{G})}(d(v) - 3)$ since the cubic case is solved in Theorem 4.2.3. Let S be a spanning even subgraph of G such that there are only two components of S containing an odd number of odd-vertices of G. Apply Thorem A.1.14 to each vertex v with $d(v) \geq 4$. If $d_S(v) \geq 4$ then split some pair of edges of $S \cap E(v)$ away from v by preserving the connectedness of this S-component. If $d_{G-E(S)}(v) \geq 3$ then split some pair of edges of $E(v) - E(S)$ away from v by preserving the property of bridgelessness. If $d_{G-E(S)}(v) = d_S(v) = 2$ then split a pair of edges of away from v and add a new edge between split vertices by preserving the connectedness of this S-component. The final resulting graph is a cubic graph with the same oddness as that of G. Then apply Theorem 4.2.3.

Chapter 5

Hint for Exercise 5.4. It is $G \oplus_3 K_{3,3}$.

Hint for Exercise 5.7. It is easy to check $K_{3,3} = M_3$. And M_{2k+1} is obtained from M_{2k-1} by Method 5.4 (Case 2).

Hint for Exercise 5.9. Apply Exercise 5.8. It is easy to find a disconnected even 2-factor.

Hint for Exercise 5.10. First show each of them is a strong Kotzig graph. Applying Exercise 5.8, we are to show that none of them has a disconnected even 2-factor.

It is trivial for $M_3 = K_{3,3}$.

For the Heawood graph (Figure 5.2), the shortest even circuit is of length 6. Hence, if the Heawood graph (of order 14) has a disconnected even 2-factor, it must be the union of a 6-circuit and an 8-circuit. However, for every 6-circuit C, the complement of C does not contain an 8-circuit.

For the dodecahedron graph (Figure 5.3), the shortest even circuit is of length 8. Hence, if the dodecahedron graph (of order 20) has a disconnected even 2-factor, it must be the union of an 8-circuit and a 12-circuit, or a pair of 10-circuits. Note that, for every 8-circuit C, the complement of C does not contain a 12-circuit. And similarly, for every 10-circuit C, the complement of C is not connected.

None of them is uniquely edge-3-colorable due to their highly symmetry.

Hint for Exercise 5.12. Let K be a spanning Kotzig graph of the contracted graph G/H. Let K^+ be the graph obtained from K by applying the $(Y \to \triangle)$-operation at every vertex v of K if v corresponds to a contracted vertex of G/H. Then K^+ is a spanning Kotzig minor of G.

Hint for Exercise 5.13. By Corollary A.1.10, G has a 2-factor F such that G/F is 4-edge-connected. By Theorem A.1.6, G/F contains two edge-disjoint spanning trees T_1 and T_2. By Lemma A.2.15, T_1 contains a parity subgraph P of G/F. After suppressing all degree 2 vertices of $G - E(P)$, the graph $\overline{G - E(P)}$ is 3-edge-colorable and connected since $G/F - E(P)$ is even and $T_2 \subset G/F - E(P)$.

Chapter 6

Hint for Exercise 6.1. Use Exercise 3.4.

Hint for Exercise 6.2. Let $v_1 \ldots v_n$ be a Hamilton path of G with $v_a v_1 \in E(G)$ for some $a > 2$, let $C = v_1 \ldots v_a v_1$. Apply Proposition 6.4.5.

Hint for Exercise 6.3. Similar to (a part of) the proof of Proposition 6.1.4.

Hint for Exercise 6.4. Let C be a Hamilton circuit of $G - v$ where $v \in V(G) - V(S)$. Then, by Exercise 6.3, $G' = \overline{G - (E(S) - E(C))}$ is bridgeless. By Theorem 6.2.4, G' has a circuit double cover $\mathcal{F}'$ containing C. Thus, $\mathcal{F}' - \{C\} \cup \{(S \triangle C), S\}$ is an even subgraph double cover of G containing S.

Hint for Exercise 6.5. By Lemma 6.4.2, the suppressed cubic graph $G_1 = \overline{G - (E(S) - E(C))}$ has a circuit double cover $\mathcal{F}_1$ with $C \in \mathcal{F}_1$.

Since C is a Hamilton circuit in $G_2 = G[E(C) \cup E(S)]$, G_2 has an even subgraph double cover $\mathcal{F}_2 = \{C\} \cup \{S\} \cup \{S \triangle C\}$.

Hence, $(\mathcal{F}_1 - \{C\}) \cup (\mathcal{F}_2 - \{C\}$ is an even subgraph double cover of G.

Hint for Exercise 6.6. By Exercise 2.3, $E_{w=1}$ induces an even subgraph C_0. If C_0 has more than one component, then it is a 2-factor consisting of two 5-circuits which is the excluded case of (P_{10}, w_{10}) (by Proposition B.2.26). Hence, C_0 is a circuit. By Proposition B.2.4, C_0

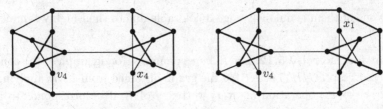

(a) Stable circuit C (bold edges)   (b) Semi-extension D (bold edges)
   missing v_4, x_4 missing v_4, x_1

Figure D.4 *A semi-extension D of the stable circuit C*

is of length $r = 5, 6, 8$ or 9. Apply Proposition 6.4.5 to each case since $P_{10} - V(C_0)$ must be a small-end Hamilton path or a small-end Y-tree (for a 6-circuit, one needs Proposition B.2.8; for an 8-circuit, $P_{10} - V(C_0)$ is a single edge, see Figure B.10).

Hint for Exercise 6.7. Let $(G, 2) = (G_1, w_1) + (G_2, w_2)$ be an eulerian weight decomposition where $E(G_1) \cap E(G_2) = E(C)$ and, for each $j = 1, 2$, $w_j(e) = 1$ if and only if $e \in E(C)$. Here, G_1 consists of C, all chords of C and one vertex of $V(G) - V(C)$, while G_2 consists of C and the remaining part $V(G) - V(G_1)$. By Corollary 6.2.5, $(\overline{G_1}, w_1)$ has a faithful cover $\mathcal{F}_1$.

Note that $|V(\overline{G_2})| \leq 16$. By Exercise 2.7, let $\mathcal{F}_2$ be a faithful cover of (G_2, w_2). Then $\mathcal{F}_1 \cup \mathcal{F}_2$ is a CDC of G.

Hint for Exercise 6.8. Let $w : E(G) \mapsto \{1, 2\}$ such that $E_{w=1}$ induces the stable circuit C. Let T be the cyclical 3-edge-cut with components P_1 and P_2. By Proposition B.2.27, let $\mathcal{F}_\mu$ be a faithful circuit cover of $(G/P_\mu, w)$. Then we are able to construct a faithful cover for (G, w) from $\mathcal{F}_1, \mathcal{F}_2$ by merging some members of them along the cut T.

Hint for Exercise 6.9. See Figure D.4.

Hint for Exercise 6.10. By Proposition 6.4.5, it is sufficient to show that $H = G - V(C)$ has a small-end Hamilton path.

Since G is cubic and 2-edge-connected, H has at least two degree ≤ 2 vertices.

Suppose that H has some Hamilton path, but none of them is small-end.

So, H has no Hamilton circuit, for otherwise, a small-end Hamilton path can be obtained from a Hamilton circuit starting at some small degree vertex.

Let $Q = x_1 \ldots x_t$ be a Hamilton path of H with $d_H(x_1) = d_H(x_t) = 3$ and let $N(x_1) = \{x_2, x_{i_1}, x_{i_2}\}$ $(2 < i_1 < i_2 < t)$ and $N(x_t) = \{x_{t-1}, x_{j_1}, x_{j_2}\}$ $(1 < j_1 < j_2 < t-1)$. Since G is cubic, $\{x_{i_1}, x_{i_2}, x_{j_1}, x_{j_2}\}$ is a set of four distinct degree 3 vertices. Note that x_{i_1-1} is an end of a Hamilton path $x_{i_1-1} \ldots x_1 x_{i_1} \ldots x_t$. Hence, (1) x_{i_1-1} is of degree 3 in H and (2) $x_{i_1-1} \notin N(x_t)$, for otherwise, H has a Hamilton circuit. Similar for x_{j_2+1}. Thus, $\{x_1, x_{i_1-1}, x_{i_1}, x_{i_2}, x_{j_1}, x_{j_2}, x_{j_2+1}, x_t\}$ is a set of eight degree-3 vertices of H. This contradicts that H has at least two degree 2 vertices.

Hint for Exercise 6.11. By Proposition 6.4.5, it is sufficient to show that $H = G - V(C)$ has either a small-end Hamilton path or a small-end Y-tree. Let $x_1 \ldots x_{t-1} + x_{t-2} x_t$ be a Y-tree of H. By Exercise 6.10, we may further assume that H has no Hamilton path. So, $x_1 x_{t-1}, x_1 x_t \notin E(G)$.

Since G is 2-edge-connected, the edge $x_{t-2} x_{t-3}$ is not a cut-edge of G. Hence, there is a vertex $x_h \in \{x_2, \ldots, x_{t-3}\}$ such that x_h is adjacent to some vertex of $\{x_{t-1}, x_{t-2}\} \cup V(C)$. Thus, $N(x_1) \subseteq \{x_2, \ldots, x_{t-3}\} - \{x_h\}$.

The remaining part is similar to the proof of Exercise 6.10. Let $N(x_1) = \{x_2, x_i, x_j\}$. Note that x_{i-1} is also an end of a Y-tree $x_{i-1} \ldots x_1 x_i \ldots x_{t-1} + x_{t-2} x_t$. Hence, $\{x_{i-1}, x_i, x_j\}$ is a set of three degree-3 vertices contained in $\{x_2, \ldots, x_{t-3}\} - \{x_h\}$. That is, $t - 5 \geq 3$. This contradicts that $t \leq 7$.

Hint for Exercise 6.12. Let R be a spanning tree of $H = G - V(C)$. Since $|V(H)| \leq 5$, every spanning tree is either a Hamilton path or a Y-tree. Then, apply Exercise 6.11.

Hint for Exercise 6.13. See Proposition B.2.32.

Hint for Exercise 6.14. Let G be a 2-edge-connected graph and C be a connected even subgraph of G. Suppose that the pair (G, C) is a counterexample to the strong circuit double cover conjecture such that

(1) $|E(G)|$ is as small as possible,

(2) subject to (1), $|E(C)|$ is as large as possible.

The remaining part follows the proof of Theorem 6.6.4.

Chapter 7

Hint for Exercise 7.3. Apply Theorem A.1.14.

Hint for Exercise 7.4. By Lemma A.2.4.

Hint for Exercise 7.5. Let $\mathcal{F}$ be an r-even subgraph cover of G with $ced_{\mathcal{F}}(e)$ as small as possible. If $ced_{\mathcal{F}}(e) = h > 1$, then let $C_1, C_2 \in \mathcal{F}$ be two even subgraphs containing e. G has another r-even subgraph cover $\mathcal{F}' = \mathcal{F} - \{C_2\} + \{C_1 \triangle C_2\}$ with $ced_{\mathcal{F}'}(e) = ced_{\mathcal{F}}(e) - 1$. This contradicts our choice of $\mathcal{F}$.

Hint for Exercise 7.6. Let $\mathcal{F}_i$ be an r-even subgraph cover of H_i. By Exercise 7.5, let $ced_{\mathcal{F}_i}(e) = 1$ where e is the new edge in $H_i - M_i$. By properly joining the even subgraphs containing the new edge, we obtain an r-even subgraph cover of G.

Hint for Exercise 7.7. The triangle is collapsible (apply the definition directly). By Lemma 7.4.9, the entire graph is collapsible. Then apply Theorems 7.4.3 and 7.1.2.

Hint for Exercise 7.8. Apply Exercise 7.7.

Hint for Exercise 7.9. Contract all cliques of $L(G)$ by applying Exercise 7.7.

Hint for Exercise 7.10. Let G' be the graph obtained from G by inserting a vertex into every edge. It is obvious that $L(G)$ has a CDC if $L(G')$ has a CDC. Contract all cliques of $L(G')$ (by applying Exercise 7.7). The resulting graph (C-reduced) is the same as G. Hence, $L(G')$ has a CDC since G has.

Hint for Exercise 7.11. By Lemma 7.4.9, let $\mathcal{P} = \{V_1, \ldots, V_t\}$ be a partition of $V(G)$ such that $G[V_i]$ is a maximal collapsible subgraph of G (for each $i = 1, \ldots, t$). Let $G^r = G/\mathcal{P}$ be the C-reduced graph of G with the vertex set $\{v_1, \ldots, v_t\}$ where v_i is obtained by contracting V_i. Assume that H^r is a non-trivial collapsible subgraph of G^r with the vertex set $\{v_1, \ldots, v_h\}$. We are to show that *the induced subgraph* $G[V_1 \cup \cdots \cup V_h]$ *is collapsible*.

Let $X \subseteq V_1 \cup \cdots \cup V_h$ of even order and $X_i = V_i \cap X$ for each $i = 1, \ldots, h$. Let $X^r = \{v_i : |X_i| \text{ is odd }\}$. Since H^r is collapsible, let Q^r be a connected spanning subgraph of H^r with $O(Q^r) = X^r$. For each $i \in \{1, \ldots, h\}$, let $Y_i = \{y \in V_i : |E_{Q^r}(y)| \text{ is odd }\}$ and $Z_i = Y_i \triangle X_i$. Since each $G[V_i]$ is collapsible, let Q_i be a connected spanning subgraph of $G[V_i]$ with $O(Q_i) = Z_i$. Let Q be the subgraph of G induced by edges of Q^r and all Q_i. It is not hard to see that Q is a connected spanning subgraph of $G[V_1 \cup \cdots \cup V_h]$ and $O(Q) = X$. Thus, each $G[V_i]$ is not a maximal collapsible subgraph of G for $i = 1, \ldots, h$. This proves that the induced subgraph $G[V_1 \cup \cdots \cup V_h]$ is collapsible.

Hint for Exercise 7.12. "⇒": Assume that G/H is $\mathcal{P}$-contractible. We show that G is $\mathcal{P}$-contractible.

Let J' be a supergraph of G. By the definition of $\mathcal{P}$-contractible configuration, we need to show the following statement is always true:

(α)   *J' has the property $\mathcal{P}$ ⇔ J'/G has the property $\mathcal{P}$.*

Note that J'/H is a supergraph of G/H. Since G/H is $\mathcal{P}$-contractible, we have the true statement:

(α_1)   *J'/H has the property $\mathcal{P}$ ⇔ $(J'/H)/(G/H)$ has the property $\mathcal{P}$.*

Note that $J'/G = (J'/H)/(G/H)$. The statement (α_1) is equivalent to the following true statement:

(α_2)   *J'/H has the property $\mathcal{P}$ ⇔ J'/G has the property $\mathcal{P}$.*

Since H is $\mathcal{P}$-contractible, we have the following true statement:

(α_3)   *J'/H has the property $\mathcal{P}$ ⇔ J' has the property $\mathcal{P}$.*

The combination of statements (α_2) and (α_3) is the statement (α).

"⇐": Assume that G is $\mathcal{P}$-contractible. We show that G/H is also $\mathcal{P}$-contractible. Let J be a supergraph of G/H, by definition, we need to show that the following statement is always true:

(β)   *J has the property $\mathcal{P}$ ⇔ $J/(G/H)$ has the property $\mathcal{P}$.*

It is easy to find a graph J' such that J' is a supergraph of G, $J'/H = J$ and $J'/G = J/(G/H)$. Since G is $\mathcal{P}$-contractible, we have the following true statement:

(β_1)   *J' has the property $\mathcal{P}$ ⇔ J'/G has the property $\mathcal{P}$.*

Note that $J'/G = J/(G/H)$. The statement (β_1) is equivalent to the following statement:

(β_2)   *J' has the property $\mathcal{P}$ ⇔ $J/(G/H)$ has the property $\mathcal{P}$.*

Furthermore, since H is $\mathcal{P}$-contractible, we have the following true statement:

(β_3)   *J' has the property $\mathcal{P}$ ⇔ J'/H has the property $\mathcal{P}$.*

Note that $J'/H = J$. The statement (β_3) is equivalent to the following statement:

(β_4) *J' has the property* $\mathcal{P}$ $\Leftrightarrow$ *J has the property* $\mathcal{P}$.

The combination of true statements (β_2) and (β_4) yields the statement (β).

Chapter 8

Hint for Exercise 8.2. Apply Theorem 8.1.2 after contracting one edge of each 2-edge-cut.

Hint for Exercise 8.4. Corollary of Lemmas A.2.10 and A.2.12.

Hint for Exercise 8.5. We assume that G is not an even graph. If a non-trivial parity subgraph decomposition $\mathcal{P}$ of G has $t > 3$ members then replacing $(t-2)$ members of $\mathcal{P}$ with their union yields a non-trivial parity subgraph decomposition with precisely three members (note that t is odd, see Exercise 8.3). Thus, this theorem is an immediate corollary of Theorem 8.2.2-(iv) and Exercise 8.4.

Hint for Exercise 8.6. By Exercise 8.5.

Hint for Exercise 8.7. "$\Leftarrow$": Let P_1, P_2 be two edge-disjoint parity subgraphs of G. It is obvious that $P_1 \cup P_2$ is an even subgraph of G, by Lemma A.2.11. Since a vertex is of odd degree in P_1 if and only if it is of odd degree in G, by Lemma A.2.7, each component of P_1 contains an even number of odd degree vertices of G. Since the vertex set of each component C of S is the union of the vertex sets of several components of P_1, the component C of S also contains an even number of odd degree vertices of G. So, $S = P_1 \cup P_2$ is an evenly spanning even subgraph of G.

"$\Rightarrow$": Let S be an evenly spanning even subgraph of G and let $O(G) = \{v_1, \ldots, v_{2t}\}$ be the set of all odd degree vertices of G such that v_{2i-1} and v_{2i} are contained in the same component of S for each $i = 1, \ldots, t$. Let P_i be a path joining v_{2i-1} and v_{2i} and contained in S. By Lemma A.2.5, the symmetric difference Q of all P_i for $i = 1, \ldots, t$ is a parity subgraph of G. Thus, the even subgraph S is the union of two parity subgraphs Q and $S - E(Q)$ of G.

Hint for Exercise 8.8 (Proof 1). An immediate corollary of Exercise 8.7 and Theorem 8.2.4.

Hint for Exercise 8.8 (Proof 2). "⇒": Let $O(G)$ be the set of all odd vertices of G. Get a mod-4-flow as follows. Orient $E(G) - E(S)$ arbitrarily and assign weight 2 to each edge in $E(G) - E(S)$. For each component C of S, let $|O(G) \cap V(C)| = 2t$ and $T = v_1 \ldots v_1$ be an eulerian tour of C. Let

$$v_1 \ldots v_{i_1}, \quad v_{i_1} \ldots v_{i_2}, \quad \ldots \quad , v_{i_{2t-1}} \ldots v_1$$

be the subsequences of T where $O(G) \cap V(C) = \{v_{i_0}, v_{i_1}, \ldots, v_{i_{2i-1}}\}$ (where $i_0 = 1$). For each $\mu = 0, \ldots, t - 1$, the edges of the subsequence $v_{i_{2\mu-1}} \ldots v_{i_{2\mu}}$ are oriented forward and the edges of the subsequence $v_{i_{2\mu}} \ldots v_{i_{2\mu+1}}$ are oriented backward and assign weight 1 to each edge of C. It is easy to see that (D, f) is a nowhere-zero mod-4-flow of G.

"⇐". By Exercise C.15, let (D, f) be a nowhere-zero mod-4-flow with $1 \le f(e) \le 2$. Show that the subgraph S of G induced by weight 1 edges is an evenly spanning even subgraph. Let C be a component of S, then, by Exercise C.4,

$$\sum_{e \in (V(C), V(G) - V(C))} f(e) \equiv 0 \pmod 4.$$

This implies that $|(V(C), V(G) - V(C))|$ is even and therefore C contains an even number of odd vertices of G.

Hint for Exercise 8.11. (2) ⇒ (1). Let $\mathcal{P} = \{P_1, P_2, P_3\}$ be a parity subgraph decomposition of G with $E_{w=2} = E(P_1)$. Then, obviously, $\{P_1 \cup P_2, P_1 \cup P_3\}$ is a faithful even subgraph cover of G with respect to w.

(1) ⇒ (2). Let $\{C_1, C_2\}$ be a faithful even subgraph cover of G with respect to w. Note that $E_{w=1} = C_1 \triangle C_2$ is an even subgraph of G. Therefore,

$$G - E(C_1 \triangle C_2) = G[E(C_1) \cap E(C_2)] = G[E_{w=2}]$$

is a parity subgraph of G and so,

$$\{G[E_{w=2}], \quad C_1 - E_{w=2}, \quad C_2 - E_{w=2}\}$$

is a parity subgraph decomposition of G.

The equivalence of (1) and (4) is trivial, the equivalence of (2) and (3) is proved by a similar argument as Exercise 8.7.

Hint for Exercise 8.12. Apply Theorem 8.2.8.

Hint for Exercise 8.13. Apply Theorem 8.2.8.

Hint for Exercise 8.14. Since $G - \{e\}$ admits a nowhere-zero 4-flow, $G - \{e\}$ is bridgeless. Let C be the union of two edge-disjoint paths joining the endvertices of e. By Theorem 8.2.8, let $\mathcal{F}$ be a 4-even subgraph double cover of $G - \{e\}$ such that $C \in \mathcal{F}$. Let C_1, C_2 be circuits of G with $C_1 \triangle C_2 = C$ and $E(C_1) \cap E(C_2) = \{e\}$. Then $(\mathcal{F} - \{C\}) \cup \{C_1, C_2\}$ is a 5-even subgraph double cover of G.

Hint for Exercise 8.16. By Exercise A.4, the girth is at most 5.

Hint for Exercise 8.17. Since G admits a 3-flow (D, f) with $supp(f) \supseteq E(G) - E(S)$, color the faces of G according to the 3-flow. Then re-color the faces bounded by each component of S with the fourth color.

Hint for Exercise 8.18. Follow the proof of Theorem 8.5.4. $(D, \frac{f_1}{2} + \frac{3f_2}{2})$ is a nowhere-zero 5-flow of G.

Chapter 9

Hint for Exercise 9.1. Let H be a counterexample with $|E(H)|$ as small as possible.

I. We claim that H *is 2-connected*. Suppose that H is not 2-connected. Let T be its block tree which has at least two leaf blocks B_1 and B_2. Since no leaf block of H consists of only two vertices, let C_j be a circuit of B_j of length ≥ 3 $(j = 1, 2)$. This contradicts that H is a counterexample.

II. We claim that H must have some digon. For otherwise, let $\mathcal{F}$ be a circuit decomposition of H and any pair of members of $\mathcal{F}$ is what we needed.

III. By II, let $\{e_1, e_2\}$ be a pair of parallel edges with endvertices v_1 and v_2. Let $H' = H - \{e_1, e_2\}$. Since H' is connected and even (by I), let P_1, P_2 be a pair of edge-disjoint paths joining v_1 and v_2 in H'.

Since the multiplicity of H is at most 2, both P_1, P_2 are of lengths at least 2. Then $\{P_1 \cup e_1, P_2 \cup e_2\}$ is a pair of circuits that we needed. This contradicts that H is a counterexample and completes the proof.

Hint for Exercise 9.2. (This is a corollary of Exercise 9.1.) We only need to work on a face f that is not homeomorphic to a closed disk. Let W be a closed walk around the boundary of f. Construct an even graph H from W by duplicating every edge that W passes through twice. It is easy to see that H satisfies Exercise 9.1. That is, the closed walk W (the boundary of f) is of length $\geq 2g$.

Hint for Exercise 9.3. We have the inequality

$$2|E(G)| \geq g\,|F|$$

if the embedding is circular, and

$$2|E(G)| \geq g\,(|F|-1) + 2g$$

if the embedding is not circular (apply Exercise 9.2 for a non-circular face).

Chapter 10

Hint for Exercise 10.3. Suppose that $(K_5, \mathcal{P}_5)$ has a compatible circuit decomposition $\mathcal{C}$, then each member C of $\mathcal{C}$ must be of length precisely 4 (two in $C_1 = v_1v_2v_3v_4v_5v_1$ and two in $C_2 = v_1v_3v_5v_2v_4v_1$, alternately along the compatible circuit C). But $|E(K_5)| = 10$ is not a multiple of 4.

Another proof. Note that $(K_5, \mathcal{P}_5)$ corresponds to the eulerian weighted Petersen graph (P_{10}, w_{10}). See Proposition B.2.26.

Hint for Exercise 10.4. Let $G \in \Lambda_{\overline{x}}$ be bridgeless and w be an admissible eulerian weight of G. Construct an eulerian graph H by replacing each edge e of G by $w(e)$ parallel edges which therefore forms a forbidden part of a forbidden system $\mathcal{P}$ of H (Operation 3). It is obvious that G has a faithful cover if and only if H has a circuit decomposition compatible to $\mathcal{P}$.

Hint for Exercise 10.5. Draw a K_5 on the plane with one crossing. By considering the crossing point as a vertex, we obtain a 4-regular planar graph H with six vertices. For some forbidden system of H, the corresponding weighted graph has a P_{10}-minor. A family of such graphs can be constructed by properly connecting copies of H at their vertices.

Hint for Exercise 10.7. Color each component of $E_{w=1}$ with red and blue alternately. Then the set of all mono-colored circuits in the original eulerian graph H is a CCD of $(H, \mathcal{P})$.

Hint for Exercise 10.8. Color each circuit of the decomposition $\mathcal{F}_e$ alternately with red and blue. Then the subgraph induced by red edges (blue edges) is an even subgraph compatible to $\mathcal{F}_e$. Then the union of the circuit decompositions of the red-even subgraph and the blue-even subgraph is a circuit decomposition $\mathcal{F}_c$ of G compatible to $\mathcal{F}_e$.

Hint for Exercise 10.9. Apply Theorem 10.3.3.

Hint for Exercise 10.10. Let H' be an eulerian graph obtained from H by splitting each vertex without breaking the Euler tour so that the resulting graph H' is 4-regular. Then, by applying Operation 2 to H', we obtain a cubic graph G with a Hamilton circuit C. So G is 3-edge-colorable. By Lemma 2.2.1, G has a circuit cover that covers all edges of C once and all chords twice. By Lemma 10.2.1, it corresponds to a circuit decomposition of H compatible with the Euler tour.

Hint for Exercise 10.11. Let H' be an eulerian graph obtained from H by splitting each vertex without breaking the Euler tour so that the resulting graph H' is 6-regular and remains bipartite. Then, by applying Operation 1 to H', we obtain a cubic bipartite graph G with a dominating circuit C. G admits a nowhere-zero 3-flow since it is bipartite (by Exercise C.31). So, by Theorem 8.2.2, G has a circuit cover that covers all edges of C once and all other edges twice, which, by Lemma 10.2.1, corresponds to a circuit decomposition of H compatible with the Euler tour.

Hint for Exercise 10.12. Let $\mathcal{F}$ be a circuit double cover of a K_5-minor-free graph $G = (V, E)$ ($\mathcal{F}$ exists because of Theorem 3.3.1). Construct an auxiliary graph F_v for each vertex v of G. The vertex set of F_v is $E(v)$ and the edge set of F_v is the set of circuits of $\mathcal{F}$ containing v, and an edge C is incident with two vertices e' and e'' if and only if the edges e' and e'' of G are contained in the circuit C of $\mathcal{F}$. Obviously, each component of F_v is a circuit. Denote $\omega(F_v)$ be the number of components of the graph F_v. Thus, it is obvious that *a graph G has a circular 2-cell embedding on a 2-manifold surface if and only if G has a circuit double cover $\mathcal{F}$ such that $\omega(F_v) = 1$ for each vertex v of G.*

Let $\mathcal{F}$ be a circuit double cover of G such that

$$\sum_{v \in V(G)} \omega(F_v)$$

is as small as possible. Assume that $\omega(F_z) > 1$ for some vertex z of G. Let $N_1, N_2, \ldots, N_r$ be the components of the graph F_z, and $\mathcal{F}_\mu \ (\subseteq \mathcal{F})$ be the set of edges in N_μ for $\mu = 1, \ldots, r$. Construct an auxiliary graph H such that $V(H) = \mathcal{F}$ and $C', C'' \in \mathcal{F}$ are adjacent in H if and only if $[V(C') \cap V(C'')] - \{z\} \neq \emptyset$. Since $G - \{z\}$ is connected, so is H. Let $\mathcal{Q}$ be a shortest path of H joining a pair of distinct $\mathcal{F}_{\mu'}$ and $\mathcal{F}_{\mu''}$. Let $\mathcal{Q} = C_1 \ldots C_s$ and $C_1 \in \mathcal{F}_{\mu'}, C_s \in \mathcal{F}_{\mu''}$. (Here, $\mathcal{Q}$ is a circuit chain in

G.) By the choice of Q,

$$\left(\bigcup_{i=2}^{s-1} V(C_i)\right) \cap \{z\} = \emptyset \quad \text{and} \quad \left(\bigcup_{i=2}^{s-1} E(C_i)\right) \cap E(z) = \emptyset.$$

Let G' be a subgraph of G induced by the set of all edges in $\bigcup_{i=1}^{s} E(C_i)$. Construct G'' as the following, $V(G'') = V(G')$ and replacing each edge of G' contained in two circuits of Q by two parallel edges. G'' is therefore eulerian with $d_{G''}(v) = 2$ or 4 for each vertex v of G''. Properly define an admissible forbidden system $\mathcal{P}$ on G'' so that (by Theorem 10.3.3) $(G'', \mathcal{P})$ has a CCD $\mathcal{F}^*$. Thus, $\mathcal{F}' = (\mathcal{F} - Q) + \mathcal{F}^*$ is a circuit double cover of G with

$$\sum_{v \in V(G)} \omega(F_v') < \sum_{v \in V(G)} \omega(F_v).$$

This contradicts the choice of $\mathcal{F}$.

Chapter 11

Hint for Exercise 11.1. Each bi-colored circuit corresponds to an even circuit in a circuit decomposition of the line graph.

Hint for Exercise 11.2. By Theorem 11.1.3, H has an even circuit decomposition. Then apply Exercise 10.8.

Hint for Exercise 11.3. Apply Propositions B.2.25 and B.2.4.

Chapter 12

Hint for Exercise 12.1. "$\Rightarrow$": Let $\mathcal{F}^S$ be a faithful cover of (G^S, ϕ) and $C \in \mathcal{F}^S$ containing the new edge e_0'. If C is a digon, then $\mathcal{F}^S - \{C\}$ is a faithful cover of (G, ϕ^{-D}). If C is not a digon, then $\mathcal{F}^S$ is a faithful cover of (G, ϕ^M).

"$\Leftarrow$": Similar.

Hint for Exercise 12.2. "$\Rightarrow$": Suppose neither ϕ^M nor ϕ^{-D} is admissible in G. Then, in G, the edge e_0 is contained in an edge-cut T_1 in which $\phi^M(e_0) > \frac{1}{2}\phi^M(T_1)$, and e_0 is also contained in another edge-cut T_2 in which there is an edge e_1 (other than e_0) with $\phi^{-D}(e_1) > \frac{1}{2}\phi^{-D}(T_2)$.

Note that the symmetric difference $T_1 \triangle T_2 = T$ is an edge-cut of both G and G^S containing the edge e_1 but not e_0 and e_0', in which

$$\phi(e_1) > \frac{1}{2}\phi(T). \tag{D.1}$$

This contradicts that ϕ is admissible in G^S.

The following is the detailed calculation of the inequality (D.1). As eulerian weights, both $\phi^M(T_1)$ and $\phi^{-D}(T_2)$ are even. Hence, we have that

$$\phi^M(e_0) \geq \phi^M(T_1 - e_0) + 2 \tag{D.2}$$

and

$$\phi^{-D}(e_1) \geq \phi^{-D}(T_2 - e_1) + 2. \tag{D.3}$$

Thus, the combination of (D.2) and (D.3) yields that

$$
\begin{aligned}
\phi(e_1) \ &= \phi^{-D}(e_1) \\
&\geq \phi^{-D}(T_2 - e_1) + 2 && \text{by (D.3)} \\
&= \phi^{-D}(T_2 - e_1 - e_0) + \phi^{-D}(e_0) + 2 \\
&= \phi^{-D}(T_2 - e_1 - e_0) + \phi^M(e_0) \\
&\geq \phi(T_2 - e_1 - e_0 - e_0') + \phi^M(T_1 - e_0) + 2 && \text{by (D.2)} \\
&\geq \phi(T_2 - e_1 - e_0 - e_0') + \phi(T_1 - e_0 - e_0') + 2 \\
&\geq \phi(T_1 \triangle T_2 - e_1) + 2.
\end{aligned}
$$

"$\Leftarrow$": Trivial.

Chapter 13

Hint for Exercise 13.1. Let $\{C_0, \ldots, C_{t+1}\}$ be a circular $(t+2)$-even subgraph double cover of G. Then $\{C_0, \ldots, C_{t-3}, C_{t-2} \cup C_t, C_{t-1} \cup C_{t+1}\}$ is a circular t-even subgraph double cover of G.

Hint for Exercise 13.2. Let G be a 3-edge-connected graph and D be an arbitrary orientation of G. Let $3G$ be the graph obtained from G by replacing each edge by three parallel edges, and let the direction of each edge of $D(3G)$ remain the same as in $D(G)$. The new graph $3G$ is 9-edge-connected and therefore, by the assumption, admits a mod-5-flow (D, f) with $f(e) \in \{1, -1\}$. Now (D, f') is a mod-5-flow of G where, for each $e \in E(G)$ which was replaced with three parallel edges $e_1, e_2, e_3 \in E(3G)$,

$$f'(e) = f(e_1) + f(e_2) + f(e_3).$$

Hint for Exercise 13.4. By Theorem 13.2.8-(4) and similar to the proof of Exercise 13.1.

Chapter 14

Hint for Exercise 14.1. Let G be a graph admitting a nowhere-zero 4-flow. By Theorem 8.2.4, let $\{P_1, P_2, P_3\}$ be a parity subgraph decomposition of G and let $|E(P_1)| \leq \frac{|E(G)|}{3}$. Then, $\{P_1 \cup P_2, P_1 \cup P_3\}$ is a 2-even subgraph cover with total length at most $\frac{4|E(G)|}{3}$.

Hint for Exercise 14.2. By Lemma A.2.19, let P be a parity subgraph of G with $|E(P)| \leq \frac{|E(G)|}{2h+1}$. Define $f : E(G) \mapsto \{1, 2\}$ as follows:

$$f(e) = \begin{cases} 2 & \text{if } e \in E(P) \\ 1 & \text{if } e \notin E(P). \end{cases} \tag{D.4}$$

It is not hard to see that f is an eulerian weight of G. Since the odd-edge-connectivity of G is at least 5, by Exercise 8.2 (a corollary of Theorem 8.1.2), G admits a nowhere-zero 4-flow. By Theorem 8.2.8, the graph G has a faithful even subgraph cover $\mathcal{F}$ with respect to the eulerian $(1, 2)$-weight f defined in (D.4), and furthermore, $|\mathcal{F}| \leq 3$.

Hint for Exercise 14.3. Assume that $\mathcal{F} = \{C_1, \ldots, C_k\}$ is a shortest k-even subgraph cover of G such that G has a circuit C with $ced_{\mathcal{F}}(e) > \frac{k}{2}$ for each $e \in E(C)$. Since

$$\sum_{i=1}^{k} |E(C_i) \cap E(C)| = \sum_{e \in E(C)} ced_{\mathcal{F}}(e) > \frac{k}{2}|E(C)|,$$

there is a $\mu \in \{1, \ldots, k\}$ such that $|E(C_\mu) \cap E(C)| > \frac{|E(C)|}{2}$. Then

$$\mathcal{F}' = \{C_\mu \triangle C\} \cup \{C_i : i \neq \mu\}$$

is a k-even subgraph cover of G with $\mathcal{L}(\mathcal{F}') < \mathcal{L}(\mathcal{F})$.

Hint for Exercise 14.4. Use Theorem A.1.16 for vertex splitting (the 4-edge-connectivity can be relaxed to 5-odd-edge-connectivity), and the weight w is considered as the lengths of induced paths after splitting operation.

Hint for Exercise 14.5. "$\Rightarrow$": The edge-cut T is covered by C_2 and C_3 (not C_1). Since T is an odd edge-cut and $|T \cap C_3|$ is even, $T \cap C_2 \neq \emptyset$.
    "$\Leftarrow$": Find a parity subgraph P of G/C_1 that $P \subseteq C_2$. Then $G/C_1 - P$

is even and can be extended to an even subgraph C_3 of G (by including some edges of C_1).

Hint for Exercise 14.6. By Definition 14.3.20 and Proposition 14.3.17, it is sufficient to show that G has a 3-even subgraph cover $\{C_1, C_2, C_3\}$ with C_1 as a 2-factor. By Corollary A.1.10, let M be a perfect matching of G such that $|M \cap T| = 1$ for every 3-edge-cut T of G. Let $C_1 = G - M$ which is a 2-factor of G. The contracted graph G/C_1 is of odd-edge-connectivity at least 5 since C_1 passes through every 3-edge-cut. By Exercise 8.2, G/C_1 admits a nowhere-zero 4-flow. By Theorem 8.2.2, the contracted graph G/C_1 has a 2-even subgraph cover $\{C_2, C_3\}$. Properly adding some edges of C_1 into C_2 and C_3, each of which can be extended to an even subgraph of G. Here, $\{C_1, C_2, C_3\}$ is a 3-even subgraph cover of G.

Chapter 15

Hint for Exercise 15.1. Let T be a 2-edge-cut with components Q_1, Q_2. Find a faithful cover for each G/Q_i ($i = 1, 2$) and merge them together at edges of T.

Hint for Exercise 15.4. After some iterations of this algorithm, the resulting graph G_i may have a bridge e^* (although no endvertex of e^* is of degree 1).

Hint for Exercise 15.5. Let C be the even subgraph induced by $E_{w=h_o}$. By Theorem 7.3.3, let $\mathcal{F}$ be an h_e-cover of G. Add $(h_o - h_e)$ copies of C into $\mathcal{F}$.

Hint for Exercise 15.6. By Lemma 15.3.9, it is sufficient to show two smallest cases: $w : E(G) \mapsto \{h_o, h_o + 1\}$ with $h_o = 3$ or 5.

Let C be the even subgraph induced by $E_{w=h_o}$.

By Theorem 7.3.4, let $\mathcal{F}_4 = \{C_1, \ldots, C_7\}$ be a 7-even subgraph 4-cover of G. Then $\mathcal{F}_4^* = \{C_1 \triangle C, \ldots, C_7 \triangle C\}$ is a 7-even subgraph faithful cover of (G, w_4) for $w_4 : E(G) \mapsto \{3, 4\}$.

Similarly, by Theorem 7.3.5, let $\mathcal{F}_6 = \{C_1, \ldots, C_{10}\}$ be a 10-even subgraph 6-cover of G. And $\mathcal{F}_6' = \{C_1 \triangle C, \ldots, C_{10} \triangle C\}$ is a 10-even subgraph faithful cover of (G, w_6') for $w_6' : E(G) \mapsto \{4, 6\}$. Then $\mathcal{F}_6^* = \mathcal{F}_6' \cup \{C\}$ is an 11-even subgraph faithful cover of (G, w_6) for $w_6 : E(G) \mapsto \{5, 6\}$.

Remark. Let $\mathcal{F}$ be the union of one copy of $\mathcal{F}_4^*$ (or $\mathcal{F}_6^*$) and several copies of $\mathcal{F}_6'$ (which is a $(4,6)$-cover) and possibly some copies of $\mathcal{F}_4$ or $\mathcal{F}_6$. The circuit family $\mathcal{F}$ may also be a faithful cover for some weight w where $w(G) = \{h_o, h_e\}$ with a larger difference of $h_e - h_o$.

Chapter 16

Hint for Exercise 16.1. By Lemma 2.2.1 (and Equation (2.1)),

$$\{(M_i \cup M_j) \triangle E_{w=1} : \{i,j\} \subset \{1,2,3\}\}$$

is a faithful even subgraph cover of (G, w). Since (G, w) is Hamilton weighted, every faithful cover of (G, w) has only two circuits. Hence, one member of $\{(M_i \cup M_j) \triangle E_{w=1} : \{i,j\} \subset \{1,2,3\}\}$ must be empty. That is, $(M_i \cup M_j) = E_{w=1}$ for some $\{i,j\} \subset \{1,2,3\}\}$. Thus, $E_{w=2} = M_h$ for $\{h\} = \{1,2,3\} - \{i,j\}$.

Hint for Exercise 16.2. (i) $\Rightarrow$ (ii): Let $c : E(G) \mapsto \mathbb{Z}_3$ be the unique 3-edge-coloring of G. It is easy to prove that the 2-factor induced by $c^{-1}(i) \cup c^{-1}(j)$ has only one component, for otherwise, alternating the colors along one component yields another 3-edge-coloring. So, there are three Hamilton circuits, each of which is bi-colored. If there is a fourth Hamilton circuit, it induces a different coloring: bi-colored Hamilton circuit and mono-colored chord set. This contradicts that G is uniquely 3-edge-colorable.

(ii) $\Rightarrow$ (iii): Since (G, w) is a Hamilton weighted graph, each faithful cover consists of a pair of Hamilton circuits. If (G, w) has at least two distinct faithful covers, then G has at least four distinct Hamilton circuits which contradicts (ii).

(iii) $\Rightarrow$ (iv): Let $\{H_1, H_2\}$ be the unique faithful cover of (G, w) (where each H_j is a Hamilton circuit). Let C be a component of $E_{w=1}$. If C is not a Hamilton circuit of G, then $\{H_1 \triangle C, H_2 \triangle C\}$ is another faithful even subgraph cover of (G, w) and distinct from $\{H_1, H_2\}$. This contradicts that w is a Hamilton weight with only one faithful circuit cover.

(iv) $\Rightarrow$ (i): Let $\{M_1, M_2, M_3\}$ be an arbitrary 1-factorization of G. Since (G, w) is Hamilton weighted, by Exercise 16.1, let $E_{w=2} = M_1$. Hence, $M_2 \cup M_3 = E_{w=1}$. Since $E_{w=1}$ is a Hamilton circuit, the partition of $E_{w=1}$ into a pair of 1-factors $\{M_2, M_3\}$ is unique.

Hint for Exercise 16.3. Let w and w' be two distinct Hamilton weights of G, and $\{M_1, M_2, M_3\}$ be an arbitrary 1-factorization of G. By Exercise 16.1, Both $E_{w=2}$ and $E_{w'=2} \in \{M_1, M_2, M_3\}$. Note that $E_{w=2} \neq E_{w'=2}$ since w and w' are distinct. Hence, let $E_{w=2} = M_1$ and $E_{w'=2} = M_2$. Thus, every 1-factorization of G must be in the form $\{E_{w=2}, E_{w'=2}, E(G) - E_{w=2} - E_{w'=2}\}$. It proves the uniqueness of 1-factorization of G.

Hint for Exercise 16.4. In Figure 16.25, label the exterior circuit

$$C_E = v_0 v_1 \ldots v_8 v_0,$$

and the interior circuit

$$C_I = u_0 u_4 u_8 u_3 u_7 u_2 u_6 u_1 u_5 u_0,$$

the matching between C_E and C_I is $\{v_\mu u_\mu : \mu = 0, \ldots, 8\}$.

The unique 1–factorization $\mathcal{M} = \{M_1, M_2, M_3\}$ of $P(9, 2)$ is

$$
\begin{aligned}
M_1 &= \{v_0 u_0, v_3 u_3, v_6 u_6, v_1 v_2, v_4 v_5, v_7 v_8, u_1 u_5, u_2 u_7, u_4 u_8\}, \\
M_2 &= \{v_1 u_1, v_4 u_4, v_7 u_7, v_2 v_3, v_5 v_6, v_8 v_0, u_2 u_6, u_3 u_8, u_5 u_0\}, \\
M_3 &= \{v_2 u_2, v_5 u_5, v_8 u_8, v_3 v_4, v_6 v_7, v_0 v_1, u_3 u_7, u_4 u_0, u_6 u_1\}.
\end{aligned}
$$

Note that $P(9, 2)$ has an automorphism α that $\alpha(u_i) = u_{i+1}$ and $\alpha(v_i) = v_{i+1}$ (mod 9). Furthermore,

$$(\star) \qquad \alpha(M_i) = M_{i+1} \text{ (mod 3)}.$$

Assume that $P(9, 2)$ admits a Hamilton weight w. Since $P(9, 2)$ is uniquely 3-edge-colorable, by Exercise 16.2 ((i) and (iv)), $E_{w=1}$ induces a Hamilton circuit of $P(9, 2)$. Thus, $E_{w=2}$, which is the set of all chords of the Hamilton circuit, must be one member of $\mathcal{M}$, and $E_{w=1}$ is the union of two members of $\mathcal{M}$. By $(\star)$, without loss of generality, let $M_2 = E_{w=2}$ (bold lines in Figure 16.25). That is,

$$w(e) = \begin{cases} 2 & \text{if } e \in M_2, \\ 1 & \text{if } e \in M_1 \cup M_3. \end{cases}$$

However, the graph $P(9, 2)$ has a nonhamiltonian faithful cover $\mathcal{F} = \{C_1, C_2, C_3, C_4\}$ where

$$
\begin{aligned}
C_1 &= v_2 v_3 v_4 u_4 u_0 u_5 v_5 v_6 u_6 u_2 v_2, \\
C_2 &= v_0 u_0 u_5 u_1 v_1 v_2 v_3 u_3 u_8 v_8 v_0, \\
C_3 &= v_0 v_1 u_1 u_6 u_2 u_7 v_7 v_8 v_0, \\
C_4 &= v_7 u_7 u_3 u_8 u_4 v_4 v_5 v_6 v_7.
\end{aligned}
$$

Hint for Exercise 16.5. Induction on $|V(G)|$ by contracting triangles.

Hint for Exercise 16.6. Apply Exercise 16.5.

Hint for Exercise 16.7. *Proof of (1).* It is a corollary of Theorem 3.2.1.

Proof of (2). If (G, w) has a removable circuit C with $e_0 \notin E(C)$, then, by applying Theorem 3.2.1 to $(G, w) - (C, 1)$, (G, w) has a faithful cover $\mathcal{F}$ consisting of ≥ 3 circuits, one of which is C. This contradicts that w is a Hamilton weight of G.

Proof of (3). So, every faithful cover of $(G - e_0, w)$ must be a circuit chain joining the endvertices of e_0 (otherwise, any circuit not in the chain is removable).

Hint for Exercise 16.8. By Exercise 16.7-(3), every faithful cover of $(G - e, w)$ must be a circuit chain joining the endvertices of e. Furthermore, each faithful cover (circuit chain) induces a 3-edge-coloring of $\overline{G - e}$: edges of $(E_{w=2} - e)$ are colored with yellow, while $E_{w=1}$ are alternately colored with red and blue (where each circuit of $\mathcal{F}$ is either red-yellow bi-colored or blue-yellow bi-colored).

Proof of (1). If $(G - e, w)$ has two distinct faithful covers $\mathcal{F}_1, \mathcal{F}_2$, then there are two distinct 3-edge-colorings. Hence, $E_{w=1}$ must have at least two components (otherwise, the colorings must be the same). This contradicts Exercise 16.5 that $E_{w=1}$ induces a Hamilton circuit.

Proof of (2). This is trivial.

Proof of (3). $|\mathcal{F}| = 2$ if and only if w is a Hamilton weight of $(G - e, w)$. By Theorem 16.4.1, $(\overline{G - e}, w) \in \langle \mathcal{K}_4 \rangle^{HW}$. By Lemma 16.3.12-(2) and (4), e is untouched.

Proof of (4). By Lemma 16.3.12-(2) and (4) again, e is touched if and only if $(\overline{G - e}, w) \notin \langle \mathcal{K}_4 \rangle^{HW}$. And by Theorem 16.4.1 again, $(\overline{G - e}, w) \notin \langle \mathcal{K}_4 \rangle^{HW}$ if and only if w is not a Hamilton weight of $\overline{G - e}$.

Hint for Exercise 16.9. By Proposition B.2.15, both G and $\overline{G - e}$ are Petersen minor free as long as one of them is planar (note, $\langle \mathcal{K}_4 \rangle$-graphs are planar). Hence, by Theorem 3.2.1, both (G, w) and $(\overline{G - e}, w)$ have faithful circuit covers.

(1) $\Rightarrow$ (2): Since $(\overline{G - e}, w)$ is Hamilton weighted and Petersen minor free, by Theorem 16.4.1 (or by Exercise 16.8-(3), Lemma 16.3.12-(2) and (4)), $(\overline{G - e}, w) \in \langle \mathcal{K}_4 \rangle_2^{HW}$.

(2) $\Rightarrow$ (1): If (G, w) has a faithful cover consisting of ≥ 3 circuits, then the one not containing e is removable in (G, w). So, (G, w) is Hamilton weighted. By Theorem 16.4.1 (or by Lemma 16.3.13), $(G, w) \in \langle \mathcal{K}_4 \rangle_2^{HW}$.

Hint for Exercise 16.10. (1) is obvious because of the maximality of $|\mathcal{F}|$. (2) is a corollary of (1) by applying Exercise 3.3-(b) and Exercise 16.2.

Hint for Exercise 16.11. By Theorem 1.2.4-(1), G is cubic. Let $(J, w_J) = (\overline{(C_1 \cup C_2)}, w_{\{C_1, C_2\}})$. By the choice of $\mathcal{F}$, (J, w_J) is a Hamilton weighted graph. If $(J, w_J) \in \langle \mathcal{K}_4 \rangle^{HW}$, then (1) holds. So, assume that neither (1) nor (2) holds, that is, $(J, w_J) \notin \langle \mathcal{K}_4 \rangle^{HW}$ and is 3-connected. By Theorem 16.4.1, J must contain a Petersen minor.

Appendix A

Hint for Exercise A.4. Use the Euler formula (Theorem A.1.1).

Hint for Exercise A.5. Apply Theorem A.1.12.

Hint for Exercise A.6. The case of $\lambda_o = 3$ is trivial. Let $\lambda_o = 5$ and G be a smallest counterexample to the exercise. First show that G has no 2-edge-cut (by the fact that G is a smallest counterexample, and, apply Lemma A.2.14). Then, by Theorem A.1.6 and Lemma A.2.15, G has two edge-disjoint parity subgraphs.

Hint for Exercise A.8. "$\Rightarrow$": Assume that G is an r-graph. Let T be an odd edge-cut of G. Let X and Y be two components of $G - T$. That is, $T = (X, Y)$. Contracting one component Y of $G - T$, we obtain a graph G' consisting of all odd degree vertices since one of them, created by the contraction, is of degree $|T|$, while all others, vertices of X, remain of degree r. Since the number of odd degree vertices in G' must be even, we have that $|X|$ is odd. By the definition of r-graph, the edge-cut T must be of order at least r. Hence, every odd edge-cut of G must be of size at least r.

"$\Leftarrow$": Assume that the odd-edge-connectivity of G is r. Let $X \subseteq V(G)$ be of odd order. With a similar argument as aboves by considering the parity of the number of odd vertices in a graph, we can prove that the edge-cut $(X, V(G) - X)$ is of odd size. Hence, by the definition of odd-edge-connectivity, the edge-cut $(X, V(G) - X)$ must be of size at least r. Hence, G is an r-graph.

Hint for Exercise A.9. By Corollary A.1.10, G has a 2-factor F such that $F \cap T \neq \emptyset$ for every 3-edge-cut T of G. Thus, G/F has no 3-edge-cut, and, therefore, is of odd-edge-connectivity at least 5.

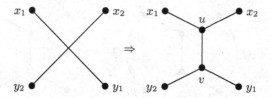

Figure D.5 *Exercise B.2, a pair of crossing edges e_1, e_2*

Appendix B

Hint for Exercise B.2. Draw G on the plane with only one pair of crossing edges $e_1 = x_1 y_1$ and $e_2 = x_2 y_2$ (see Figure D.5). Construct a planar graph G' by adding two new vertices u and v, and replacing e_1, e_2 with five new edges $x_1 u, x_2 u, uv, vy_1, vy_2$. By the 4-color theorem (Theorem B.1.2), G' has a 3-edge-coloring $c' : E(G') \mapsto \mathbb{Z}_3$. Let $c'(uv) = 0$.

Case 1. $c'(x_1 u) = c'(y_1 v) = 1$, $c'(x_2 u) = c'(y_2 v) = 2$. For this case, The coloring c' is also a proper 3-edge-coloring of G if one colors e_1 with 1 and e_2 with 2.

Case 2. $c'(x_1 u) = c'(y_2 v) = 1$, $c'(x_2 u) = c'(y_1 v) = 2$. Let Q be the component (circuit) of $c'^{-1}(1) \cup c'^{-1}(2)$ containing the vertex u. (See Figure D.6.)

If $v \in Q$, let P be the segment of Q joining x_1 and y_2, but not containing u and v (note that G' is planar). Let c be the coloring of G obtained from c' such that $c(e) = c'(e)$ if $e \notin E(P) \cup \{e_1, e_2\}$, $c(e) = 3 - c'(e)$ if $e \in E(P)$, $c(e_1) = c(e_2) = 2$.

If $v \notin Q$, let $c'' : E(G') \mapsto \mathbb{Z}_3$ obtained from c' by alternating the colors along the circuit Q. The new coloring c'' is Case 1 (solved).

Hint for Exercise B.3. It is not hard to see that G is either planar or of crossing number 1. Thus, this exercise is a corollary of Exercise B.2.

Hint for Exercise B.4. Let $c \mapsto \{0, 1, 2\}$ and assume that $c(e) \in \{1, 2\}$ for each $e \in E(M)$. It is sufficient to show that $G - E(M)$ is an even 2-factor. For each component C of $G - E(M)$, the edge-cut $T = (V(C), V(G) - V(C))$ contains no 0-colored edges. Hence, by Lemma B.1.3, the edge-cut T is of even size. Therefore, the circuit C is of even length.

Hint for Exercise B.6. Construct H from G by identifying all degree 1

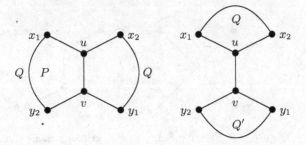

Figure D.6 *Exercise B.2, changing the coloring c'*

vertices as a single vertex w. By Lemma A.2.7, for each $\alpha \in \mathbb{Z}_2^2 - \{(0,0)\}$, $H - c^{-1}(\alpha)$ is an even subgraph of H and $c^{-1}(\alpha)$ is a parity subgraph of H. Then apply Lemma A.2.14 to the edge-cut $E(w)$.

Hint for Exercise B.7. By Exercise B.5, let G be a snark and $T = \{e_1, \ldots, e_4\}$ be a smallest cyclic edge-cut. Let Q_1, Q_2 be the components of $G - T$. For each $\{i, j\} = \{1, 2\}$, let $G_i = G - E(Q_j)$. (See Figure D.7-(a).)

A 3-edge-coloring $c : E(G_i) \mapsto \mathbb{Z}_3$ is denoted by a 4-tuple: $\alpha_1 \alpha_2 \alpha_3 \alpha_4$ where e_μ is colored α_μ. By Exercise B.6, there are only four possible types of 3-edge-colorings of G_i ($i = 1, 2$):

$$1111, 1122, 1212, 1221$$

(under permutations of colors). Since G is a snark, G_1 and G_2 cannot have the same type of 3-edge-coloring.

Assume that G_1 has at most one *type* of 3-edge-coloring, say, 1111 or 1122 (under permutations of the subscripts of $\{e_1, e_2, e_3, e_4\}$). Let P', P'' be two edge-disjoint paths in G_2 joining $\{e_1, e_2\}$ and $\{e_3, e_4\}$ (since G has no non-trivial 3-edge-cut). (See Figure D.7-(b).) If the 3-edge-coloring of G_1 is the type 1122 or G_1 is not 3-edge-colorable, then $\overline{G_1 \cup P' \cup P''}$ is a snark. So, assume that the 3-edge-coloring of G_1 is the type 1111. Since the graph G has no non-trivial edge-cut of order less than four, let R be a path in Q_2 joining P' and P''. Thus, $\overline{G_1 \cup P' \cup P'' \cup R}$ is a snark. (See Figure D.7-(c).)

So, we assume that each of G_1 and G_2 has precisely two types of 3-edge-colorings. Without loss of generality, let 1122, 1212 be the types of the 3-edge-colorings of G_1 and let 1221, 1111 be the types of the 3-edge-colorings of G_2. By the fact that G_1, G_2 do not have the same type of coloring, it is not hard to see that, under the coloring of the type 1111

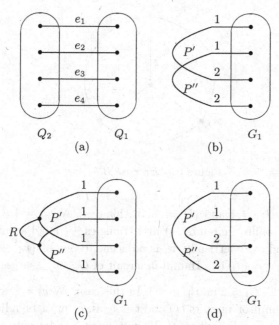

Figure D.7 *Exercise B.7*

of G_2, the (1,2)-bi-colored paths P' and P'' in G_2 must join e_1 and e_4, e_2 and e_3, respectively. Thus, $\overline{G_1 \cup P' \cup P''}$ is a snark since G_1 has only two types of 3-edge-colorings: 1122 and 1212. (See Figure D.7-(d).)

Hint for Exercise B.8. Consider a smallest counterexample to the statement. It is obvious that G is triangle-free, for otherwise, a Hamilton circuit of the contracted graph G/T (T is a triangle) can be easily extended to a Hamilton circuit of G.

Let C be a longest circuit of G and $W = V(G) - V(C)$. We only consider the case of $|V(G)| = 10$ and $|V(C)| = 9$, all other cases are similar and simpler. Let $C = v_0 \ldots v_8 v_0$ and $\{w\} = W$. Let

$$N(w) = \{v_{i_0}, v_{i_1}, v_{i_2}\} \text{ with } 0 = i_0 < i_1 < i_2 < 8.$$

(See Figure D.8.)

It is obvious that, for every pair $\mu, \nu \in \mathbb{Z}_3$,

($\star$) $\qquad\qquad v_{i_\mu+1}v_{i_\nu+1}, \ v_{i_\mu-1}v_{i_\nu-1} \notin E(G)$

for otherwise $v_{i_\mu+1}v_{i_\nu+1} \ldots v_{i_\mu}wv_{i_\nu} \ldots v_{i_\mu+1}$ is a Hamilton circuit.

Let $I_\mu = \{v_j : i_\mu < j < i_{\mu+1}\}$ for each $\mu \in \mathbb{Z}_3$. Without loss of generality, let $|I_0| \leq |I_1| \leq |I_2|$.

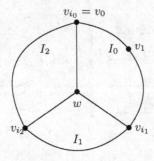

Figure D.8 *Exercise B.8*

Case 1. $|I_0| = |I_1| = 1$ and $|I_2| = 4$. In this case, $N(w) = \{v_0, v_2, v_4\}$. By avoiding Hamilton circuits $(\star)$ and triangles, v_1 is adjacent to one of $\{v_6, v_7\}$. And symmetrically, v_3 is adjacent to one of $\{v_6, v_7\}$. Then $v_0 v_8 v_5 v_4 v_3 v_6 v_7 v_1 v_2 w v_0$ is a Hamilton circuit of G.

Case 2. $|I_0| = 1$, $|I_1| = 2$ and $|I_2| = 3$. In this case, $N(w) = \{v_0, v_2, v_5\}$. By avoiding Hamilton circuits $(\star)$ and triangles, v_1 must be adjacent to v_7, and, v_8 must be adjacent to v_3. Then v_4 must be adjacent to v_6. But this creates a triangle.

Case 3. $|I_0| = |I_1| = |I_2| = 2$. In this case, $N(w) = \{v_0, v_3, v_6\}$. By avoiding Hamilton circuits $(\star)$ and triangles, $v_1 v_5 \in E(G)$. Symmetrically, $v_2 v_7, v_4 v_6 \in E(G)$. So, $G = P_{10}$ (see Figure B.12).

Appendix C

Hint for Exercise C.1. Contract one edge of a 2-edge-cut. A nowhere-zero k-flow of the contracted graph can be extended to the original graph.

Hint for Exercise C.13. Reverse the direction, change the sign (by Exercise C.12), and add k to each edge on the circuit. That is, $f'(e) = f(e)$ if $e \notin E(C)$, and $= k - f(e)$ otherwise.

Hint for Exercise C.15. By Exercise C.14.

Hint for Exercise C.16. By Lemma C.1.7 (or Exercise C.8), let C be the even subgraph that $E(C) = \{e : f(e) \equiv 1 \pmod 2\}$. Let (D, g) be a 2-flow of G with $supp(g) = E(C)$ (by Theorem C.1.8 or Exercise C.6).

Then, define

$$f_1 = \frac{f - g}{2} \text{ and } f_2 = \frac{f + g}{2}.$$

Hint for Exercise C.17. The graph K_4.

Hint for Exercise C.19. By Theorem C.2.6.

Hint for Exercise C.20. Let (D, f) be a positive k-flow of G with $supp(f) = E(G) - \{e\}$. By Exercise C.19, $D(G)$ is strongly connected. Let P be a directed path of $D(G)$ from x to y where $e = xy$ and let (D, g) be a non-negative 2-flow of G with $supp(g) = E(P) \cup \{e\}$. Thus $(D, g + f)$ is a positive $(k + 1)$-flow of G.

Hint for Exercise C.21. Let D be an orientation of G. For each $e \in E(G) - E(T)$, let C_e be the unique circuit of G contained in $T \cup \{e\}$, and let (D, f_e) be a 2-flow of G with $supp(f_e) = E(C_e)$ and $f_e(e) = 1$. Then $(D, \sum_{e \in E(G) - E(T)} c(e) f_e)$ is the Γ-flow that we wanted.

Hint for Exercise C.22. Consider $T = \{e_1, e_2, e_3\}$ with $|T| = 3$. Let (D, f) be a positive k-flow of G. The opposite orientation of D is denoted by $\tilde{D}$. Assume that neither D nor $\tilde{D}$ agrees with D_T at all edges of T. Without loss of generality, assume that D agrees with D_T at only one edge e_3 of T and but not with other two, $e_1, e_2 \in T$. Since (D, f) is a positive flow of G and neither $[Q_1, Q_2]_{D_T}$ nor $[Q_2, Q_1]_{D_T}$ is empty, it is not hard to see that e_1 and e_2 have opposite directions under D. Thus, either $f(e_1) = f(e_2) + f(e_3)$ or $f(e_2) = f(e_1) + f(e_3)$. By Theorem C.2.4, there is a directed circuit C containing both e_1, e_2. Complete the proof by applying Exercise C.13.

Hint for Exercise C.23. The "if" part: using Exercise C.15, each vertex is adjacent to precisely one weight 2 edge and two weight 1 edges, and is either the heads of both weight 1 arcs or the tails of both weight 1 arcs. Thus, each component of the subgraph H_1 of G induced by weight 1 edges is an even length circuit. Alternately color the edges in each component of H_1 and color the weight 2 edges with the third color.

The "only if" part: the subgraph induced by red and yellow edges is an even subgraph C_1 of G and the subgraph induced by red and blue edges is an even subgraph C_2 of G. By Theorem C.1.8, let (D, f_i) be a 2-flow of G with $supp(f_i) = E(C_i)$ for $i = 1, 2$. Then $(D, 2f_1 + f_2)$ is a nowhere-zero 4-flow of G.

Hint for Exercise C.24. Two alternative proofs.
Proof 1. Properly split the vertices with degrees greater than three

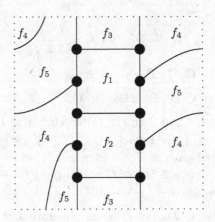

Figure D.9 *The dual (on the torus) of this graph contains a K_5*

so that the suppressed graph of the resulting graph is hamiltonian and cubic, hence, is 3-edge-colorable. Apply Exercise C.23.

Proof 2. Let C_1 be a Hamilton circuit of G and P be a Hamilton path contained in C_1. For each $e \in E(G) - E(P)$, let C_e be the unique circuit contained in $G[P \cup \{e\}]$. Let C_2 be the symmetric difference of the circuits C_e for all $e \in E(G) - E(C_1)$. Let (D, f_i) $(i = 1, 2)$ be a 2-flow of G with $supp(f_i) = E(C_i)$. Then $(D, f_1 + 2f_2)$ is a nowhere-zero 4-flow of G.

Hint for Exercise C.25. Let H be a Hamilton circuit of G. Let F_i be the set of all faces inside of H and F_o be the set of all faces outside of H. In the dual graph $G^\star$, each F_i, F_o induces a tree which is 2-vertex-colorable in $G^\star$. So, $G^\star$ is 4-vertex-colorable, and therefore, G is 4-face-colorable.

Hint for Exercise C.26. See Figure D.9. The graph embedded on the torus. The graph itself is a permutation graph containing a 4-circuit (hence it has a Hamilton circuit). Note that, on the torus, its dual contains a K_5 which is not 4-vertex-colorable.

Hint for Exercise C.27. "$\Rightarrow$": Let $\{C_1, C_2\}$ be a 2-even subgraph cover of G. Let $\mathcal{F}_i$ be a circuit decomposition of C_i $(i = 1, 2)$ so that each pair of circuits of $\mathcal{F}_i$ may intersect each other but not cross each other on the sphere. For each $i = 1, 2$, let $f_i : F(G) \mapsto \mathbb{Z}_2$ where $f_i(F) = 1$ if F ($\in F(G)$) is a face of G contained in the interior of an odd number of circuits of $\mathcal{F}_i$, and, $f_i(F) = 0$ otherwise. Then $f = (f_1, f_2) : F(G) \mapsto \mathbb{Z}_2 \otimes \mathbb{Z}_2$ is a 4-face coloring of G.

"$\Leftarrow$": Let $F(G)$ be colored red, blue, yellow and white. The symmetric difference of the boundary of all red and blue (or red and yellow) faces is an even subgraph C_{RB} (or C_{RY}, respectively). It is not hard to see that $\{C_{RB}, C_{RY}\}$ is a 2-even subgraph cover of G.

Hint for Exercise C.28. Since G admits a nowhere-zero 4-flow, $E(G)$ has a partition into three parity subgraphs P_1, P_2, P_3. Without loss of generality, let $e_1 \in P_1$ and let $e_2 \in P_1 \cup P_2$. Let $C_1 = P_1 \cup P_2$ and $C_2 = P_2 \cup P_3$. Now, e_1 is covered only by the even subgraph C_1 and e_2 is also covered by C_1 (and probably C_2). Let (D, f_1) be a 2-flow with support $E(C_1)$ and (D, f_2) be a 2-flow with support $E(C_2)$. Then one of $(D, f = f_1 + 2f_2)$, $(D, f = -f_1 + 2f_2)$ is a nowhere-zero 4-flow of G with $f(e_1), f(e_2) \in \{1, -1\}$. Apply Exercise C.12 to change the signs of possible negative weights of e_1 and e_2.

Hint for Exercise C.29. We can obtain a bridgeless cubic graph with at most eight vertices by splitting the vertices with degree greater than three (by Theorem A.1.14) and replacing subdivided edges by single edges. Note that the smallest snark is the Petersen graph (Proposition B.1.11).

Hint for Exercise C.30. (1) $\Rightarrow$ (2): By Exercise C.15, let (D, f) be a nowhere-zero mod-3-flow with $f(e) = 1$ for every edge $e \in E(G)$. The orientation D is a mod-3-orientation.

(2) $\Rightarrow$ (1): Let $f' : E(G) \mapsto \{1\}$. Then (D, f) is a nowhere-zero mod-3-flow.

Hint for Exercise C.31. (A proof using Exercise C.30) (1) $\Rightarrow$ (2): By Exercise C.30, let D be a mod-3-orientation of G. Since G is cubic, it is obvious that G has a bi-partition: the set of all heads and the set of all tails.

(2) $\Rightarrow$ (1): Let (A, B) be a bi-partition of G. Let D be the orientation of all edges from A to B and $f : E(G) \mapsto \{1\}$. Then (D, f) is a nowhere-zero mod-3-flow.

Hint for Exercise C.32. This is a dual problem of Exercise C.31 for planar graphs. The following is an alternative proof without using any definition or technique of integer flows.

"$\Rightarrow$": Let $v \in V(G)$ and v be colored 0. Then all neighbors of v are colored either 1 or 2. Since G is planar and triangulated, the neighbors

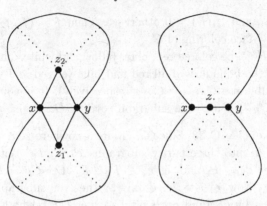

Figure D.10 *3-vertex-coloring of an even planar triangulation*

of v induce a circuit around v. Note that a bi-colored circuit is of even length. Thus, v is of even degree and, therefore, G is an even graph.

"$\Leftarrow$": Induction on $|V(G)|$. Let xyz_1x and xyz_2x be two adjacent triangles with xy as the edge on their common boundaries. (See Figure D.10.) Construct G' by deleting the edge xy, xz_2, yz_2 and identifying z_1, z_2 as a single vertex z. The new graph G' remains as an even planar triangulation. By induction, G' is 3-vertex-colorable. Let z be colored 0. The circuit C of G' consisting of neighbors of z is alternately bi-colored $1, 2$. Since G is even, the lengths of the two subpaths of C joining x, y are all of odd length. That is, x and y are colored differently, say 1, 2, respectively. Color z_1, z_2 with 0 and the colors of the other vertices of G' remain the same. The coloring we obtained is a 3-vertex-coloring of G.

Hint for Exercise C.33. For a facial circuit $x_1 \ldots x_{3k}x_1$ with $k > 1$, add a triangle $x_1x_3x_5x_1$. Repeat this procedure until the graph becomes triangulated. Then apply Exercise C.32.

Hint for Exercise C.34. "$\Leftarrow$": Apply Exercise C.32. "$\Rightarrow$": For a facial circuit $x_1 \ldots x_rx_1$ with $r \geq 4$, add a chord x_ix_j if $i \neq j \pm 1 \pmod{r}$ and x_i, x_j are colored differently. For a bi-colored facial circuit of length 4, add a new vertex inside the face and four new edges joining the new vertex and the vertices of the circuit. Repeat this procedure until G is triangulated. Apply Exercise C.32.

Hint for Exercise C.35. "$\Leftarrow$": By Exercise C.31. "$\Rightarrow$": Let D be a mod-3-orientation of G. If neither $E^+(v)$ nor $E^-(v)$ is empty, for some vertex v, then split a pair of edges $e' \in E^+(v)$ and $e'' \in E^-(v)$ away from v (by maintaining the connectivity). Repeat this procedure until

$E^+(v) = \emptyset$ or $E^-(v) = \emptyset$ for each v. For each vertex v with degree greater than three, split three edges away from v (by maintaining the connectivity). Repeat this procedure until G becomes a subdivision of a bipartite cubic graph.

Hint for Exercise C.36. Assume that G has a vertex x with $d(x) \geq 6$ or $d(x) = 4$. By Theorem A.1.16, two edges of $E(x)$ can be split away from x so that the suppressed graph of the resulting graph remains odd 5-edge-connected. Then apply Exercise C.1 for resulting 2-cuts.

Hint for Exercise C.38. Use the same argument of the second proof of Theorem 8.1.3.

Hint for Exercise C.39. By Theorem 8.2.2, a graph admits a nowhere-zero 4-flow if and only if it can be covered by two even subgraphs, say C_1 and C_2. Thus, by Theorem C.1.8, we have two 2-flows (D, f_1) and (D, f_2) with supports on $E(C_1)$ and $E(C_2)$. If the edge e is contained in only one even subgraph C_1 (or C_2, symmetrically), then $(D, hf_1 + h'f_2)$ is a nowhere-zero mod-4-flow of G where $h' \in \mathbb{Z}_4 - \{0, h, -h\}$. Assume that e is contained in both C_1 and C_2. Let $C_3 = C_1 \triangle C_2$. Now we get an even subgraph cover $\{C_3, C_1\}$ where $e \in E(C_1)$ and $e \notin E(C_3)$ which is the case we have done already.

Hint for Exercise C.40. Let x and y be the endvertices of a Hamilton path of G and let $G' = G \cup \{e = xy\}$. By Exercise C.39, let (D, f) be a positive 4-flow of G' with $f(e) = 1$. Without loss of generality, we assume that e is an arc from x to y in $D(G)$. Note that each edge-cut of G' containing e must contain at least three edges since G is bridgeless. By Equation (C.2) of Theorem C.2.6, since $f(e) = 1$, there is a directed path P from x to y in $D(G)$. Thus, $(D, f + f')$ is a nowhere-zero 5-flow of G where (D, f') is a 2-flow of G' with support $E(P) \cup \{e\}$ and non-negative at each edge of P.

Hint for Exercise C.41. Similar to Exercise C.40.

Hint for Exercise C.42. Let (D, f) be a positive 4-flow of $G - \{e\}$. Then, by Equation (C.2) of Theorem C.2.6, and Theorem A.1.3, let P be a directed path joining the endvertices of e in $D(G - \{e\})$. Let (D, f') be a non-negative 2-flow with $supp(f') = E(P) \cup \{e\}$. Then $(D, f + f')$ is a nowhere-zero 5-flow of G.

Glossary of terms and symbols

All graphs considered in this book may contain parallel edges or loops.

For most basic terminology and notation in graph theory, we follow some popularly used textbooks, such as, [18], [19], [34], [41], [92], [242], etc.

acyclic: A graph is acyclic if it contains no circuit.

admissible forbidden system: A forbidden system $\mathcal{P}$ is admissible if $|P \cap T| \leq \frac{1}{2}|T|$ for every forbidden part $P \in \mathcal{P}$ and every edge-cut T of G.

admissible weight: A weight $w : E(G) \mapsto \mathbb{Z}^*$ is admissible if, for each edge-cut T and each edge $e \in T$, $w(T) \geq 2w(e)$.

attachment: See *edge-attachment* and *Tutte bridge attachment*

arc: A directed edge is called an arc.

block, *block tree, leaf block* and *trivial block*: A block is a maximal 2-connected subgraph of a graph G. A trivial block is a block consisting of one edge. The block tree $\mathcal{T}$ of a graph G is a bipartite graph with the vertex set $\mathcal{B} \cup C$ where $\mathcal{B}$ is the set of all blocks of G and C is the set of all cut-vertices of G. For $c \in C$ and $B \in \mathcal{B}$, c is adjacent to B in $\mathcal{T}$ if the cut-vertex c is contained in the block B. The block corresponding to a leaf in the block tree is called a leaf block (the block containing at most one cut-vertex).

bond: See edge-cut.

boundary of a face: Let G be embedded on a surface. The set of all edges of G lying on the topological boundary of a face of G is called the boundary of the face.

bridge and *cut-edge*: A bridge of a graph G is an edge $e \in E(G)$ such that e is not contained in any circuit of G. A bridge is also sometimes called a cut-edge.

Cayley graph: Let Γ be a group and $S \subset \Gamma$ such that $1 \notin S$ and $\alpha \in S$ if and only if $\alpha^{-1} \in S$. The Cayley graph $G(\Gamma, S)$ is the graph with the vertex set Γ such that two vertices x, y are adjacent in $G(\Gamma, S)$ if and only if $x = y\alpha$ for some $\alpha \in S$.

CCD: CCD is the abbreviation of compatible circuit decomposition.

Chinese postman tour: See *postman tour*.

chord: Let $C = v_1 \ldots v_r v_1$ be a circuit in a graph G. An edge e of $G - E(C)$ with both endvertices in $V(C)$ is called a chord of the circuit C.

circuit and *k-circuit*: A circuit is a connected 2-regular graph. A circuit of length k is sometimes called a k-circuit.

circuit chain (for *circuit cover problems*): Let x and y be two vertices of a graph G. A family $\mathcal{C}$ of circuits of G is called a circuit chain joining x, y if $\mathcal{C} = \{C_1, \ldots, C_p\}$ such that
  (1) $x \in V(C_1)$ and $y \in V(C_p)$,
  (2) $V(C_i) \cap V(C_j) \neq \emptyset$ if and only if $i = j \pm 1$.
The integer p is the length of the chain $\mathcal{C} = \{C_1, \ldots, C_p\}$.

circuit chain (for *circuit decomposition problems*): Let $v \in V(G)$. A sequence of edge-disjoint circuits $\mathcal{C} = \{C_1, \ldots, C_k\}$ $(k \geq 2)$ is a *circuit chain closed at v* if
  (1) for each $i, j \in \{1, \ldots, k\}$ with $i \neq j$, $[V(C_i) \cap V(C_j)] - \{v\} \neq \emptyset$ if and only if $j - i = \pm 1$,
  (2) $v \in V(C_1) \cap V(C_k)$.
The integer k is the *length* of the circuit chain $\{C_1, \ldots, C_k\}$.

circuit cover, *cycle cover* and *even subgraph cover*: A family $\mathcal{F}$ of even subgraphs (or circuits) of a graph G is an even subgraph cover (or a circuit cover, respectively) of G if every edge of G is contained in some member of $\mathcal{F}$. An even subgraph cover (or circuit cover) consisting of k elements is called a k-even subgraph cover (or k-circuit cover, respectively).

circuit double cover and *even subgraph double cover*: A circuit cover (or even subgraph cover) which covers each edge of a graph precisely twice is called a circuit double cover (or even subgraph double cover, respectively).

circuit k-cover and *even subgraph k-cover*: An even subgraph cover (or circuit cover) covers each edge of a graph G precisely k-times is called an even subgraph k-cover (or circuit k-cover, respectively).

circuit $(1, 2)$-cover and *even subgraph $(1, 2)$-cover*: A circuit cover or even subgraph cover covers each edge of a graph once or twice.

circular embedding: An embedding of a graph G on a surface Σ is circular if the boundary of every face is a circuit. (A circular 2-cell embedding is also called a *closed 2-cell embedding* in many articles.)

circumference: The circumference of a graph is the length of the longest circuit in a graph.

coloring: A *vertex-coloring* (or *edge-coloring*, *face-coloring*) is a mapping $c : V(G) \mapsto \{1, 2, \ldots, k\}$ (or $E(G) \mapsto \{1, 2, \ldots, k\}$, $F(G) \mapsto \{1, 2, \ldots, k\}$, respectively) such that each pair of adjacent (or incident) elements are colored differently. (This kind of coloring is called *proper coloring*. If no confusion occurs, most colorings, for faces, edges or vertices, mentioned in this book are proper colorings.) A graph G is *k-vertex-colorable* (or *k-edge-colorable*, *k-face-colorable*) if $V(G)$ (or $E(G)$, $F(G)$, respectively) can be colored with at most k colors.

complete directed graph: A directed graph is complete if, for every pair of vertices x and y, there is an arc from x to y and an arc from y to x.

compatible circuit decomposition and *compatible even subgraph decomposition*: Let G be an eulerian graph and $\mathcal{P}$ be a forbidden system of G. A circuit (even subgraph) decomposition $\mathcal{F}$ is compatible with respect to $\mathcal{P}$ if $|E(C) \cap P| \leq 1$ for every $C \in \mathcal{F}$ and every $P \in \mathcal{P}$. A compatible *circuit* decomposition is abbreviated as CCD.

component: A component of a graph G is a maximal connected subgraph of G.

connected, *k-connected* and *k-edge-connected*: A graph G is connected if there is a path joining each pair of vertices of G. A graph G is k-connected (or k-edge-connected) if each vertex-cut (or edge-cut, respectively) of G contains at least k elements.

contra pair for *faithful cover*: Let w be an admissible, eulerian weight of a graph G. If G does not have a faithful circuit cover with respect to w, then (G, w) is called a *contra pair*.

An admissible, eulerian $(1, 2)$-weighted graph is a *minimal contra pair* if it is a contra pair but $(G - e, w)$ has a faithful cover for every weight 2 edge e and (G, w) has no removable circuit.

An admissible, eulerian $(1, 2)$-weighted graph is a *critical contra pair* if it is a contra pair and (G, w) has no removable circuit.

contra pair for *compatible decomposition*: Let $\mathcal{P}$ be an admissible forbidden system of an eulerian graph G. $(G, \mathcal{P})$ is called a *contra pair* if it has no compatible circuit decompositions.

contractible: See *contraction*.

contraction and *edge-contraction*: The operation of contraction (in order to distinguish it from the vertex-shrinking, it is sometimes called *edge-contraction*) is an operation of a graph G by contracting all edges of a subset E' of $E(G)$. The graph obtained from G by the above operation is denoted by G/E'. (Note that the contraction does not create a new loop.)

Let G and H be two graphs. We say G is *contractible* to H if there is a subset $E' \subseteq E(G)$ such that $H = G/E'$.

crossing number: Let G be a graph drawn on a sphere (or plane). A pair of edges intersecting with each other not at their endvertices is called a crossing pair of the drawing. The crossing number of G, denoted by $c(G)$, is the smallest integer k such that there is a drawing of G on the sphere with at most k crossing pairs.

critical edge-cut: Let G be a graph and w be an admissible, eulerian weight of G. A critical edge-cut T of (G, w) is an edge-cut T with an edge $e \in T$ such that $w(e) = w(T - \{e\})$.

cubic graph: A 3-regular graph is also called a cubic graph.

cut-edge: See *bridge*.

cut-vertex: A vertex-cut consisting of only one vertex is called a cut-vertex.

cycle: An even subgraph is a graph such that the degree of each vertex is even. In this book, it is called an *even subgraph*.

cyclic edge-cut: An edge-cut T of G is cyclic if no component of $G - T$ is acyclic.

cyclically k-edge-connected: A graph G is cyclically-k-edge-connected if every cyclic edge-cut of G contains at least k edges.

degree of a face: A graph G is embedded on a surface (a 2-cell embedding). The degree of a face F of G is the length of the closed walk around the boundary. (For a circular 2-cell embedding, the degree of a face is simply the number of edges around the boundary.)

degree of a vertex: Let H be a subgraph of a graph G and v be a vertex of G. The degree of v in H, denoted by $d_H(v)$, is the number of edges of H incident with v. If $v \notin V(H)$, we simply say $d_H(v) = 0$. If $H = G$, we sometimes simply write $d_G(v) = d(v)$ if no confusion occurs.

directed circuit: See *directed path*.

directed even subgraph: See *directed path* or *even subgraph*.

directed path, *directed circuit* and *directed even subgraph*: A directed path P of a directed graph D is a sequence of vertices $v_0 \ldots v_p$ such that $v_i v_{i+1}$ is an arc of D and no vertex is repeated in the sequence. A directed circuit is a closed directed path $v_0 \ldots v_p v_0$. A directed even subgraph H is the union of several arc-disjoint directed circuits.

dominating circuit: A circuit C of a graph G is a dominating circuit if $G - V(C)$ is an independent set.

dual: Let a graph G be embedded on a surface Σ. The dual of G on Σ, denoted by G^*, is the graph with vertex set $F(G)$ (which is the face set of G) and edge set $E(G)$ (the same): an edge e of G^* is incident with vertices $F_i, F_j \in V(G^*)$ if and only if the corresponding faces F_i, F_j in G lie on two sides of the edge e.

edge-attachment: Let G_1, G_2 be two subgraphs of G. An edge e of $\overline{G_1}$ is called *an edge-attachment of G_2 in G_1* if e corresponds to an induced path $P = v_0 \ldots v_p$ in G_1 such that $V(P) - \{v_0, v_p\}$ contains some vertex of G_2.

edge-coloring, proper edge-coloring: $c : E(G) \mapsto S$ is an edge-coloring (or proper edge-coloring) if $c(e) \neq c(e^*)$ for every pair of incident edges e and e^*.

edge-connected: See *connected*.

edge-contraction: See *contraction*.

edge-cut, *bond* and *k-edge-cut*: Let $\{V_1, V_2\}$ be a partition of $V(G)$. The set of edges joining V_1 and V_2, denoted by $[V_1, V_2]$ or $\delta(V_1)$, is called an edge-cut of G. An edge-cut consisting of k edges is called a k-edge-cut.

An edge-cut T of a graph G is *minimal* if no proper subset of T is an edge-cut of G. An edge-cut T of a graph G is *trivial* if one component of $G - T$ is a single vertex. A minimal edge-cut is called a *bond*.

edge-depth (of a circuit cover): For an even subgraph cover $\mathcal{F}$ of G and an edge $e \in E(G)$, $ced_{\mathcal{F}}(e)$ is the number of members of $\mathcal{F}$ containing e. $ced_{\mathcal{F}}(G)$ is the maximum of $ced_{\mathcal{F}}(e)$ for all edges $e \in E(G)$, and is called the edge-depth of G with respect to $\mathcal{F}$.

edge transitive: A graph $G = (V, E)$ is edge transitive if, for every pair of edges e_1, e_2 of G, there is an automorphism $\pi \in Auto(G)$ such that $\pi(e_1) = e_2$.

embedding: Let Σ be a surface and G be a graph. The graph G is *embedded* on the surface Σ if G is drawn on the surface Σ such that the edges only meet at their common ends.

essentially k-edge-connected: A graph G is essentially k-edge connected if every edge-cut T with $|T| < k$ is trivial.

Euler tour: A closed trail T is called an Euler tour of a graph G if T passes through each edge of G precisely once.

eulerian graph: An eulerian graph is a connected even subgraph.

eulerian weight: A weight $w : E(G) \mapsto \mathbb{Z}^*$ is eulerian if the total weight of each edge-cut is even.

even circuit decomposition: Let G be a 2-connected eulerian graph. A circuit decomposition $\mathcal{F}$ of G is called an even circuit decomposition if each member of $\mathcal{F}$ is of even length.

even component: See *odd component*.

even 2-factor: A 2-factor of a cubic graph is even if every component is an even length circuit.

even subgraph and *directed even subgraph*: An even subgraph is a subgraph such that the degree of each vertex is *even*. In much of the literature on this topic, it is also called a *cycle*. A directed subgraph is *even* if the indegree equal to the outdegree at every vertex.

even vertex: See *odd vertex*.

evenly spanning even subgraph: An evenly spanning even subgraph S of G is an even subgraph such that each component of S contains an even number of odd vertices of G and each vertex not contained in S is an even vertex of G.

face and *adjacency of faces*: Let G be a graph embedded on a surface Σ. A face of G (on Σ) is a connected region of $S - G$ (where the graph G is considered as a topological subset of the surface Σ). Two faces F' and F'' are *adjacent* to each other if there is an edge e such that F' and F'' are on either side of e. (Note that if the embedding of G is not circular, then it is possible that a face is adjacent to itself since there might be an edge both sides of which are the same face.)

face-coloring, proper face-coloring: Let G be embedded on a surface with the face set $F(G)$. $c : F(G) \mapsto S$ is a face-coloring (or proper face-coloring) if $c(f) \neq c(f^*)$ for every pair of adjacent faces f and f^*.

facial circuit: A facial circuit is a circuit which is the boundary of a face of an embedded graph G.

factor and *r-factor*: A subgraph H of a graph G is an r-factor of G if H is a spanning, r-regular subgraph of G.

faithful even subgraph cover and *faithful circuit cover*: Let w be a weight: $E(G) \mapsto \mathbb{Z}^*$. An even subgraph (or circuit) cover $\mathcal{F}$ of G is faithful with respect to w if each edge of G is contained in precisely $w(e)$ members of $\mathcal{F}$.

flow: Let G be a graph and Γ be an additive group with "0" as its identity. A flow (or *group flow*) is an ordered pair (D, f) of a graph G where D is an orientation of G and $f : E(G) \mapsto \Gamma$ such that

$$\sum_{e \in E^+(v)} f(e) = \sum_{e \in E^-(v)} f(e)$$

for every vertex $v \in V(G)$. If $\Gamma = \mathbb{Z}$, then (D, f) is called an *integer flow*. If $\Gamma = \mathbb{Z}_k$, then (D, f) is called a *mod-k-flow*. A *nowhere-zero flow* (D, f) of a graph G is a flow with $supp(f) = E(G)$. A *positive integer flow* is a flow (D, f) with $f : E(G) \mapsto \mathbb{Z}$ and $f(e) > 0$ for each edge $e \in E(G)$.

forbidden part, *forbidden set* and *forbidden system*: Let G be an eulerian graph. For a vertex v in G, a forbidden set incident with v, denoted by $\mathcal{P}(v)$, is a partition of $E(v)$ (the set of edges incident with v). A member of $\mathcal{P}(v)$ is called a forbidden part (incident with v). The set $\mathcal{P} = \bigcup_{v \in V(G)} \mathcal{P}(v)$ is called a forbidden system of G.

fractional faithful even subgraph cover: Let G be a graph and $w : E(G) \mapsto \mathbb{Q}^*$ be a weight of G. An ordered pair $(\mathcal{F}, \chi)$ is called a fractional faithful even subgraph cover of (G, w) if $\mathcal{F}$ is a family of even subgraphs of G and $\chi : \mathcal{F} \mapsto \mathbb{Q}^*$ is a function such that, for each edge $e \in E(G)$,

$$w(e) = \sum_{e \in C \in \mathcal{F}} \chi(C).$$

frame: Let G be a cubic graph. A spanning subgraph H of G is called a *frame* of G if G/H is an even graph.

generalized Petersen graph $P(n, k)$: A graph $P(n, k)$, called a *generalized Petersen graph*, is defined as follows: $P(n, k)$ has $2n$ vertices, namely $v_0, \ldots, v_{n-1}, u_0, \ldots, u_{n-1}$, the vertex v_i is adjacent to three vertices $\{v_{i+1}, v_{i-1}, u_i\}$, and u_i is further adjacent to u_{i+k} and u_{i-k} (where addition is mod n).

girth: The length of a shortest circuit in a graph G is called the girth of G and is denoted by $g(G)$.

group flow: See *flow*.

Hamilton circuit and *Hamilton path*: A circuit (or a path) of a graph G containing all vertices of G is called *Hamilton circuit* (or *Hamilton path*, respectively).

Hamilton cover and *Hamilton weight*: Let w be an eulerian $(1, 2)$-weight of a cubic graph G. A faithful circuit cover $\mathcal{F}$ of (G, w) is hamiltonian if $\mathcal{F}$ is a set of two Hamilton circuits. An eulerian $(1, 2)$-weight w of G is hamiltonian if *every* faithful circuit cover of (G, w) is hamiltonian.

Hamilton weight: See *Hamilton cover*.

handle bridge: A graph G is embedded on a surface. An edge that lies on the boundary of only one face is called a handle bridge (that is, both sides of the edge are the same face).

hypohamiltonian: A graph G is hypohamiltonian if, for every $v \in V(G)$, $G - v$ contains a Hamilton circuit.

indegree and *outdegree*: The indegree (or outdegree) of a vertex v in a directed graph is the number of arcs with their heads (or tails, respectively) at the vertex v.

independent set: A subset $U \subseteq V(G)$ is independent in G if there is no edge of G joining any pair of vertices of U.

induced subgraph: Let $A \subseteq E(G)$ (or $A \subseteq V(G)$). The subgraph of G induced by A, denoted by $G[A]$, is the subgraph of G with the edge set A and the vertex set consisting of all vertices of G incident with some edges of A (or is the subgraph of G with the vertex set A and the edge set consisting of all edges of G with both endvertices in A, respectively).

induced weight: Let $\mathcal{F}$ be a set of even subgraph covers of a bridgeless graph G. The weight $w_{\mathcal{F}} : E(G) \mapsto \mathbb{Z}^*$ with $w_{\mathcal{F}}(e)$ equal to the number of members of $\mathcal{F}$ containing e, for each edge $e \in E(G)$, is called the *weight of G induced by $\mathcal{F}$*.

integer flow: See *flow*.

intersection of graphs: See *union of graphs*.

length of circuit, even subgraph and path: The length of a path (or circuit, or even subgraph) Q is the number of edges contained in Q.

loop: A loop is an edge with only one endvertex.

matching: A matching of a graph G is a 1-regular subgraph of G.

minimal edge-cut: See *edge-cut*.

minor: Let G and H be two graphs. If G contains a subgraph which is contractible to H, then H is a minor of G and we say G contains an H-minor. G is called H-minor-free if G contains no H-minor.

mod-k-flow: See *flow*.

multiplicity of an edge, and a graph: Let x and y be two distinct vertices of a graph G. The number of edges of G with endvertices x and y is the multiplicity of xy. For a graph G, the multiplicity of G is the maximum multiplicity of all edges of G.

neighbor: Let $v \in V(G)$. A vertex of G adjacent to v is called a neighbor of v in G. The set of all neighbors of a vertex v in a graph G is denoted by $N_G(v)$.

nowhere-zero flow: See *flow*.

odd component and *even component*: A component of a graph G containing an odd (or even) number of vertices is called an odd component (or an even component, respectively).

odd vertex and *even vertex*: An odd vertex (or even vertex) is a vertex of odd (or even) degree.

odd-edge-connectivity: The odd-edge-connectivity of a graph G, denoted by $\lambda_o(G)$, is the size of a smallest odd edge-cut of G.

oddness: Let G be a bridgeless graph and F be a spanning even subgraph of G. The oddness of F, denoted by $odd(F)$, is the number of components of F that contain an odd number of odd-vertices of G. The oddness of G, denoted by $odd(G)$, is the minimum of $odd(F)$ for all spanning even subgraphs F of G.

1-factor: See *factor*.

1-factorization: A set of 1-factors $\{M_1, \ldots, M_r\}$ is a 1-factorization of G if $\{E(M_1), \ldots, E(M_r)\}$ is a partition of $E(G)$.

orientable even subgraph double cover and *orientable k-even subgraph double cover*: Let $\mathcal{F} = \{C_1, \ldots, C_r\}$ be an even subgraph double cover of a graph G. The set $\mathcal{F}$ is an orientable even subgraph double cover if there is an orientation D_μ of $E(C_\mu)$, for each $\mu = 1, \ldots, r$, such that (1) $D_\mu(C_\mu)$ is a directed even subgraph; (2) for each edge e contained in two even subgraphs C_α and C_β ($\alpha, \beta \in \{1, \ldots, r\}$), the directions of $D_\alpha(C_\alpha)$ and $D_\beta(C_\beta)$ are opposite on e. An orientable k-even subgraph double cover $\mathcal{F}$ is an orientable even subgraph double cover consisting of k members.

outdegree: See *indegree*.

parallel edges: Two edges are parallel if they have the same endvertices (parallel edges are also called multiple edges).

parity 3-edge-coloring: A coloring $c : E(G) \mapsto \mathbb{Z}_3$ is a parity 3-edge-coloring if each $c^{-1}(i)$ is a parity subgraph of G.

parity subgraph: A spanning subgraph H of a graph G is called a parity subgraph of G if, for each vertex $v \in V(G)$,

$$d_H(v) \equiv d_G(v) \pmod 2.$$

parity subgraph decomposition: A decomposition of a graph G is called a parity subgraph decomposition if each member of the decomposition is a parity subgraph of G. A parity subgraph decomposition of G is *trivial* if it has only one member.

perfect matching: A 1-factor of G is also called a perfect matching of G.

permutation graph: A permutation graph is a cubic graph G such that G has a 2-factor which is the union of two chordless circuits.

planar graph: A graph that can be embedded on a sphere (or plane) is planar.

positive integer flow: See *flow*.

postman tour and *Chinese postman tour*: A closed trail containing all edges of G is called a postman tour. A shortest postman tour is called a Chinese postman tour.

proper vertex-coloring, proper edge-coloring, proper face-coloring: See *coloring*.

regular graph and *k-regular graph*: A graph is regular if the degrees of all vertices are the same. A graph is k-regular if $d(v) = k$ for every vertex $v \in V(G)$.

removable circuit: Let G be a graph associated with an admissible eulerian weight w. A circuit C of G is removable in (G, w) if (G, w) has a weight-decomposition $(G_1, w_1) + (C, 1)$ such that w_1 is admissible in G_1.

r-graph: An r-graph G is an r-regular graph such that, for each vertex subset $X \subset V(G)$ with $|X| \equiv 1 \pmod 2$ and $0 < |X| < |V(G)|$,

$$|(X, V(G) - X)_G| \geq r.$$

separating: See *vertex-cut*.

shortest cycle cover: An even subgraph cover is shortest if its total length is shortest among all even subgraph covers of the graph. The total length of a shortest cycle cover of G is denoted by $SCC(G)$.

shrinking: Let $X \subseteq V(G)$. The operation on G that identifies all vertices of X as a single vertex (and removes all resulting loops) is called a *shrinking*. The resulting graph is denoted by G/X.

This shrinking operation is extended to partitions of $V(G)$ as follows. Given a partition $\mathcal{P} = \{V_1, V_2, \ldots, V_p\}$ of $V(G)$ into non-empty parts, we shrink $\mathcal{P}$ by shrinking each subset V_i, $1 \le i \le p$, and we denote the resulting graph by $G/\mathcal{P}$. (Note that $G/\mathcal{P}$ might have multiple edges, but does not create new loops.)

The opposite operation of shrinking is vertex-splitting.

simple graph: A graph is simple if it has no loop and parallel edges.

snark: A 2-edge-connected, non-3-edge-colorable cubic graph is called a snark.

spanning: A subgraph H of a graph G is spanning if $V(H) = V(G)$.

strongly connected: A directed graph is strongly connected if, for each pair of vertices x and y, there is a directed path from x to y and a directed path from y to x.

subdivided edge: A subdivided edge of a graph G is a path $x_1 \ldots x_p$ such that $d_G(x_i) = 2$ for each $i \in \{2, \ldots, p-1\}$.

subdivision: Let G be a graph. The graph obtained by replacing each edge of G with a subdivided edge is called a subdivision of G (see Figure D.11).

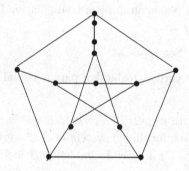

Figure D.11 *A subdivision of the Petersen graph*

supereulerian graph: A graph G is *supereulerian* if G contains a spanning, connected, even subgraph.

supergraph: If H is a subgraph of G, then G is a supergraph of H.

suppressed graph: The suppressed graph of a graph G is the graph obtained from G by replacing each maximal subdivided edge with a single edge, and is denoted by $\overline{G}$.

surface: All surfaces considered in this book are 2-manifolds.

symmetric difference: For two sets X and Y, the symmetric difference of X and Y is

$$X \triangle Y = (X \cup Y) - (X \cap Y).$$

For two subgraphs H_1 and H_2 of a graph G, the symmetric difference of H_1 and H_2, (also) denoted by $H_1 \triangle H_2$, is the subgraph of G induced by $[E(H_1) \cup E(H_2)] - [E(H_1) \cap E(H_2)]$.

tournament: A tournament is a directed graph obtained from a complete graph by assigning a direction to each edge.

trail: An alternating sequence $v_0 e_0 v_1 \ldots v_{n-1} e_{n-1} v_n$ is a trail if, for each $i \in \{0, \ldots, n-1\}$, e_i is an edge joining the vertices v_i and v_{i+1}. The *length* of the trail is n. (Note that vertices and edges in a trail might be repeated).

transitive tournament: A transitive tournament is a tournament whose vertex set can be labeled as $\{v_1, \ldots, v_n\}$ such that v_i dominates v_j if and only if $i < j$.

triangulation: A planar graph is a triangulation if all of its facial circuits are 3-circuits.

trivial edge-cut: See *edge-cut*.

trivial forbidden part: A forbidden part is trivial if it consists of a single edge.

trivial vertex (with respect to *a forbidden system* $\mathcal{P}$): Let G be an eulerian graph and $\mathcal{P}$ be a forbidden system of G. A vertex v in $(G, \mathcal{P})$ is trivial (with respect to $\mathcal{P}$) if every forbidden part incident with v is trivial.

Tutte bridge: Let H be a subgraph of G. A *Tutte bridge* of H is either a chord e of H ($e = xy \notin E(H)$ with both $x, y \in V(H)$) or a subgraph of G consisting of one component Q of $G - V(H)$ and all edges joining Q and H (and, of course, all vertices of H adjacent to Q). A chord e of H is a *trivial Tutte bridge*.

For a Tutte bridge B_i of H, the vertex subset $V(B_i) \cap V(H)$ is called the *attachment of* B_i, and is denoted by $A(B_i)$.

2-cell embedding: An embedding of a graph G on a surface Σ is called a 2-cell embedding if every face is homeomorphic to an open unit disk. A 2-cell embedding is *circular* (or *strong*) if the boundary of every face is a circuit.

2-manifold: A Hausdorff space Σ is called a 2-dimensional-manifold if each point of Σ has a neighborhood that is homeomorphic to the 2-dimensional open disk $U^2 = \{x \in \mathbb{R}^2 : |x| \leq 1\}$ (see [173], [45]). 2-manifold is an abbreviation of 2-dimensional-manifold.

union of graphs and *intersection of graphs*: Let G_1 and G_2 be two graphs. The graph $G_1 \cup G_2$ (or $G_1 \cap G_2$) is the graph with the vertex set $V(G_1) \cup V(G_2)$ and the edge set $E(G_1) \cup E(G_2)$ (or $V(G_1) \cap V(G_2)$ and $E(G_1) \cap E(G_2)$, respectively).

unique 3-edge-coloring: A uniquely 3-edge-colorable graph is a cubic graph having precisely one 1-factorization.

vertex-coloring, proper vertex-coloring: $c : V(G) \mapsto S$ is a vertex-coloring (or proper vertex-coloring) if $c(u) \neq c(v)$ for every pair of adjacent vertices u and v.

vertex-cut, *cut* and *separating*: A vertex subset T separating G into two parts G_1 and G_2 means that G_1 and G_2 are two subgraphs of G and $T = G_1 \cap G_2$ and $G_1 \cup G_2 = G$ and $d_{G_i}(v) > 0$ for each $i \in \{1, 2\}$ and each $v \in T$. A vertex subset T of a graph G separating G into two parts is called a vertex-cut (or simply, a cut).

vertex identification (vertex shrinking): Let $X \subseteq V(G)$. The graph obtained from G by identifying/merging all vertices of X as a single vertex and deleting all resulting loops is denoted by G/X. And this operation is called *vertex identification* or *vertex shrinking*. The opposite operation of shrinking is vertex-splitting if X is an independent set.

vertex shrinking: See *vertex identification*.

vertex-splitting: Let G be a graph and v be a vertex of G and $F \subset E(v)$. The graph $G_{[v;F]}$ obtained from G by splitting the edges of F away from v. (That is, adding a new vertex v' and changing the end v of the edges of F to be v'. See Figure A.1.)

vertex transitive: A graph $G = (V, E)$ is vertex transitive if, for every pair of vertices x, y of G, there is an automorphism $\pi \in Auto(G)$ such that $\pi(x) = y$.

weight decomposition: Let (G, w) be a weighted graph. $\{(G_1, w_1), (G_2, w_2)\}$ is a weight decomposition of (G, w) if $G_1 \cup G_2 = G$ and $w(e) = w_1(e) + w_2(e)$ for every edge $e \in E(G)$ (note $w_i(e) = 0$ if $e \notin E(G_i)$).

weight induced by even subgraphs: See *induced weight*.

Some symbols

$Auto(G)$: the automorphism group of a graph.

$E(G)$: the edge set of a (sub)graph G.

$E(v)$: the set of edges incident with v.

$E_D^+(v)$: the set of arcs with tails at v.

$E_D^-(v)$: the set of arcs with heads at v.

$F(G)$: the face set of an embedded graph G.

G/H or G/X: the graph obtained from an edge contraction or vertex shrinking.

$G_{[v;F]}$: vertex splitting.

$H_1 \triangle H_2$: the symmetric difference of two sets or two subgraphs H_1 and H_2.

λ_o: odd-edge-connectivity.

$N(v)$: the set of neighbors of v.

$O(G)$: the set of odd degree vertices of G.

P_{10}: the Petersen graph.

$\mathbb{Q}$: the set of all rational numbers.

$\mathbb{Q}^\star$: the set of all non-negative rational numbers.

$\mathbb{Q}^+$: the set of all positive rational numbers.

$V(G)$: the vertex set of a (sub)graph G.

$[X,Y]$: the edges between X and Y where $X, Y \subset V(G)$ and $X \cap Y = \emptyset$.

$\xi(\Sigma)$: the Euler characteristic of a surface Σ

$\mathbb{Z}$: the set of all integers.

$\mathbb{Z}^\star$: the set of all non-negative integers.

$\mathbb{Z}^+$: the set of all positive integers.

References

[1] Alon, N., and Tarsi, M. 1985. Covering multigraphs by simple circuits. *SIAM J. Algebraic Discrete Methods*, **6**, 345–350.

[2] Alspach, B., and Godsil, C. 1985. Unsolved problems. Pages 461–467 of: *Cycles in Graphs, Ann. Discrete Math.*, vol. 27. Amsterdam: North-Holland.

[3] Alspach, B., and Zhang, C.-Q. 1993. Cycle covers of cubic multigraphs. *Discrete Math.*, **111**, 11–17.

[4] Alspach, B., Goddyn, L. A., and Zhang, C.-Q. 1994. Graphs with the circuit cover property. *Trans. Amer. Math. Soc.*, **344**, 131–154.

[5] Appel, K., and Haken, W. 1977. Every map is four colorable, Part I: Discharging. *Illinois J. Math.*, **21**, 429–490.

[6] Appel, K., and Haken, W. 1989. *Every Map is Four Colorable. Contemp. Math. AMS*, vol. 98. Providence, RI: American Mathematical Society.

[7] Appel, K., Haken, W., and Koch, J. 1977. Every map is four colorable, Part II: Reducibility. *Illinois J. Math.*, **21**, 491–567.

[8] Archdeacon, D. 1984. Face coloring of embedded graphs. *J. Graph Theory*, **8**, 387–398.

[9] Ash, P., and Jackson, B. 1984. Dominating cycles in bipartite graph. Pages 81–87 of: Bondy, J.A., and Murty, U.S.R. (eds), *Progress in Graph Theory*. New York: Academic Press.

[10] Barnette, D. W. 1996. Cycle covers of planar 3-connected graphs. *J. Combin. Math. Combin. Comput.*, **20**, 245–253.

[11] Berge, C. 1973. *Graph and Hypergraph*. New York: North-Holland. (translated by E. Minieka).

[12] Bermond, J. C., Jackson, B., and Jaeger, F. 1983. Shortest coverings of graphs with cycles. *J. Combin. Theory Ser. B*, **35**, 297–308.

[13] Biggs, N. L., Lloyd, E. K., and Wilson, R. J. 1976. *Graph Theory 1736–1936*. Oxford: Clarendon Press.

[14] Blanuša, D. 1946. Problem ceteriju boja (The problem of four colors). *Hrvatsko Prirodoslovno Društvo Glasnik Mat-Fiz. Astr, Ser. II*, **1**, 31–42.

[15] Bollobás, B. 1978. *Extremal Graph Theory*. London: Academic Press.

[16] Bondy, J. A. 1990. Small cycle double covers of graphs. Pages 21–40 of: Hahn, G., Sabidussi, G., and Woodrow, R. (eds), *Cycles and Rays*. NATO ASI Ser. C. Dordrecht: Kluwer Academic Publishers.

[17] Bondy, J. A., and Hell, P. 1990. A note on the star chromatic number. *J. Graph Theory*, **14**, 479–482.

[18] Bondy, J. A., and Murty, U. S. R. 1976. *Graph Theory with Applications*. London: Macmillan.

[19] Bondy, J. A., and Murty, U. S. R. 2008. *Graph Theory*. Springer.

[20] Brinkmann, G., and Steffen, E. 1998. Snarks and reducibility. *Ars Combin.*, **50**, 292–296.

[21] Brinkmann, G., Goedgebeur, J., Hägglund, J., and Markström, K. 2011. *Generation and properties of snarks*. Preprint.

[22] Cai, L., and Corneil, D. 1992. On cycle double covers of line graphs. *Discrete Math.*, **102**, 103–106.

[23] Carroll, L. 1874. *(C. L. Dodgson) The Hunting of the Snark (An Agony in 8 Fits)*. London: Macmillan.

[24] Catlin, P. A. 1988. A reduction method to find spanning eulerian subgraph. *J. Graph Theory*, **12**, 29–45.

[25] Catlin, P. A. 1989. Double cycle covers and the Petersen graph. *J. Graph Theory*, **13**, 465–483.

[26] Catlin, P. A. 1990. Double cycle covers and the Petersen graph, II. *Congr. Numer.*, **76**, 173–181.

[27] Catlin, P. A. 1992. Supereulerian graphs: a survey. *J. Graph Theory*, **16**, 177–196.

[28] Catlin, P. A., Han, Z.-Y., and Lai, H.-J. 1996. Graphs without spanning closed trails. *Discrete Math.*, **160**, 81–91.

[29] Cavicchioli, A., Meschiari, M., Ruini, B., and Spaggiari, F. 1998. A survey on snarks and new results: products, reducibility and a computer Search. *J. Graph Theory*, **28**, 57–86.

[30] Cavicchioli, A., Murgolo, T. E., Ruini, B., and Spaggiari, F. 2003. Special classes of snarks. *Acta Applicandae Mathematicae*, **76**, 57–88.

[31] Celmins, U. A. 1984. *On cubic graphs that do not have an edge 3-coloring*. Ph.D. thesis, University of Waterloo, Ontario, Canada.

[32] Celmins, U. A., Fouquet, J. L., and Swart, E. R. 1980. *Construction and characterization of snarks*. Research Report, University of Waterloo, Ontario, Canada.

[33] Chan, M. 2009. *A survey of the cycle double cover conjecture*. Preprint, Princeton University.

[34] Chartrand, G., and Lesniak, L. 1986. *Graphs and Digraphs*. Second edn. Belmont, CA: Wadsworth and Brooks/Cole.

[35] Chen, C. C., and Quimpo, N. F. 1981. On strongly hamiltonian abelian group graphs. Pages 23–34 of: McAvaney, K. L. (ed), *Combinatorial Mathematics VIII*. Lecture Notes in Math., vol. 884. Berlin: Springer-Verlag.

[36] Chen, Z.-H., and Lai, H.-J. 1995. Reductions techniques for supereulerian graphs and related topics – a survey. Pages 53–69 of: Ku, T.-H. (ed), *Combinatorics and Graph Theory 95*. Singapore: World Scientific.

[37] Chetwynd, A. G., and Wilson, R. J. 1981. Snarks and supersnarks. Pages 215–241 of: *The Theory and Applications of Graphs.* New York: Wiley.

[38] Cutler, J., and Häggkvist, R. 2004. *Cycle double covers of graphs with disconnected frames.* Research report 6, Department of Mathematics, Umeå University, Sweden. ·

[39] Descartes, B. 1948. Network-colourings. *Math. Gazette*, **32**, 67–69.

[40] DeVos, M., Johnson, T., and Seymour, P.D. *Cut Coloring and Circuit Covering.* Submitted for publication, http://www.math.princeton.edu/~pds/papers/cutcolouring/.

[41] Diestel, R. 2010. *Graph Theory.* Fourth edn. Springer-Verlag.

[42] Ding, S.-K., Hoede, C., and Vestergaard, P. D. 1990. Strong cycle covers. *Ars Combin.*, **29C**, 130–139.

[43] Edmonds, J. 1965. Maximum matching and a polyhedron with $(0, 1)$-vertices. *J. Res. Nat. Bur. Standards* B, **69**, 125–130.

[44] Edmonds, J., and Johnson, E. L. 1973. Matching, Euler tours and the Chinese postman. *Mathematical Programming*, **5**, 88–124.

[45] Eisenberg, M. 1974. *Topology.* New York: Holt, Rinehart and Winston, Inc.

[46] Ellingham, M. N. 1984. Petersen subdivisions in some regular graphs. *Congr. Numer.*, **44**, 33–40.

[47] Ellingham, M. N., and Zha, X.-Y. 2011. Orientable embeddings and orientable cycle double covers of projective-planar graphs. *European J. Combin.*, **32**, 495–509.

[48] Esteva, E. G. M., and Jensen, T. R. 2007. On semiextensions and circuit double covers. *J. Combin. Theory Ser. B*, **97**, 474–482.

[49] Esteva, E. G. M., and Jensen, T. R. 2009. A note on semiextensions of stable circuits. *Discrete Math.*, **309**, 4952–4954.

[50] Fan, G.-H. 1992. Covering graphs by cycles. *SIAM J. Discrete Math.*, **5**, 491–496.

[51] Fan, G.-H. 1994. Short cycle covers of cubic graphs. *J. Graph Theory*, **18**, 131–141.

[52] Fan, G.-H. 1998. Proofs of two minimum circuit cover conjectures. *J. Combin. Theory Ser. B*, **74**, 353–367.

[53] Fan, G.-H., and Raspaud, A. 1994. Fulkerson's conjecture and circuits covers. *J. Combin. Theory Ser. B*, **61**, 133–138.

[54] Fan, G.-H., and Zhang, C.-Q. 2000. Circuit decompositions of eulerian graphs. *J. Combin. Theory Ser. B*, **78**, 1–23.

[55] Fiorini, S., and Wilson, R. J. 1977. *Edge colourings of graphs.* Research Notes in Mathematics, vol. 16. Pitman.

[56] Fiorini, S., and Wilson, R. J. 1978. Edge colourings of graphs. Pages 103–126 of: Beineke, L. W., and Wilson, R. J. (eds), *Selected Topics in Graph Theory.* London: Academic Press.

[57] Fish, J. M., Klimmek, R., and Seyffarth, K. 2002. Line graphs of complete multipartite graphs have small cycle double covers. *Discrete Math.*, **257**, 39–61.

[58] Fleischner, H. 1976. Eine gemeinsame Basis für die Theorie der eulerschen Graphen und den Satz von Petersen. *Monatsh. Math.*, **81**, 267–278.

[59] Fleischner, H. 1980. Eulersche Linien und Kreisuberdeckungen die vorgegebene Duurchgange inden Kanten vermeiden. *J. Combin. Theory Ser. B*, **29**, 145–167.

[60] Fleischner, H. 1983. Eulerian Graph. Pages 17–53 of: Beineke, L. W., and Wilson, R. J. (eds), *Selected Topics in Graph Theory (2)*. London: Academic Press.

[61] Fleischner, H. 1984. Cycle decompositions, 2-coverings, removable cycles and the four-color disease. Pages 233–246 of: Bondy, J. A., and Murty, U. S. R. (eds), *Progress in Graph Theory*. New York: Academic Press.

[62] Fleischner, H. 1986. Proof of the strong 2-cover conjecture for planar graphs. *J. Combin. Theory Ser. B*, **40**, 229–230.

[63] Fleischner, H. 1988. Some blood, sweat, but no tears in eulerian graph theory. *Congr. Numer.*, **63**, 9–48.

[64] Fleischner, H. 1990. Communication at Cycle Double Cover Conjecture Workshop, Barbados, February 25–March 4.

[65] Fleischner, H. 1991. *Eulerian Graphs and Related Topics, Part 1, Vol. 2. Ann. Discrete Math.*, vol. 50. North-Holland.

[66] Fleischner, H. 1994. Uniqueness of maximal dominating cycles in 3-regular graphs and Hamiltonian cycles in 4-regular graphs. *J. Graph Theory*, **18**, 449–459.

[67] Fleischner, H. 2002. Bipartizing matchings and Sabidussi's compatibility conjecture. *Discrete Math.*, **244**, 77–82.

[68] Fleischner, H. 2010. *Uniquely hamiltonian (simple) graphs of minimum degree four*. 8th French Combinatorial Conference, June 30th 2010, University Paris XI - Sud, Orsay, France.

[69] Fleischner, H. 2011. *Personal communication*, Vienna.

[70] Fleischner, H., and Frank, A. 1990. On cycle decomposition of eulerian graph. *J. Combin. Theory Ser. B*, **50**, 245–253.

[71] Fleischner, H., and Fulmek, M. 1990. $P(D)$-compatible eulerian trails in digraphs and a new splitting lemma. Pages 291–303 of: Bodendiek, R. (ed), *Contemporary Methods in Graph Theory*.

[72] Fleischner, H., and Häggkvist, R. 2009. Circuit double covers in special types of cubic graphs. *Discrete Math.*, **309**, 5724–5728.

[73] Fleischner, H., and Kochol, M. 2002. A note about the dominating circuit conjecture. *Discrete Math.*, **259**, 307–309.

[74] Fleischner, H., Hilton, A. J. W., and Jackson, B. 1990. On the maximum number of pairwise compatible Euler cycles. *J. Graph Theory*, **14**, 51–63.

[75] Fleischner, H., Genest, F., and Jackson, B. 2007. Compatible circuit decompositions of 4-regular graphs. *J. Graph Theory*, **56**, 227–240.

[76] Ford, L. R., and Fulkerson, D. R. 1962. *Flows in Networks*. Princeton, NJ: Princeton University Press.

[77] Fouquet, J. 1982. Note sur la non existence d'un snark d'ordre 16. *Discrete Math.*, **38**, 163–171.

[78] Fowler, T. G. 1998. *Unique Coloring of Planar Graphs*. Ph.D. thesis, Georgia Tech.

[79] Franklin, P. 1941. *The Four Color Problem*. New York: Scripta Mathematica, Yeshiva College.

[80] Fulkerson, D. R. 1971. Blocking and antiblocking pairs of polyhedral. *Math. Programming*, **1**, 168–194.

[81] Gardner, M. 1976. Mathematical Games. *Scientific American*, **4**, 126–130.

[82] Goddyn, L. A. 1985. A girth requirement for the double cycle cover conjecture. Pages 13–26 of: Alspach, B., and Godsil, C. (eds), *Cycles in Graphs. Ann. Discrete Math.*, vol. 27. Amsterdam: North-Holland.

[83] Goddyn, L. A. 1988. *Cycle covers of graphs*. Ph.D. thesis, University of Waterloo, Ontario, Canada.

[84] Goddyn, L. A. 1989. Cycle double covers of graphs with Hamilton paths. *J. Combin. Theory Ser. B*, **46**, 253–254.

[85] Goddyn, L. A. 1991. *Cycle double covers–current status and new approaches*. Contributed lecture at Cycle Double Cover Conjecture Workshop, IINFORM, Vienna, January 1991.

[86] Goddyn, L. A. 1993. Cones, lattices and Hilbert bases of circuits and perfect matching. *Contemporary Mathematics*, **147**, 419–440.

[87] Goddyn, L. A., van den Heuvel, J., and McGuinness, S. 1997. Removable circuits in multigraphs. *J. Combin. Theory Ser. B*, **71**, 130–143.

[88] Goddyn, L. A., Tarsi, M., and Zhang, C.-Q. 1998. On (k, d)-colorings and fractional nowhere zero flows. *J. Graph Theory*, **28**, 155–161.

[89] Goldwasser, J. L., and Zhang, C.-Q. 1996. On minimal counterexamples to a conjecture about unique edge-3-coloring. *Congr. Numer.*, **113**, 143–152.

[90] Goldwasser, J. L., and Zhang, C.-Q. 1999. Permutation graphs and Petersen graph. *Ars Combin.*, **51**, 240–248.

[91] Goldwasser, J. L., and Zhang, C.-Q. 2000. Uniquely edge-3-colorable graphs and snarks. *Graph and Combinatorics*, **16**, 257–267.

[92] Gould, R. 1988. *Graph Theory*. Menlo Park, CA: Benjamin/Cummings Publishing Company, Inc.

[93] Greenwell, D., and Kronk, H. V. 1973. Uniquely line-colorable graphs. *Canad. Math. Bull.*, **16**, 525–529.

[94] Gross, J. L., and Tucker, T. W. 1987. *Topological Graph Theory*. New York: John Willey & Sons.

[95] Guan, M.-G., and Fleischner, H. 1985. On the minimum weighted cycle covering problem for planar graphs. *Ars Combin.*, **20**, 61–68.

[96] Gusfield, D. 1983. Connectivity and edge-disjoint spanning trees. *Inform. Process. Lett.*, **16**, 87–89.

[97] Haggard, G. 1977. Edmonds Characterization of disc embedding. Pages 291–302 of: *Proceeding of the 8th Southeastern Conference of Combinatorics, Graph Theory and Computing, Utilitas Mathematica.*

[98] Häggkvist, R. 2009. *Lollipop Andrew strikes again (abstract)*. 22nd British Combinatorial Conference, July 5–10, 2009, University of St Andrews, UK.

342   *References*

[99] Häggkvist, R., and Markström, K. 2006a. Cycle double covers and spanning minors I. *J. Combin. Theory Ser. B*, **96**, 183–206.

[100] Häggkvist, R., and Markström, K. 2006b. Cycle double covers and spanning minors II. *Discrete Math.*, **306**, 762–778.

[101] Häggkvist, R., and McGuinness, S. 2005. Double covers of cubic graphs with oddness 4. *J. Combin. Theory Ser. B*, **93**, 251–277.

[102] Hägglund, J. 2011. *Personal communication.*

[103] Hägglund, J., and Markström, K. 2011. On stable cycles and the cycle double covers of graphs with large circumference. *Discrete Math..* doi:10.1016/j.disc.2011.08.024.

[104] Heawood, P. J. 1898. On the four-color map theorem. *Quarterly J. Pure Math. Applied Math.*, **29**, 270–285.

[105] Heinrich, K., Liu, J.-P., and Zhang, C.-Q. 1998. Triangle-free circuit decompositions and Petersen-minor. *J. Combin. Theory Ser. B*, **72**, 197–207.

[106] Hind, H. R. 1988. *Restricted edge-colourings, Chapter 5.* Ph.D. thesis, Cambridge University, UK.

[107] Hoffman, A. J. 1958. Page 80 of *Théorie des Graph* (by C. Berge).

[108] Hoffman, A. J. 1960. Some recent applications of the theorem of linear inequalities to extremal combinatorial analysis. *Proc. Symp. Appl. Math.*, **10**, 113–127.

[109] Hoffman, F., Locke, S. C., and Meyerowitz, A. D. 1991. A note on cycle double cover in Cayley graphs. *Mathematica Pannonica*, **2**, 63–66.

[110] Hoffmann-Ostenhof, A. 2007. A counterexample to the bipartizing matching conjecture. *Discrete Math.*, **307**, 2723–2733.

[111] Hoffmann-Ostenhof, A. 2012. *Nowhere-zero flows and structures in cubic graphs.* Ph.D. thesis, University of Vienna, Austria.

[112] Holton, D. A., and Sheehan, J. 1993. *The Petersen Graph.* Australian Mathematical Society Lecture Series, vol. 7. Cambridge University Press.

[113] Holyer, I. 1981. The NP-completeness of edge-coloring. *SIAM J. Comput.*, **10**, 718–720.

[114] Huck, A. 1993. *On cycle-double covers of bridgeless graphs with hamiltonian paths.* Tech. rept. 254. Institute of Mathematics, University of Hannover, Germany.

[115] Huck, A. 2000. Reducible configurations for the cycle double cover conjecture. *Discrete Appl. Math.*, **99**, 71–90.

[116] Huck, A. 2001. On cycle-double covers of graphs of small oddness. *Discrete Math.*, **229**, 125–165.

[117] Huck, A., and Kochol, M. 1995. Five cycle double covers of some cubic graphs. *J. Combin. Theory Ser. B*, **64**, 119–125.

[118] Isaacs, R. 1975. Infinite families of non-trivial trivalent graphs which are not Tait colorable. *Amer. Math. Monthly*, **82**, 221–239.

[119] Itai, A., and Rodeh, M. 1978. Covering a graph by circuits. Pages 289–299 of: *Automata, Languages and Programming.* Lecture Notes in Computer Science, vol. 62. Berlin: Springer-Verlag.

[120] Jackson, B. 1990. Shortest circuit covers and postman tours of graphs with a nowhere-zero 4-flow. *SIAM J. Comput.*, **19**, 659–665.

[121] Jackson, B. 1993. On circuit covers, circuit decompositions and Euler tours of graphs. Pages 191–210 of: Walker, K. (ed), *Surveys in Combinatorics*. London Math. Soc. Lecture Note Series, vol. 187. Cambridge: Cambridge University Press.

[122] Jackson, B. 1994. Shortest circuit covers of cubic graphs. *J. Combin. Theory Ser. B*, **60**, 299–307.

[123] Jaeger, F. 1975. On nowhere-zero flows in multigraphs. *Proceedings of the Fifth British Combinatorial Conference 1975, Congr. Numer.*, **XV**, 373–378.

[124] Jaeger, F. 1976. Balanced valuations and flows in multigraphs. *Proc. Amer. Math. Soc.*, **55**, 237–242.

[125] Jaeger, F. 1978a. *On interval hypergraphs and nowhere-zero flow in graphs*. Research Report of Mathematics Application and Information, Universite Scientifique et Medicale et Institut National Polytechnique de Grenoble, No. 126, Juillet.

[126] Jaeger, F. 1978b. Sue les flots dans les graphes et certaines valuations dans les hypergraphes d'intervalles. Pages 189–193 of: Benzaken, C. (ed), *Proc. Colloque Algèbre Appliquèe et Combinatoire, Grenoble*.

[127] Jaeger, F. 1979. Flows and generalized coloring theorems in graphs. *J. Combin. Theory Ser. B*, **26**, 205–216.

[128] Jaeger, F. 1980. Tait's theorem for graphs with crossing number at most one. *Ars Combin.*, **9**, 283–287.

[129] Jaeger, F. 1984. On circular flows in graphs. Pages 391–402 of: *Finite and Infinite Sets, Vol. I, II (Eger, 1981), Colloquia Mathematica Societatis János Bolyai 37*. Amsterdam: North Holland.

[130] Jaeger, F. 1985. A survey of the cycle double cover conjecture. Pages 1–12 of: Alspach, B., and Godsil, C. (eds), *Cycles in Graphs. Ann. Discrete Math.*, vol. 27. Amsterdam: North-Holland.

[131] Jaeger, F. 1988. Nowhere-zero flow problems. Pages 71–95 of: Beineke, L. W., and Wilson, R. J. (eds), *Selected Topics in Graph Theory (3)*. London: Academic Press.

[132] Jaeger, F., and Swart, T. 1980. Conjecture 1. Pages 304–305 of: Deza, M., and Rosenberg, I.G. (eds), *Combinatorics 79. Ann. Discrete Math.*, vol. 9. Amsterdam: North-Holland.

[133] Jamshy, U., and Tarsi, M. 1992. Shortest cycle covers and the cycle double cover conjecture. *J. Combin. Theory Ser. B*, **56**, 197–204.

[134] Jamshy, U., Raspaud, A., and Tarsi, M. 1987. Short circuit covers for regular matroids with nowhere-zero, 5-flow. *J. Combin. Theory Ser. B*, **43**, 354–357.

[135] Jensen, T. R. 2010. Splits of circuits. *Discrete Math.*, **310**, 3026–3029.

[136] Jensen, T. R., and Toft, B. 1994. *Graph Coloring Problems*. John Wiley & Sons.

[137] Kahn, J., Robertson, N., and Seymour, P. D. 1987. Communication at Bellcore.

344 *References*

[138] Kaiser, T., and Raspaud, A. 2010. Perfect matchings with restricted intersection in cubic graphs. *European J. Combin.*, **31**, 1307–1315.

[139] Kaiser, T., and Škrekovski, R. 2008. Cycles intersecting edges-cuts of prescribed sizes. *SIAM J. Discrete Math.*, **22**, 861–874.

[140] Kaiser, T., Král, D., Lidický, B., and Nejedlý, P. 2010. Short Cycle Covers of Graphs with Minimum Degree Three. *SIAM J. Discrete Math.*, **24**, 330–355.

[141] Kilpatrick, P. A. 1975. *Tutte's first colour-cycle conjecture.* Ph.D. thesis, Cape Town, South Africa.

[142] Knuth, D. E. 2008. *The Art of Computer Programming.* Vol. 4, Fascicle 0, Introduction to Combinatorial Algorithms and Boolean Function. Addison-Wesley.

[143] Kochol, M. 1993a. *Construction of cyclically 6-edge-connected snarks.* Technical Report TR-II-SAS-07/93-5, Institute for Information, Slovak Academy of Sciences, Bratislava, Slovakia.

[144] Kochol, M. 1993b. *Cycle double covering of graphs.* Technical Report TR-II-SAS-08/93-7, Institute for Informatics, Slovak Academy of Sciences, Bratislava, Slovakia.

[145] Kochol, M. 1996a. A cyclically 6-edge-connected snark of order 118. *Discrete Math.*, **161**, 297–300.

[146] Kochol, M. 1996b. Snarks without small cycles. *J. Combin. Theory Ser. B*, **67**, 34–47.

[147] Kochol, M. 2000. Equivalence of Fleischner's and Thomassen's conjectures. *J. Combin. Theory Ser. B*, **78**, 277–279.

[148] Kochol, M. 2001. Stable dominating circuits in snarks. *Discrete Math.*, **233**, 247–256.

[149] Kochol, M. 2004. Reduction of the 5-flow conjecture to cyclically 6-edge-connected snarks. *J. Combin. Theory Ser. B*, **90**, 139–145.

[150] Kochol, M. 2010. Smallest counterexample to the 5-flow conjecture has girth at least eleven. *J. Combin. Theory Ser. B*, **100**, 381–389.

[151] Kostochka, A. V. 1995. The 7/5-conjecture strengthens itself. *J. Graph Theory*, **19**, 65–67.

[152] Kotzig, A. 1958. Bemerkung zu den faktorenzerlegungen der endlichen paaren regulren graphen. *Časopis Pěst. Mat.*, **83**, 348–354.

[153] Kotzig, A. 1962. Construction of third-order Hamiltonian graphs. *Časopis Pěst. Mat.*, **87**, 148–168.

[154] Kotzig, A. 1964. Hamilton graphs and Hamilton circuits. Pages 63–82 of: *Theory of Graphs and its Applications, Proceedings of the Symposium of Smolenice 1963.* Prague: Publ. House Czechoslovak Acad. Sci.

[155] Kotzig, A., and Labelle, J. 1978. Strongly Hamiltonian graphs. *Utilitas Mathematica*, **14**, 99–116.

[156] Král, D., Nejedlý, P., and Šámal, R. 2008. *Short cycle covers of cubic graphs.* KAM-DIMATIA Series 2008 (2008–846) Department of Applied Mathematics, Charles University, Prague, Czech.

[157] Král, D., Máčajová, E., Pangrác, O., Raspaud, A., Sereni, J.-S., and Škoviera, M. 2009. Projective, affine, and abelian colourings of cubic graphs. *European J. Combin.*, **30**, 53–69.

[158] Kriesell, M. 2006. Contractions, cycle double covers, and cyclic colorings in locally connected graphs. *J. Combin. Theory Ser. B*, **96**, 881–900.

[159] Kundu, S. 1974. Bounds on the number of disjoint spanning trees. *J. Combin. Theory Ser. B*, **17**, 199–203.

[160] Kuratowski, C. 1930. Sur le problème des courbes gauches en topologie. *Fund. Math.*, **15**, 271–283.

[161] Lai, H.-J. 1994. *Extension of a 3-coloring result of planar graphs*. Unpublished manuscript.

[162] Lai, H.-J. 1995. The size of graphs without nowhere-zero 4-flows. *J. Graph Theory*, **19**, 385–395.

[163] Lai, H.-J., and Lai, H.-Y. 1991a. Cycle covering of plane triangulations. *J. Combin. Math. Combin. Comput.*, **10**, 3–21.

[164] Lai, H.-J., and Lai, H.-Y. 1991b. Small cycle covers of planar graphs. *Congr. Numer.*, **85**, 203–209.

[165] Lai, H.-J., and Zhang, C.-Q. 2001. Hamilton weight and Petersen minor. *J. Graph Theory*, **38**, 197–219.

[166] Lai, H.-J., Yu, X.-X., and Zhang, C.-Q. 1994. Small circuit double covering of cubic graphs. *J. Combin. Theory Ser. B*, **60**, 177–194.

[167] Little, C. H. C., and Ringeisen, R. D. 1978. On the strong graph embedding conjecture. Pages 479–487 of: *Proceeding of the 9th Southeastern Conference on Combinatorics, Graph Theory and Computing, Utilitas Mathematica*.

[168] Little, C. H. C., Tutte, W. T., and Younger, D. H. 1988. A theorem on integer flows. *Ars Combin.*, **26A**, 109–112.

[169] Loupekine, F., and Watkins, J. J. 1985. Cubic graphs and the four-color theorem. Pages 519–530 of: Alavi, Y., Chartrand, G., Lesniak, L., Lick, D.R., and Wall, C.E. (eds), *Graph Theory and its Application to Algorithms and Computer Science*. New York: John Wiley & Sons, Inc.

[170] Lovász, L. 1978. Kneser's conjecture chromatic number, and homotopy. *J. Combin. Theory Ser. A*, **25**, 319–324.

[171] MacGillivray, G., and Seyffarth, K. 2001. Classes of line graphs with small cycle double covers. *Austral. J. Combin.*, **24**, 91–114.

[172] Markström, K. 2011. Even cycle decompositions of 4-regular graphs and line graphs. *Discrete Math.*, doi:10.1016/j.disc.2011.12.007.

[173] Massey, W. S. 1967. *Algebraic Topology: An Introduction*. New York: Springer-Verlag.

[174] Matthews, K. R. 1978. On the eulericity of a graph. *J. Graph Theory*, **2**, 143–148.

[175] Máčajová, E., and Škoviera, M. 2005. Fano colourings of cubic graphs and the Fulkerson conjecture. *Theor. Comput. Sci.*, **349**, 112–120.

[176] Máčajová, E., and Škoviera, M. 2009. On a Conjecture of Fan and Raspaud. *Electronic Notes in Discrete Mathematics*, **34**, 237–241.

[177] Máčajová, E., Raspaud, A., and Škoviera, M. 2005. Abelian colourings of cubic graphs. *Electronic Notes in Discrete Mathematics*, **22**, 333–339.

[178] Máčajová, E., Raspaud, A., Tarsi, M., and Zhu, X.-D. 2011. Short cycle covers of graphs and nowhere-zero flows. *J. Graph Theory*, **68**, 340–348.

[179] McGuinness, S. 1984. *The double cover conjecture.* Ph.D. thesis, Queen's University, Kingston, Ontario, Canada.

[180] Menger, K. 1927. Zur allgemeinen Kurventheorie. *Fund. Math.*, **10**, 96–115.

[181] Mohar, B. 2010. Strong embeddings of minimum genus. *Discrete Math.*, **310**, 2595–2599.

[182] Mohar, B., and Thomassen, C. 2001. *Graphs on Surfaces.* Baltimore, MD: The Johns Hopkins University Press.

[183] Naserasr, R., and Škrekovski, R. 2003. The Petersen graph is not 3-edge-colorable – a new proof. *Discrete Math.*, **268**, 325–326.

[184] Nash-Williams, C. St. J. A. 1961. Edge-disjoint spanning trees of finite graphs. *J. London Math. Soc.*, **s1–36**, 445–450.

[185] Nelson, D., Plummer, M. D., Robertson, N., and Zha, X.-Y. 2011. On a conjecture concerning the Petersen graph. *Electronic Journal of Combinatorics*, **18**, P20.

[186] Nowakowski, R. J., and Seyffarth, K. 2008. Small cycle double covers of products I: Lexicographic product with paths and cycles. *J. Graph Theory*, **57**, 99–123.

[187] Nowakowski, R. J., and Seyffarth, K. 2009. Small cycle double covers of products II: Categorical and strong products with paths and cycles. *Graph and Combinatorics*, **25**, 385–400.

[188] Ore, O. 1967. *The Four-Color Problem.* New York: Academic Press.

[189] Petersen, J. 1891. Die Theorie der Regulären Graphen. *Acta Math.*, **15**, 193–220.

[190] Petersen, J. 1898. Sur le théoreme de Tait. *Intermed. Math.*, **15**, 225–227.

[191] Polesskii, V. P. 1971. A lower bound for the reliability of information network. *Probl. Peredachi Inf.*, **7:2**, 88–96.

[192] Preissmann, M. 1981. *Sur les colorations des arêtes des graphes cubiques,* Thèse de Doctorat de 3eme. Ph.D. thesis, Université de Grenoble, France.

[193] Preissmann, M. 1982. Snarks of order 18. *Discrete Math.*, **42**, 125–126.

[194] Rizzi, R. 2001. On 4-connected graphs without even cycle decompositions. *Discrete Math.*, **234**, 181–186.

[195] Robertson, N., Seymour, P.D., and Thomas, R. *Cyclically 5-connected cubic graphs.* in preparation.

[196] Robertson, N., Seymour, P.D., and Thomas, R. *Excluded minors in cubic graphs.* in preparation.

[197] Robertson, N., Sanders, D., Seymour, P. D., and Thomas, R. 1997a. The 4-color theorem. *J. Combin. Theory Ser. B*, **70**, 2–44.

[198] Robertson, N., Seymour, P. D., and Thomas, R. 1997b. The Tutte's 3-edge-coloring conjecture. *J. Combin. Theory Ser. B*, **70**, 166–183.

[199] Sanders, D. P., and Thomas, R. *Edge 3-coloring cubic apex graphs.* in preparation.

[200] Sanders, D. P., Seymour, P.D., and Thomas, R. *Edge 3-coloring cubic doublecross graphs.* in preparation.

[201] Seyffarth, K. 1989. *Cycle and Path Covers of Graphs.* Ph.D. thesis, University of Waterloo, Ontario, Canada.

[202] Seyffarth, K. 1992. Hajós' conjecture and small cycle double covers of planar graphs. *Discrete Math.*, **101**, 291–306.

[203] Seyffarth, K. 1993. Small cycle double covers of 4-connected planer. *Combinatorica*, **13**, 477–482.

[204] Seymour, P. D. 1979a. On multi-colorings of cubic graphs and the conjecture of Fulkerson and Tutte. *Proc. London Math. Soc.*, **s3–38**, 423–460.

[205] Seymour, P. D. 1979b. Sums of circuits. Pages 342–355 of: Bondy, J.A., and Murty, U.S.R. (eds), *Graph Theory and Related Topics.* New York: Academic Press.

[206] Seymour, P. D. 1981a. Even circuits in planar graphs. *J. Combin. Theory Ser. B*, **31**, 327–338.

[207] Seymour, P. D. 1981b. Nowhere-zero 6-flows. *J. Combin. Theory Ser. B*, **30**, 130–135.

[208] Seymour, P. D. 1981c. On Tutte's extension of the four-color problem. *J. Combin. Theory Ser. B*, **31**, 82–94.

[209] Seymour, P. D. 1990. Communication at Cycle Double Cover Conjecture Workshop, Barbados, February 25–March 4.

[210] Seymour, P. D. 1995. *Personal communication.*

[211] Seymour, P. D. 2012. *Personal communication.*

[212] Seymour, P. D., and Truemper, K. 1998. A Petersen on a Pentagon. *J. Combin. Theory Ser. B*, **72**, 63–79.

[213] Shu, J., and Zhang, C.-Q. 2005. A note about shortest cycle covers. *Discrete Math.*, **301**, 232–238.

[214] Shu, J., Zhang, C.-Q., and Zhang, T.-Y. 2012. Flows and parity subgraphs of graphs with large odd edge connectivity. *J. Combin. Theory Ser. B*, (to appear).

[215] Stahl, S. 1998. The multichromatic numbers of some Kneser graphs. *Discrete Math.*, **185**, 287–291.

[216] Steinberg, R. 1976. *Grötzsch's Theorem dualized.* M.Phil. thesis, University of Waterloo, Ontario, Canada.

[217] Steinberg, R. 1984. Tutte's 5-flow conjecture for projective plane. *J. Graph Theory*, **8**, 277–285.

[218] Stephens, D. C., Tucker, T. W., and Zha, X.-Y. 2007. *Representativity of Cayley maps.* Preprint.

[219] Szekeres, G. 1973. Polyhedral decompositions of cubic graphs. *Bull. Austral. Math. Soc.*, **8**, 367–387.

[220] Tait, P. G. 1880. Remarks of the coloring of maps. *Proc. R. Soc. Edinburgh*, **10**, 729.

[221] Tarsi, M. 1986. Semi-duality and the cycle double cover conjecture. *J. Combin. Theory Ser. B*, **41**, 332–340.

[222] Tarsi, M. 2010. *Personal communication.*

[223] Thomas, R. *Generalizations of The Four Color Theorem.* http://people.math.gatech.edu/~thomas/FC/generalize.html.

[224] Thomas, R. 1998. An update on the four-color theorem. *Notices of the AMS*, **45**, 848–859.

[225] Thomason, A. 1978. Hamiltonian Cycles and uniquely edge colorable graphs. *Ann. Discrete Math.*, **3**, 259–268.

[226] Thomason, A. 1982. Cubic graphs with three hamiltonian cycles are not always uniquely edge colorable. *J. Graph Theory*, **6**, 219–221.

[227] Thomassen, C. 1997. On the complexity of finding a minimum cycle covers of graphs. *SIAM J. Comput.*, **26**, 675–677.

[228] Tutte, W. T. 1946. On Hamilton circuits. *J. London Math. Soc.*, **s1–21**, 98–101.

[229] Tutte, W. T. 1949. On the imbedding of linear graphs in surfaces. *Proc. London Math. Soc.*, **s2–51**, 474–483.

[230] Tutte, W. T. 1954. A contribution on the theory of chromatic polynomial. *Canad. J. Math.*, **6**, 80–91.

[231] Tutte, W. T. 1956. A class of Abelian groups. *Canad. J. Math.*, **8**, 13–28.

[232] Tutte, W. T. 1961. On the problem of decompositing a graph into n connected factors. *J. London Math. Soc.*, **s1–36**, 221–230.

[233] Tutte, W. T. 1966. On the algebraic theory of graph colourings. *J. Combin. Theory*, **1**, 15–50.

[234] Tutte, W. T. 1967. A geometrical version of the four color problem. In: Bose, R.C., and Dowling, T.A. (eds), *Combinatorial Mathematics and its Applications*. Chapel Hill, NC: University of North Carolina Press.

[235] Tutte, W. T. 1976. Hamiltonian circuits. Pages 193–199 of: *Colloquio Internazional sulle Teorie Combinatorics, Atti dei Convegni Lincei 17, Accad. Naz. Lincei, Roma I.*

[236] Tutte, W. T. 1984. *Graph Theory*. Encyclopedia of Mathematics and Its Applications, vol. 21. Cambridge Mathematical Library.

[237] Tutte, W. T. 1987. *Personal correspondence with H. Fleischner* (July 22, 1987).

[238] Veblen, O. 1912–1913. An application of modular equations in analysis situs. *Ann. Math.*, **12**, 86–94.

[239] Vince, A. 1988. Star chromatic number. *J. Graph Theory*, **12**, 551–559.

[240] Watkins, J. J. 1989. Snarks. In: Capobianco, M., Guan, M., Hsu, D.F., and Tian, F. (eds), *Graph Theory and Its Applications: East and West, Proceeding of the First China–USA International Graph Theory Conference*. New York: New York Academy of Sciences.

[241] Watkins, J. J., and Wilson, R. J. 1991. A survey of snarks. Pages 1129–1144 of: Alavi, Y., Chartrand, G., Oellermann, O.R., and Schwenk, A.J. (eds), *Graph Theory, Combinatorics, and Applications, Proceeding of the Sixth Quadrennial International Conference on the Theory and Applications of Graphs*. New York: John Wiley & Sons, Inc.

[242] West, D. B. 1996. *Introduction to Graph Theory*. Upper Saddle River, NJ: Prentice Hall.

[243] White, A. T. 1984. *Graphs, Groups and Surfaces*. Revised edn. North-Holland Mathematics Studies, vol. 8. Amsterdam: North-Holland.

[244] Wilson, R. J. 1976. Problem 2. In: *Proc. 5th British Comb. Conf.*, *Utilitas Mathematica*.

[245] Xie, D.-Z., and Zhang, C.-Q. 2009. Flows, flow-pair covers, cycle double covers. *Discrete Math.*, **309**, 4682–4689.

[246] Xu, R. 2009. Note on cycle double covers of graphs. *Discrete Math.*, **309**, 1041–1042.

[247] Ye, D. 2010. *Personal communication.*

[248] Ye, D., and Zhang, C.-Q. 2009. *Circumference and Circuit Double Covers.* Preprint.

[249] Ye, D., and Zhang, C.-Q. 2012. Cycle Double Covers and Semi-Kotzig Frame. *European J. Combin.*, **33**, 624–631.

[250] Younger, D. H. 1983. Integer flows. *J. Graph Theory*, **7**, 349–357.

[251] Zha, X.-Y. 1995. The closed 2-cell embeddings of 2-connected doubly toroidal graphs. *Discrete Math.*, **145**, 259–271.

[252] Zha, X.-Y. 1996. Closed 2-cell embeddings of 4 cross-cap embeddable graphs. *Discrete Math.*, **162**, 251–266.

[253] Zha, X.-Y. 1997. Closed 2-cell embeddings of 5-crosscap embeddable graphs. *European J. Combin.*, **18**, 461–477.

[254] Zhang, C.-Q. 1990. Minimum cycle coverings and integer flows. *J. Graph Theory*, **14**, 537–546.

[255] Zhang, C.-Q. 1994. On even circuit decompositions of eulerian graphs. *J. Graph Theory*, **18**, 51–57.

[256] Zhang, C.-Q. 1995. Hamiltonian weights and unique 3-edge-colorings of cubic graphs. *J. Graph Theory*, **20**, 91–99.

[257] Zhang, C.-Q. 1996a. Nowhere-zero 4-flows and cycle double covers. *Discrete Math.*, **154**, 245–253.

[258] Zhang, C.-Q. 1996b. On embeddings of graphs containing no K_5-minor. *J. Graph Theory*, **21**, 401–404.

[259] Zhang, C.-Q. 1997. *Integer Flows and Cycle Covers of Graphs.* New York: Marcel Dekker.

[260] Zhang, C.-Q. 2002. Circular flows of nearly eulerian graphs and vertex-splitting. *J. Graph Theory*, **40**, 147–161.

[261] Zhang, C.-Q. 2010. Cycle covers (I) – minimal contra pairs and Hamilton weights. *J. Combin. Theory Ser. B*, **100**, 419–438.

[262] Zhang, X.-D., and Zhang, C.-Q. 2012. Kotzig frames and circuit double covers. *Discrete Math.*, **312**, 174–180.

[263] Zhu, X.-D. 2001. Circular chromatic number: a survey. *Discrete Math.*, **229**, 371–410.

[264] Zhu, X.-D. 2006. Recent developments in circular colouring of graphs. Pages 497–550 of: Klazar, M., Kratochvil, J., Matousek, J., Thomas, R., and Valtr, P. (eds), *Topics in Discrete Mathematics.* Springer.

Author index

Subject index

Printed in the United States
by Baker & Taylor Publisher Services